THE SQUARE OF OPP

Jean-Yves Beziau and Gillman Payette (eds)

THE SQUARE OF OPPOSITION
A GENERAL FRAMEWORK FOR COGNITION

PETER LANG
Bern • Berlin • Bruxelles • Frankfurt am Main • New York • Oxford • Wien

Bibliographic information published by die Deutsche Nationalbibliothek
Die Deutsche Nationalbibliothek lists this publication in the Deutsche Nationalbibliografie; detailed bibliographic data is available on the Internet at ‹http://dnb.d-nb.de›.

British Library Cataloguing-in-Publication Data: A catalogue record for this book is available from The British Library, Great Britain

Library of Congress Cataloging-in-Publication Data

World Congress on the Square of Opposition (1st : 2007 : Montreux, Switzerland)
The square of opposition : a general framework for cognition / Jean-Yves Béziau and Gillman Payette (eds.).
p. cm.
ISBN 978-3-0343-0537-2
1. Opposition, Theory of–Congresses. I. Béziau, Jean-Yves. II. Payette, Gillman. III. Title.
BC199.06W67 2007
165–dc23

2011053170

Publication supported by a grant of the Swiss National Science Foundation.

Layout: forte, Thomas Stark, Nova Gora, Austria

ISBN 978-3-0343-0537-2 ISBN 978-3-0352-0163-5 (eBook)

Hochfeldstrasse 32, CH-3012 Bern, Switzerland
info@peterlang.com, www.peterlang.com

Table of Contents

Part 3: Philosophy of Language

Part 4: Mathematical Logic

Part 5: Linguistics and Semiotics

Part 6: Cognitive and Social Sciences

Preface

JEAN-YVES BEZIAU and GILLMAN PAYETTE

1. Introduction to the Square

The square of opposition is a simple geometrical figure expressing some fundamental ideas that were the foundation of logic for two thousand years. In the following we give a brief introduction to some of the basic ideas in the traditional square of opposition, and some of its extensions, for the uninitiated.

Although Aristotle, as far as we know, never wrote of the square of opposition as a geometrical figure in his works, he does provide the basis for the doctrine of opposition, i.e., the logical relationships presented in the traditional square of opposition.[1] Aristotle used these rather simple relationships to analyze categorical propositions. Categorical propositions are so-called since they involve categorical terms, i.e., terms that name categories (e.g., common nouns). Each categorical proposition is composed of two terms and a quantifier, the first term is the subject and the second is the predicate. For example, in the categorical proposition 'Some Dogs are Mammals': 'Some' is referred to as the quantifier, 'Dogs' is the subject term, and 'Mammals' is the predicate term. What made up the basic syntactic structure of Aristotle's logic was an abstraction that replaced subject and predicate terms by, anachronistically speaking, schematic variables. Instead of 'Some dogs are mammals' we get 'Some As are Bs' or more simply, 'Some A are B'. In the new representation the subject term was replaced by 'A' and the predicate term by 'B'.

The square elucidates the notion of negation by portraying the logical relationships between statements (or propositions) of contrary-ness, subcontrary-ness, contradictory-ness and subalternate-ness. Contrary propositions are propositions that cannot be *true* together, but could be false together.

1 For a brief history of the square and its genesis in Aristotle's works see Tarence Parsons's entry on the square in the Standford Encyclopedia of Philosophy.

Subcontraries are the dual of contraries in that they cannot be false together, but can be true together. Two propositions contradict one another when they are both contraries and subcontraries of each other. The proposition P is a subalternate of another proposition Q when Q implies P, logically speaking. The four categorical propositions in the square of opposition are: All A are B, Some A are B, No A are B, and Some A are not B. Frequently reference to these four propositions is made by calling propositions of the form 'All A are B' A-propositions, 'Some A are B' I-propositions, 'No A are B' E-propositions, and 'Some A are not B' O-propositions. These four propositions are placed in the square according to Aristotle's ideas as follows:

Figure 1

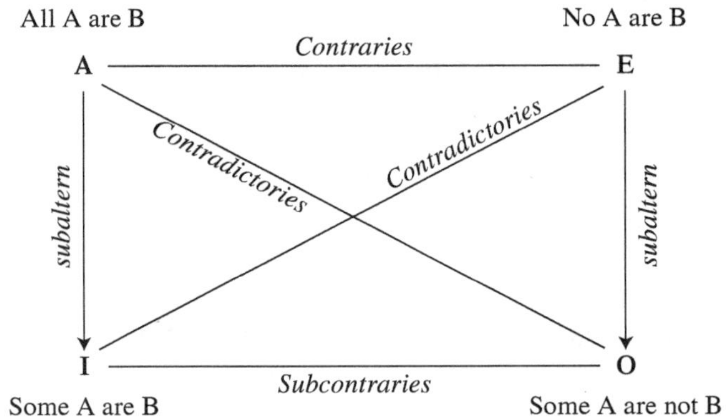

The A and E propositions are called *universal* propositions and the I and O propositions are *particular* propositions. The propositions on the left side of the square are the *affirmative* propositions, the right side are the *negative* propositions. Thus the corners are referred to, alternatively, as follows: **A** Universal Affirmative, **E** Universal Negative, **I** Particular Affirmative, and **O** Particular Negative. Universal negative and particular negative propositions are called negative because they incorporate negation. The negation in the universal case is called outer negation since the whole predication is negated. Inner negation is when the predicate is negated and occurs in the particular negative case. Such diagrams are recorded at least as early as Boethius's translation of Aristotle in the fifth century CE.

The use of the square as a tool for analysis isn't restricted to categorical propositions, it can also be used to analyze propositions concerning necessity and possibility. In that case one gets a *modal* square of opposition. The propositions involved are the various modes of a proposition P, i.e., manners in which P can be true. We can give a modal version of A, E, I and O propositions. The m(odal) A is Necessary P, that is P is necessarily true. The m(odal) I position is Possible P, i.e., it is possible that P is true. The other two are their negations mE Impossible P (or Necessary not P) and mO Possible not P (or Not Necessary P).

Figure 2

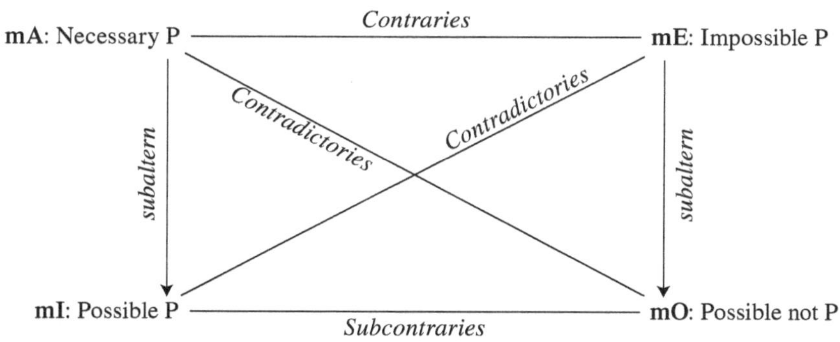

The logical relationships portrayed in the modal square are fairly intuitive, but there was debate over the relationships in the Aristotelian square for quantifiers. Particularly over instances where, in the first square above, the A term is empty or, equivalently, there are no As. An example of an A proposition using an empty term is 'All unicorns have horns'. Since 'All A are B' implies that 'some A are B', it must be that there are As when A propositions are true. That implication is referred to as existential import, and could pose a problem for the range of application of traditional logic since when reasoning using terms for imaginary or hypothetical categories, which are presupposed to be empty. In the 20th century the modern predicate calculus supplanted the square as the basis of logic, and the problem of existential import was solved by rejecting the relation of subalternation between the universals and particulars. That rejection results in an austere set of relations, i.e., just the contradictories. The square of opposition became an X of opposition.

The debate over existential import doesn't occur for the modal square since there isn't a similar problem of empty terms. When the square dealing with quantifiers was made insignificant, the interest in using opposition to analyze logical relationships for other concepts remained. During the 20th century, and in recent years, interest in the square of opposition has been extended to a number of new areas: psychology, semiotics, linguistics, ethnology, arts, argumentation theory and cognitive science, to name a few. And the generalization of the theory of opposition to more complex geometrical objects –hexagons, octagons and even hypercubes –that had begun in medieval times has been carried much further. The new research looks at the theory of oppositionas including the old logical relationships, and, in some cases, developing logic using opposition theory as the starting point.

2. The Essays

This is a collection of new research papers concerning several contemporary avenues of research using the square, written by internationally recognized scholars from various fields. The papers have been selected from a larger number of contributions most of them based on talks presented at the First World Congress on the Square of Opposition organized in Montreux in June 2007.[2]

The goal of this book is to revitalize both interest in the square of opposition for its own sake, and to make evident the square's relevance to new research projects. This volume will be of interest to advanced students, professors and researchers from many fields –it has an interdisciplinary perspective. We have grouped the essays into their general fields of study. Each essay is self contained and we give a brief idea about the topics covered in each paper below.

2 The congress was organized by Jean-Yves Beziau (Neuchâtel/SNF, Switzerland), Michael Frauchiger (Open University and Lauener Foundation, Bern, Switzerland), Katarzyna Gan-Krzywoszynska (Poznan/Poland, Nancy/France), Alessio Moretti (Neuchâtel/Switzerland, Nice/France), Gillman Payette (Dalhousie/Calgary, Canada), Fabien Schang (Archives Poincaré/University of Nancy 2,France).

2.1 History

"Quantification Theory and the Square of Opposition" is a discussion of the early days of modern logic. Volker Peckhaus argues in his essay that quantification theory should not be used as a point of difference between the algebraic and mathematical approaches to logic as initiated by Boole and Frege, respectively. The reason being that although quantification theory isn't developed in Boole's account of the square of opposition, it is developed in the work of C. S. Peirce and taken up in the work of Schröder. The work of Peirce, however, was independent from that of Frege.

Yurii Khomskii's essay "William of Sherwood, Singular Propositions and the Hexagon of Opposition" argues a point of intellectual priority. In 1955 Tadeusz Czeżowski suggests a solution to providing an analysis of particular propositions like 'Socrates is a man' within a framework like the square. Czeżowski suggests augmenting the square to a hexagon. Khomskii argues that this augmentation existed already in the work of William of Sherwood in the thirteenth century.

John Martin's essay "Existential Commitment and the Cartesian Semantics of the Port Royal Logic" discusses how the semantics in Arnaud and Nicole's seventeenth century work *La logique, ou l'art de penser* maintains a correspondence theory of truth, even though it breaks from the medieval orthodoxy in which there must be a causal relationship between concepts and the world. The connection to the square of opposition in Martin's essay is that, contrary to another author, the Port-Royal semantics validates existential import for the A and I positions in the square. This makes the semantics align with the Aristotelian doctrine of the square.

In "The Medieval Modal Octagon and The S5 Lewis Modal System" Juan Campos-Benítez looks at a medieval extension of the square to an octagon. The octagon was used to analyze propositions that combined the A, E, I, O propositions with modal propositions, resulting in propositions corresponding to sentences like: All A are Necessarily not B. Campos-Benítez uses modern formal techniques to analyze the structure of the positions that occur in the octagon, then investigates what is required of the modern modal systems to to prove the correspondences in the medieval octagon.

2.2 Diagrams

In his paper "From the Logical Square to the Logical Poly-simplexes and Beyond" Alessio Moretti outlines the history of the logical square, beginning with the ancient Greeks and ending with n-opposition theory or NOT. There are four primary stops on this route. First is the square of the middle ages that was used primarily as a pedagogical device for teaching logic. Second is the use of the square by the structuralist thinkers such as Piaget and Gottschalk in providing foundations for reasoning in various social sciences. There is also the use made of the square by analytic philosophers, in some directions providing a degenerate square as mentioned in the introduction above and using other polygons to analyze deontic or modal concepts. There is finally the discussion of NOT in its genesis from the logical hexagon and its generalization as the geometry of opposition. The paper concludes with some consequences of NOT for epistemology.

Jonathan Westphal shows how to represent syllogistic arguments using vectors in a plane. In "The Application of Vector Theory to Syllogistic Logic" he describes how to represent the A, E, I, and O propositions in the square of opposition, and uses these representations to show that the square resulting in his "vector logic" doesn't permit existential import. The vector representation also provides a method for showing validity and invalidity of syllogisms generally. The final point of the paper is to analyze syllogisms involving the quantifier 'most' in the vector representation.

S. S. Sharma constructs a new system of diagrams for giving diagrammatic demonstrations and refutations of syllogisms in "Interpreting Squares of Opposition With the Help of Diagrams". The paper has an extended discussion of the problem of existential import and the contemporary responses to it. Other systems of diagrammatic proofs are also discussed and found lacking since they can only be used to represent the Aristotelian or contemporary view of the categorical propositions, exclusively. However, the system proffered in Sharma's paper can be used to represent both views on categorical propositions.

2.3 *Philosophy of Language*

Although the square of opposition is used to analyze many concepts like necessity and obligation, Dag Westerståhl suggests in his paper "Classical Vs. Modern Squares of Opposition and Beyond" that, fundamentally, the square concerns the interaction of the quantifiers *all* and *some* with negation (inner and outer). But the square can be generalized to look at the relationship between negation and other kinds of quantifiers. Using the formal machinery of generalized quantifiers, Westerståhl explores the general structure of the modern square of opposition with many different quantifiers. He shows how to place general quantifiers in a square of opposition based on structural properties of monotonicity and symmetry. The result is looking at possessive determiners that generate a cube of opposition.

Pieter Seuren's article is part of his natural logic project which is to construct a logic that is empirically accurate with respect to human reasoning. Seuren's goal in "From Formal Intuitions to Natural Logic" is to develop logics which correctly capture humans' innate reasoning and our more reflective, well-trained reasoning. These logics are supposed to be those that underly the logic of natural language. Seuren thinks that modern predicate calculus is ill suited to that job, but older predicate calculi that arise from the traditional square of opposition and a modified version of a logic developed by William Hamilton are up to the task. The notion of implicature doesn't do the job, Seuren argues, of explaining away natural inferences that run afoul of modern predicate logic. His solution is to replace implicature with a logic of presuppositions.

Schang has developed a question-answer semantics which he uses to interpret the concept of opposition in the paper "Questions and Answers About Oppositions". From this he provides further analysis of relations of opposition to the n-ary opposition of Moretti by giving an explanation of how opposition arises. This question-answer semantics is used to understand the problem of existential import and the concept of negation.

2.4 Mathematical Logic

Peter Schroeder-Heister's essay uses a framework of reasoning with respect to clausal definitions of atoms, which are given by production rules like $a \Leftarrow B$ where B is a list of atomic propositions and a is an atomic proposition. Further on in"Definitional Reasoning in Proof-Theoretic Semantics and the Square of Opposition" he then distinguishes four forms of judgement: direct and indirect assertion, and direct and indirect denial. The direct forms are established by applying a definitional clause like the one above to generate the asserion/denial. The indirect result from showing that all possible premises (Bs) of the opposite judgement can be refuted. The deductive relationships between these four forms of judgement correspond to those represented in the square of opposition. Schroder-Heister also looks at the interrelationships of definitions which respect the different forms of opposition in the square.

Pellissier's article "2-opposition and the Topological Hexagon" contains some interesting results in n-opposition theory. First is the introduction of weak and strong opposition and the geometrical characterization of n-opposition for $n > 2$. That is, what geometrical figures are characteristic of n-opposition. What comes out of this theorem is that weak 2-opposition *is* Aristotle's square of opposition. Pellissier then goes on to develop a topological interpretation of n-opposition and explores the relationship between topology, intuitionistic negation and paraconsistent negation.

Strößner and Strobach use a little known concept in logic to construct an extension of a three valued logic in "Veridications and their Square of Opposition". Veridications are unary operations from semantics values to true and false that send true to true and false to false. Although the only veridication for classical truth functional logic is trivial –the identity – there are many that can be defined on many valued logics. In the paper various veridications are added to a three valued logic, and the polyhedra of opposition arising from the addition of the veridications are presented and discussed.

2.5 Linguistics and Semiotics

Laurence Horn undertakes an investigation of the nameless corner of the square of opposition in "Histore d'*0 – Lexical Pragmatics and the Geometry of Opposition". It is interesting that the O corner doesn't have a lexicalization, a word, that names it, usually. For the modal square there is possibility, necessity and impossibility, but no non-necessity –'unnecessary' usually has a deontic reading. Similarly with *or, and* and *nor*, there is no *nand*. This lack of lexicalization has had various responses, one of which is to jettison the square and replace it with a triangle. A triangle stipulates that the I and O corners of the square are, in fact, semantically equivalent. But that doesn't sit well with Horn who argues that the belief that some A are B implies (and vice versa) some A are not B is a pragmatic, (Neo-) Gricean implicature. This view is contrary to that taken in Seuren's article.

Badir presents the development of the semiotic square in his essay "How the Semiotic Square Came to Be". The semiotic square concerns the relationship between the meaning of various symbols. As it turns out the logical square didn't play as much of a role in the development of the semiotic square as one might think. But rather was influenced by Levi-Strauss among others.

2.6 Cognitive and Social Sciences

Bianchi and Savardi have developed, over the past ten years, a research project that looks at contrariety not from a linguistic or semantic perspective, but from a perceptual perspective. What they set out in their essay "The cognitive dimensions of contrariety" are the results of their experiments studying whether contrariness is indeed a perceptual phenomenon. They look at whether contrariness is actually one dimensiona –there only being one *logical* contrary –as the square of opposition treats it, or, rather, as N-dimensional –there being many contraries. They also consider whether contraries sit at the ends of a bounded continuum.

Beller's article outlines an experiment testing reasoning processes involving the concepts obligatory, permitted, banned or omissible. The experiment presented in "Human Deontic Reasoning and the Deontic Square of Opposition", Beller contends, provides evidence that people reason in accord with the deontic square of opposition. The deontic square of opposition has obligation in the mA position of the modal square, permission in the mI corner, prohibition (ban) in the mE corner and need not to do the action (omissibility) in the mO corner, with all of the relevant oppositions that exist in the modal square. An important part of Beller's discussion is that people interpret deontic rules in certain peculiar ways. These peculiar interpretations can account for the subject's divergence from the deontic square of opposition.

In "n-opposition theory to structure debates" the authors discuss experiments they have done using the concepts from the square of opposition to structure on-line debates. The series of experiments develops a computer program that allows users to debate problems centered around a collection of literature, and make judgments as to whether various theses have been established within that debate. The relations of contrariness, contradictoriness, subalternation and subcontrariness are used to connect the various judgments that each participant of the debate can make on the various theses. These judgments are constructed in a defeasible logic based on the work of D. Nute.

3. The DVD

The congress on the square in Montreux was an interdisciplinary meeting with some artistic components. In the two-hour DVD[3] accompanying the book there you will find: clips of talks presented at the first world congress on the square of opposition, interviews of participants, the round square table, extracts of a jazz show presented during the event with some pieces of music based on the square and the trailer of a movie based on the square made especially for this event: the Square of Salomé.

Extracts of talks. There are extracts of talks of invited speakers of the event mixed with extracts of talks of the contributed papers. These are presented using the multiscreen technique, the screen being divided into four parts inspired by the square.[4] The invited speakers were: Sieghard Beller (Psychology, Freiburg), Jean-Yves Beziau (SNF, Neuchâtel), Ivana Bianchi

3 The DVD was produced by Catherine Chantilly and Jean-Yves Beziau.

4 About multi-screen, see Sobieszczanski and Lacroix (2010).

(Education, Macerata), Pascal Engel (Philosophy, Geneva), Laurence R. Horn (Linguistics, Yale), Alessio Moretti (Psychology, Neuchâtel), Terence Parsons (Linguistics and Philosophy, UCLA), Ugo Savardi (Psychology and Cultural Anthropology, Verona), Pieter Seuren (Max Planck Inst. for Psycholinguistics, Nijmegen), Hans Smessaert (Linguistics, Leuven), Jan Woleński (Philosophy, Kraków). About 70 talks were given on contributed papers by presenters from all over the world. For a complete list see the handbook at:

"http://www.square-of-opposition.org"

What is the square of opposition for you? A one hour movie with interviews of 33 participants by Catherine Chantilly. Each participant is placed in a similar formal setting and explains the meaning of the square for him.

The round square table. Shot of the final discussion of the event discussing the different perspectives opened by this meeting and future developments.

The Square of Jazz. In this DVD there are some extracts of the Square of Jazz concert. For the conference on the square of opposition, four jazz musicians from Switzerland composed some musical pieces based on the square, and they presented those pieces during the event. The band members are Michael Beck (piano), Phil Stöckli (tenor sax), Dominik Egli (drums) and Dominique Girod (acoustic bass).

The following songs were composed and performed: Passacaglia - by Dominique Girod, Ballade pour la trichotomie subcontraire - by Dominique Girod, Square of Opposition - by Dominik Egli, Circle of Steps - by Phil Stöckli, Subcontrary Reality (or: A World of Compromises) - by Phil Stöckli, Continuum -by Phil Stöckli, Empty square - by Michael Beck, Move It - by Michael Beck.

In the following paragraph Phil Stöckli explains the motivation and methodology of the performance. Following that are some explanations of some of the songs composed by him.

Opposition as Displayed in the Square as a Means of Composition. Music can function, be perceived and analyzed on a great number of levels, namely a) its basic musical elements like rhythm, harmony, dynamics, tempo, density; b) larger musical structures like motive, melody, harmonic progression, level of dissonance, form, instrumentation as well as c) more artistic (or subjective?) criteria like style, emotional expression, energy, narrative or metaphorical content. Consequently the search for contrasting or opposing musical elements, and their respective placement within the square of opposition, can take place on a great number of levels as well. Music in virtually all styles and periods relies strongly on such contrasting elements, and musical development can be seen as a continuous "play" between them. For this project we have tried to find specific musical ideas and to examine them as to their behavior within the square of opposition, eventually constructing variations or additions to them which complete the square. This has proved to be quite an interesting source of compositional materials, leading to unexpected, yet musically meaningful results. Of course, musical material found in this purely rational way has to pass the test of the more intuitive selection of the musician in order to qualify as actual music worthy of performance in front of an audience. The program of the concert for Jazz quartet consists of a suite of original compositions and guided improvisations which are all based in certain of their parameters on the structure of the square of opposition. Some of them may serve as an acoustic illustration of the square where the structure is more transparent, while others may be heard as music "only", nonetheless owing part of their existence to the square.

Circle of Steps. Before writing the piece, I was asking myself what is the opposite of the square of opposition. Taking it literally, it is a circle of harmony, continuity or togetherness. The piece reflects this continuity mostly through its harmonic structure, which is influenced by John Coltrane's famous composition "Giant Steps" from 1961. The structure constantly repeats itself in a circular motion, comparable to the M.C. Escher painting of the ever ascending/descending steps. The 7/8 rhythmic brings out this effect even more. The piece is a real challenge to each soloist, and therefore should be an inspirational voyage.

Subcontrary Reality (or A World of Compromises). The structure of the composition builds upon a 6/8 Rhythm, that functions as a carpet, which could also be seen as our subjective reality. Through the continuous repetition of this carpet, we might tend to forget that it even exists. However, on this sound carpet, I have a written a melody that fits in with the background and pints out some more interesting aspects of the harmony. Philosophically spoken, the message of this tune is the following: Our world is not a world of dichotomy, it cannot solely be divided into universal statements of the kind "every S is a P" or "No S is a P". Rather, life as we know it is always full of compromises and subcontraries, hence the title "Subcontrary reality".

Continuum. Similar to the other piece of mine, *Circle of Steps*, this one, literally, is based on the opposite of the square of opposition. It is designed to reflect the style of the late John Coltrane / Mc Coy Tyner quartet. The opposites in this piece can be found between the two main sections: one in flowing 3/4 time, the other with contrasting accents of the rhythm section. Philosophically, "Continuum" stands for the continuity of life, for Newtons eternal energy equivalence, for the balance of the contraries which in complement to their opposite always equal a complete sum of unity.

The Square of Salomé. This DVD includes the trailer of the movie the Square of Salomé. This movie was produced especially for this event and projected in Montreux. This is a remake of the famous biblical story using the square to display the strong and ambiguous relations between the main four characters: Salomé (Joana Medeiros), Iokanaan (Jean-Yves Beziau), Herod (Alessio Moretti) and Herodias (Catherine Chantilly). The movie was shot in Paris, Morocco and Saint Gérand Castle in April/May 2007.

The story itself is from the Biblical period. Herod is Herod Antipas, tetrarch of Galilee and Peraea, and son of Herod the Great who is the King of Judea and author of the massacre of the innocents. Herod Antipas jailed Iokanaan, later on known as St. John the Baptist, who was announcing the coming of Jesus Christ. Iokanaan was, also, slandering the wife of Herod: Herodias. Herodias was formerly married to Herod's brother with whom she had a daughter, Salomé, to whom Herod was attracted. According to the story, on one of his birthdays Herod promised to give Salomé whatever she wanted. Salomé, supposedly influenced by her mother who hated Iokanaan, asked for the head of Iokanaan. The first writings about the story can be found in the Bible, but there Salomé is just named as "the daughter of Herodias". The name Salomé came from the Jewish tradition through which the story was transmitted; Salomé is a Hebraic name with the same origin as Shalom, meaning peace. This story became a legend, a myth, and has been transformed and adapted by many authors, in particular by Gustave Flaubert and Oscar Wilde. Oscar Wilde wrote a play where Salomé seduced Herod by dancing and kissed the mouth of Iokaanan once his head had been severed.

In this creation it was decided not to necessarily follow such and such version of the story –nobody knows exactly where the limit between reality and mythology lies –but rather to see how it was possible to describe the characters through their relations between each other using the square. The following square was chosen:

Figure 3

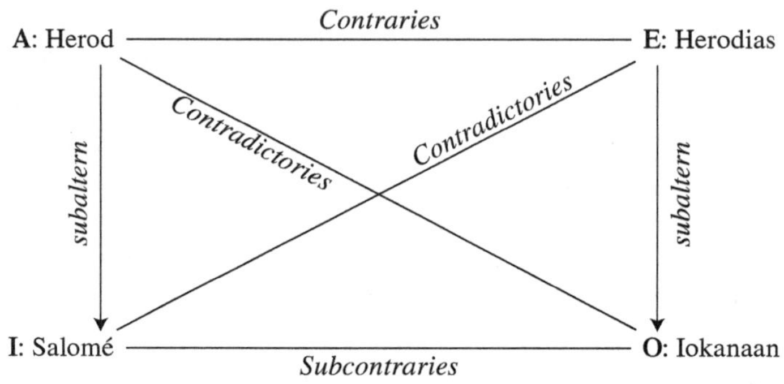

References

Sobieszczanski, M., & C. Lacroix (Eds.) (2010). *From split-screen to multi-screen*. Bern: Peter Lang.

Part 1: History

Algebra of Logic, Quantification Theory, and the Square of Opposition

VOLKER PECKHAUS[1]

Abstract

The distinction between logic as calculus and logic as language aims at explaining the main differences between the two main directions in 19th century symbolic logic, the algebra of logic originating in Boole and developed by Peirce and Schröder, and mathematical logic created by Frege and developed by Russell. The absence of modern quantification theory in the algebra of logic is seen as one of the essential points of difference between the two directions. The paper challenges this view. The different interpretations of the square of oppositions as can be found in the works of Boole, Frege and Schröder are discussed. Whereas Frege presented the modern quantificational interpretation still accepted today, no trace of quantification theory can be found in Boole and Schröder. The reason is that Boole and Schröder dealt with the square of oppositions in the context of the calculus of concepts, not of the calculus of propositions like Frege. As soon as Schröder skipped to the calculus of propositions he introduced quantification theory equivalent to Frege's, but taken over from Peirce. Quantification theory is therefore not suitable to distinguish these directions in logic.

Keywords: logic as calculus, logic as language, quantification theory, interpretations of the square of oppositions, elective symbols, calculus of classes, calculus of propositions

1 volker.peckhaus@upb.de, Universität Paderborn, Institut für Humanwissenschaften: Philosophie, Warburger Str. 100, D-33098 Paderborn, Germany. Revised English version of Peckhaus, 2005. I would like to thank Franz Fritsche (Duisburg) and Pierre Joray (Neuchâtel) for helpful comments.

1. Introduction: Logic as Calculus and Logic as Language

1967 was one of the most important years for the historiography of modern logic. In that year Jean van Heijenoort published his *From Frege to Gödel*, according to its subtitle "A Source Book in Mathematical Logic, 1879–1931" (van Heijenoort, 1967a). The seminal character of this collection is due to the fact that for the first time basic texts from the history of mathematical logic were gathered together, most of them for the first time in the English language. Its formative influence on the historiography of logic had, however, some negative effects. In concentrating on the logical tradition starting with Frege's *Begriffsschrift* (Frege, 1879) it largely ignored the logical direction with which the 19th century rebirth of the interest in formal logic started: the algebra of logic. In saying that "the second half of the nineteenth century saw a rebirth of logic" van Heijenoort was clearly aware of the pre-Fregean algebraic tradition of logic (van Heijenoort, 1967a, vi). However, he sees this tradition suffering from a number of limitations (*ibid.*):

> It tried to copy mathematics too closely, and often artificially. The multiplicity of interpretations of what became known as Boolean algebra created confusion and for a time was a hindrance rather than an advantage.

This led van Heijenoort to his final assessment that the period of the algebra of logic "would, no doubt, leave its mark upon the history of logic, but it would not count as a great epoch" (*ibid.*).

According to van Heijenoort "the great epoch in the history of logic did open in 1879, when Gottlob Frege's *Begriffsschrift* was published" (*ibid.*). The reasons were, according to his opinion, that the *Begriffsschrift* freed logic from the artificial connection with mathematics, at the same time preparing "a deeper interrelation between these two sciences". Among his criteria van Heijenoort stated that "it presented to the world, in full-fledged form, the propositional calculus and quantification theory" (*ibid.*). It seems to be obvious that this should imply the assertion that there was no quantification theory in the algebra of logic.

I feel first of all urged to a short comment on van Heijenoort's assertion that in the algebra of logic the connection between mathematics and logic was too narrow. Even if he were right, this insight hides the really interesting point, the fact namely that the development of modern structural mathematics and modern logic ran basically parallel. Most algebraists of logic were also involved in the creation of structural mathematics, and the mathematics involved in logic was in fact this new mathematics (cf. Peckhaus, 1997, chs. 5 and 6). But now back to quantification theory and van Heijenoort!

It was in the same year 1967 that van Heijenoort published his paper "Logic as Calculus and Logic as Language" (van Heijenoort, 1967b) which was even more important than his source book for our standard understanding of the early directions in modern logic. Nevertheless, despite its significance, I insist on my judgement that it is a model for sloppiness, one-sidedness, and ignorance, showing a dreadful insensibility for the complexity of the historical development.

Van Heijenoort took up Leibniz's distinction between the "calculus ratiocinator" and the "lingua characteristica" both of which provide tools for Leibniz's *scientia generalis.* But he interpreted this distinction as standing for two kinds of approaching logic. The dichotomy between logic as calculus and logic as language is used to distinguish between the two big traditions in the newer history of logic, the algebra of logic on the one hand and the tradition opened with Gottlob Frege's *Begriffsschrift* of 1879 (Frege, 1879), the latter sometimes identified with mathematical logic. In doing so, van Heijenoort opposes George Boole and Leopold Löwenheim on the algebraic side with Gottlob Frege and Bertrand Russell on the side of mathematical logic. He mentions Ernst Schröder, the most important representative of the German algebra of logic, and Charles S. Peirce (to whom we owe the term "quantifier"), his ingenious American colleague and inspirer, only in passing, but in beating Boole and Löwenheim, he also hits Schröder and Peirce, of course.

Frege's logic is described as a universal language which does not have, because of its universality, any separate semantics. "The universal formal language supplants the natural language", he writes (van Heijenoort, 1967b, 327),

> and to preserve, outside the system a notion of validity based on intuitive set theory, does not seem to fit into the scientific reconstruction of the language.

The algebraic propositional calculus uses, on the contrary, a model-theoretic approach. Algebraic systems always need interpretations of operation signs and categories, i. e., classes of concepts. They can clearly be divided into a structural, or syntactic, and a semantical side, the latter providing interpretations of the figures used.

Van Heijenoort maintained that the universality of Frege's *lingua characteristica* is founded in Frege's quantification theory, a feature lacking in Boole's propositional calculus (van Heijenoort, 1967b, 324–325). This is in fact true, but van Heijenoort withholds some other verities. He does not mention that almost at the same time when Frege published his quantification theory, an equivalent one was developed in the school of Charles S. Peirce, later taken over by Schröder, and extensively discussed in the second volume of the *Vorlesungen über die Algebra of Logik* (Schröder, 1891). Therefore it makes no sense to use quantification theory to distinguish between the algebra of logic and the Frege kind of mathematical logic.

Nevertheless, the story is quite complicated, and it is the aim of this paper, to shed some light on it. In the meantime the historiography of logic (but not the philosophy of logic and mathematics) has overcome van Heijenoort's one-sidedness. Already van Heijenoort's doctoral student and biographer Irving H. Anellis wrote that the "distinction between the 'algebraic' and the 'quantification-theoretic' traditions is artificial at best" (Anellis, 1994), 111). Anellis thought that van Heijenoort's distinction was mainly due to Bertrand Russell's influence (Anellis, 1994, 103–104; cf. Jean-Yves Béziau's review of Anellis' book Béziau, 1998). Among recent work I would like only to mention Geraldine Brady's book *From Peirce to Skolem* (Brady, 2000) which focusses on the Peirce-Schröder tradition in the development of first-order logic, and Ivor Grattan-Guinness's seminal survey *The Search for Mathematical Roots* (Grattan-Guinness, 2000) which gives deep insights into the interrelations between the different traditions in logic.

In the following I will reconsider the differences in quantification theory between the two traditions by taking a common point of reference into account, their positions toward the Aristotelian-Scholastic theory of judgement.

2. The Square of Opposition

2.1 The Aristotelian-Scholastic Model

The Aristotelian-Scholastic theory of judgement is based on a special kind of propositions, the so-called "standard forms" which contains some sort of quasi-quantification. The list of standard forms runs as follows:

All *S* are *P*	*SaP*
No *S* is *P*	*SeP*
Some *S* are *P*	*SiP*
Some *S* are not *P*	*SoP*

Such standardized judgements consist of two concepts, the subject *S* and the predicate *P*, which are combined with the help of a connector, the copula, expressed by the words "is" or "are". Two of these judgements are universal, indicated by the affirmative quantifier "all" and the negative quantifier "no", the two other judgements are particular, particular-affirmative and particular-negative. The respective quantifier "some" as well as the universal quantifiers usually refer only to the subject term. The relations between these judgements can be illustrated with the help of the square of opposition:

Figure 1

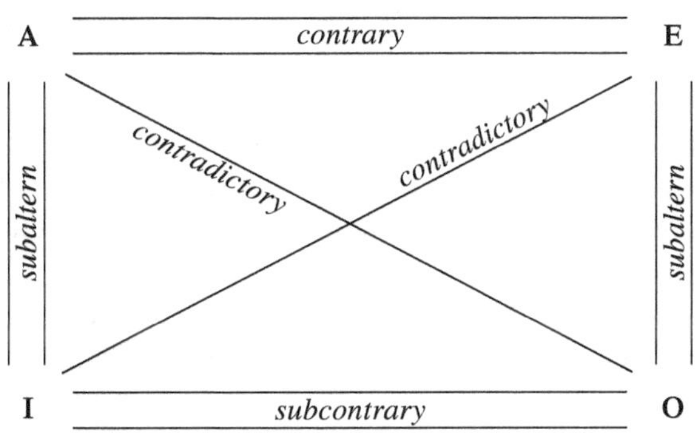

2.2 Frege's Interpretation

The modern interpretation of the standard forms is the following:

All *S* are *P*	*SaP*	$\bigwedge_x(Sx \rightarrow Px)$
No *S* is *P*	*SeP*	$\bigwedge_x(Sx \rightarrow \neg Px)$
Some *S* are *P*	*SiP*	$\neg\bigwedge_x(Sx \rightarrow \neg Px)$, or $\bigvee_x(Sx \wedge Px)$
Some *S* are not *P*	*SoP*	$\neg\bigwedge_x(Sx \rightarrow Px)$, or $\bigvee_x(Sx \wedge \neg Px)$

This interpretation goes back to Frege who used a similar one in his *Begriffsschrift*. There one can find the following square which is here reproduced with a slight modification. Frege mistakenly described the relation between the I judgement and the O judgement as "contrary" instead of "subcontrary" (Frege, 1879, 24).

Figure 2

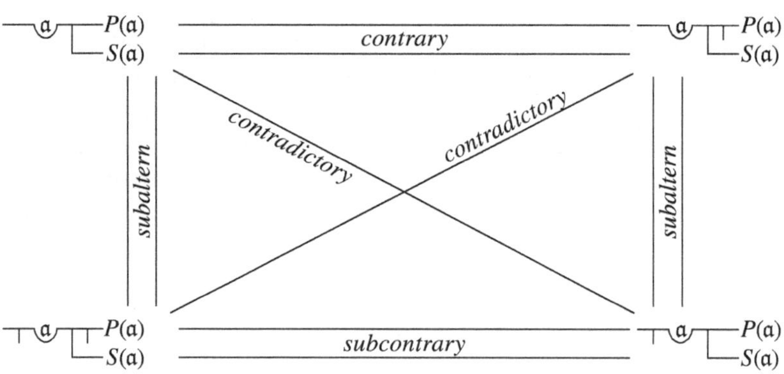

In Frege's logic only negation, implication and universal quantification are used. The quantifiers range over a domain of individuals marked by a Gothic letter written into a "Höhlung", a concavity. About its necessity Frege writes that "it delimits the scope of the generality signified by the letter. Only within its scope does a Gothic letter retain its meaning" (Frege, 1879, 20, transl. Beaney, 1997, 71). In signifying the range of a quantifier Frege made the decisive step towards modern quantification theory. This step was made possible by Frege's revision of the understanding of a judgement. It is no longer the connection of the concepts of or the terms for subject and predicate with the help of a copula, but a function of arguments. The judgement "*A* is Φ", or, more exactly "*A* has the property Φ", is therefore expressed by $\vdash \Phi(A)$ (cf. Frege, 1879, 18). A side-effect of this new understanding is that Frege's notation isolated the statements used in a complex proposition from the logical structure of this proposition. Frege's explanations given in a letter to Ludwig Darmstaedter of July 1919 are instructive in this respect (Frege, 1983, 273, transl. taken from Beaney, 1997, 362):

> So I do not begin with concepts and put them together to form a thought or judgement; I come by the parts of a thought by analyzing the thought. This marks off my *Begriffsschrift* from the similar inventions of Leibniz and his successors, despite what the name [*Begriffsschrift*] suggests; perhaps it was not a very happy choice on my part.

2.3 Boole's Interpretation

Contrary to Frege, the algebraists initially stuck to the traditional concept of the judgement as combination of concepts. In *The Mathematical Analysis of Logic* (Boole, 1847), Boole mentioned that his juvenile interest for logic had been revived by the priority struggle between the Scottish philosopher Sir William Hamilton and the London mathematician and philosopher Augustus De Morgan about the quantification of the predicate (cf. Boole, 1847, 1). This struggle concerned a modification of the traditional theory of the judgement, and although neither Hamilton nor De Morgan had any priority in this respect, it nevertheless indicated a new interest in logical research in Great Britain at that time (cf. Heath, 1966).

In respect to the topics treated Boole followed closely the model of traditional syllogistic. Using Boole's symbolism the square of opposition would look as in Figure 3.

In Boole's notation, capital letters stand for all members of a certain class. The small letters are introduced as follows (Boole, 1847, 15):

> The symbol *x* operating upon any subject comprehending individuals or classes, shall be supposed to select from that subject all the Xs which it contains. In like manner the symbol *y*, operating upon any subject, shall be supposed to select from it all individuals of the class Y which are comprised in it and so on.

Take A as the class of animals, then *x* might signify the selection of all sheep from these animals, which then can be regarded as a new class from which we select further objects, etc. This might be illustrated by the example of Figure 4.

Figure 3

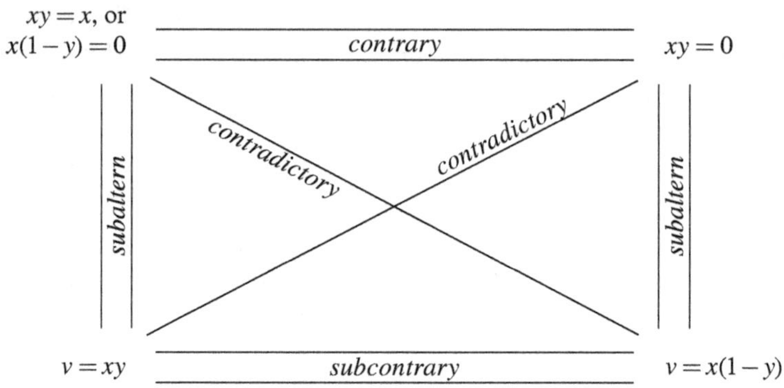

Figure 4

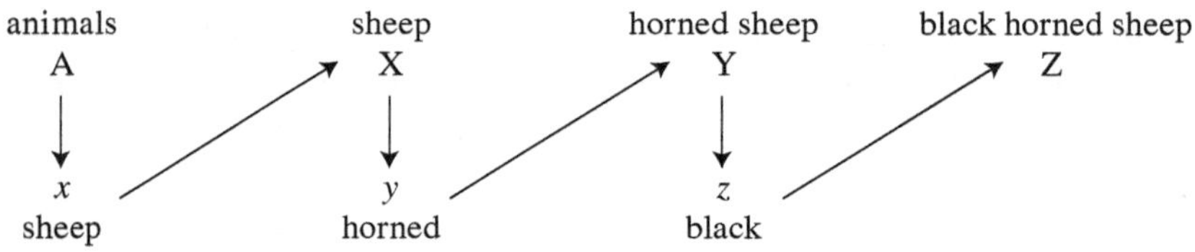

What we have done here is a successive selection which leads to individuals being common to the classes A, X, Y, and Z. xyz stands for animals that are sheep, that are horned and black. It can be regarded as the logical product of some common marks or common aspects relevant for our selection. In his major work *An Investigation of the Laws of Thought* (Boole, 1854) Boole gave up the distinction between capital and small letters, thereby significantly simplifying his notation.

The symbol 1 denotes the universe, and therefore it is consequent that if the class X is determined by the symbol x, the class not-X has to be denoted by the symbol $1-x$. $1-x$ forms the supplement to x, thus $x(1-x)=0$, where 0 symbolizes nothing or the empty class. Now we can consider Boole's interpretation of the universal-affirmative judgement. The universal-affirmative judgement "All Xs are Ys" is expressed by the equation $xy=x$ or, by simple arithmetical transformation, $x(1-y)=0$ (Boole, 1847, 22):

> As all the Xs which exist are found in the class Y, it is obvious that to select out of the Universe all Ys, and from these to select all Xs, is the same as to select at once from the Universe all Xs.

The universal-negative judgement "No Xs are Ys" asserts that there are no terms common to the classes X and Y. All individuals common would be represented by xy, but this product forms the empty class. The particular-affirmative judgement "Some Xs are Ys" says that there are some terms common to both classes forming the class V. These terms are expressed by the elective symbol v. The judgement is thus represented by $v = xy$. Boole furthermore considers to use $vx = xy$ with vx for "some X" and vy for "some Y", but observes "that this system does not express quite so much as the single equation […]" (Boole, 1847, 22–23). The particular-negative judgement "Some Xs are not Ys" can be reached by simply replacing y in the last formula with $1 - y$.

As mentioned, Boole revised his logical theory in his later logical work, especially in his seminal *An Investigation of the Laws of Thought* (Boole, 1854). Although he developed there a theory of logical functions of which first ideas can already be found in the earlier work, and although he introduced "secondary propositions" as propositions about propositions, thereby having necessary elements of the Fregean paradigm of logic at hand, no trace of a quantification theory can be found. So van Heijenoort seems to be right in denying the existence of quantification theory in the algebra of logic.

But Boole does not represent all of the algebra of logic, and so he should not be taken for all.

3. Schröder's Quantification Theory

In the following I will concentrate on the German algebraist of logic Ernst Schröder who started to adopt and revise Boole's algebra of logic 20 years after Boole. He published the first fruits of his research in his *Der Operationskreis des Logikkalkuls* of 1877 (Schröder, 1877).

3.1 Misunderstanding Frege's Quantification Theory

Schröder's *Operationskreis* gives further evidence for the assumption that it was the algebraic paradigm of regarding judgements as combinations of concepts that prevented the early algebraists of logic to interpret the old standard forms of the syllogism in terms of modern quantification theory. Notorious in this respect is Schröder's review of Frege's *Begriffsschrift*, published in 1880 (Schröder, 1880). According to Hans Sluga's influential assessment this review was infamous for its hidden polemics which was said to have been to a great extent responsible for Frege's ineffectiveness in his time: "It tore the book apart –in the politest possible way. There was a good deal of hurt vanity in Schröder's words" (Sluga, 1980, 68). A close reading of this review suggests, however, to ascribe Sluga's assessment to the logical folklore (cf. Vilkko, 1988). The review is polemical in parts, no doubt, but this style of argument may be ignored as typical for the time in which it was written. If this is done one becomes aware that the reviewer took a favorable view on Frege's book, using the advantage, however, to advertise the new subject symbolic logic as such. This made it possible to direct the interest of the readers not only on Frege's logic, but also on the algebraic direction started by Boole. So the review was some sort of comparative advertising, and it is clear that Schröder, as an algebraist, had to come to the conclusion that the algebra of logic was the better realization of common goals.

For our topic the review is of special interest because Schröder directly referred to Frege's quantification theory. He admitted some shortcomings in Boole's treatment of particular judgements which found in Boole's theory "only an inadequate, in a rigorous reading, no expression", as he said (Schröder, 1880, 91). This was harsh, but it was harsh against Boole, not against Frege. Schröder justified his assessment by the following argument (Schröder, 1880, 90–91):

> The indefinite factor v, which is used by Boole to express in the first part of the logical calculus the statement "some a are b" in the form of $va=vb$, does not serve its purpose because this equation is always identically fulfilled by the assumption $v=ab$, even in the case that no a is b. In the section on "universality" *Frege* justly gives such stipulations which allow him to express these judgements indubitably. I will not follow him slavishly in this respect, rather show that they do not justify his further deviations from *Boole*'s notation, but that the latter can also be modified and extended by analogy. Author reaches this essentially in the way that he introduced Gothic letters in the meaning of general signs and stipulated a notation to negate this universality [...].

This is not true! Frege's syntactical sign for universality is the concavity. The Gothic letter signifies the scope of the quantifier, i. e., the range of arguments which can be used in the quantified formula. This quote shows that Schröder (at least in 1880) simply did not grasp the concept of the scope of a quantifier. His easy modification of the Boolean notation consisted in plainly restricting the formula $va=vb$ by introducing the sign $\neq$ for "not equal" and stipulating that $va\neq 0$ or $ab\neq 0$ which together would also express that some a are b (*ibid.*).

3.2 Schröder's Interpretation of the Judgements

The review was published in 1880. It belongs to Schröder's early period of logical research. His mature theory is represented by the three volumes of his *Vorlesungen über die Algebra der Logik* with which he brought the algebra of logic to its provisional perfection. This seminal and influential work was published between 1890 and 1905 (Schröder, 1890). It remained, despite of its 2155 pages, unfinished.

The first volume was devoted to the calculus of domains (arbitrary elements separate from each other) and the calculus of classes. The second volume in two parts dealt with the calculus of propositions, and the final third one brought the "Algebra and Logic of Relatives", finished only in its first, algebraic, part. The algebra and logic of relatives was regarded by Schröder as the central tool for providing an algebraic scientific universal language (cf. Peckhaus, 1991; Peckhaus, 1994; Peckhaus, 1997, ch. 6).

The second volume contains among other topics considerations on "Die Syllogismen der Alten", i. e., "The Syllogisms of the Ancient" (Schröder,

1891, § 42, 221). There Schröder gave a very simple interpretation of the standard forms which can be summarized in the following square of opposition:

Figure 5

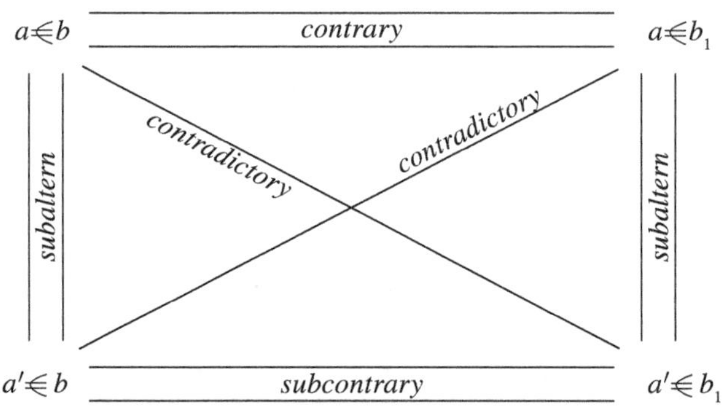

Small letters stand for universal classes, the prime indicates particularity, negation is expressed by a postponed stroke, the copula is represented by Schröder's central operator, the subsumtion, representing in the class calculus the relation "being sorted into … or being equal to …".

Earlier in the book, in a lecture dealing with the traditional division of categorical judgements in respect to quality and quantity, Schröder brings a more sophisticated discussion of the standard forms taking up his earlier criticism of Boole's treatment of particular judgements (Schröder, 1891, § 33, 90–93). He again argues that the Boolean calculus is incapable of representing particular judgements. Schröder's solution, similar to that given in 1880, comes along, however, as an expedient, for he again introduces a new operator, $\neq$ for "not equal". Schröder's modified square of opposition looks like Figure 6.

Again, like in Boole's work, we find an interpretation of the standard forms in terms of concepts. No trace of a quantification theory can be found, at least in this context.

But the same volume nevertheless disproves the assessment implied by van Heijenoort that there is no quantification theory at all in the algebra of logic. Schröder's quantification theory is, however, not connected to the theory of the syllogism.

Figure 6

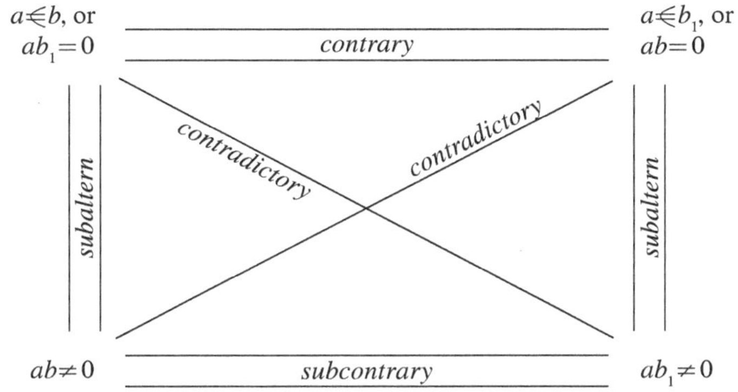

The step from the calculus of classes to the calculus of propositions is taken with the help of an alteration of the basic interpretation of the formula used. Whereas the calculus of classes was bound to a spatial interpretation especially in terms of the part–whole relation, Schröder now employed a temporal interpretation. He took this idea from Boole's *Laws of Thought* (Boole, 1854, 164–165). This may be illustrated by considering subsumtion as the basic connecting relation. In the calculus of classes $a \preccurlyeq b$ means that the class a is part of or equal to the class b. In the calculus of propositions this formula may be interpreted in the following way (Schröder, 1891, § 28, 13):

> The time during which a is true is completely contained in the time during which b is true, i. e., *whenever* [...] *a is valid, b is valid* as well. In short, we will often say: "*If a* is valid, then b is valid," "*a entails b*" [...], "from a follows b."

Related to the general calculus of domains, Schröder then introduces two new logical symbols, the "sign of products" Π, and the "sign of sums" Σ. He uses Π_x to express that propositions referring to a domain x are valid for any domain x in the basic manifold 1, and Σ_x to say that the proposition is not necessarily valid for all, but at least for a certain domain x, or for several certain domains x of our manifold 1, i. e., for at least one x (Schröder, 1891, § 29, 26–27).

For Schröder the use of Σ and Π in logic is perfectly analogous to arithmetic. Existential quantifier and universal quantifier are therefore interpreted as possibly indefinite logical addition or disjunction and logical multiplication or conjunction, respectively. This is expressed by the following definition which also shows the duality of Σ and Π (Schröder, 1891, § 30, 35).

$$\sum_{\lambda=1}^{\lambda=n} a_\lambda = a_1 + a_2 + a_3 + \ldots + a_{n-1} + a_n \mid \prod_{\lambda=1}^{\lambda=n} a_\lambda = a_1 a_2 a_3 \cdot \ldots \cdot a_{n-1} a_n .$$

Proceeding from this interpretation Schröder can be counted among the precursors of infinitary logic (cf. Moore, 1997).

4. Conclusions

It is obvious that Schröder had a real quantification theory, similar to that of Frege, but small wonder, Schröder's book was published in 1891, Frege's *Begriffsschrift* in 1879. So Schröder had 12 years to make himself familiar with modern quantification and to adopt it in his own theory. This is a simple and plausible answer, but it is false.

Schröder never claimed any priority for his quantification theory, but he did not take it from Frege. Schröder himself gives the credit for his use of Σ and Π to Charles S. Peirce and Peirce's student Oscar Howard Mitchell (Schröder, 1891, 27). About these contributions Geraldine Brady recently wrote (Brady, 2000, 6, cf. *ibid.*, 86–94) that in 1883

> one of Peirce's students, O. H. Mitchell, developed a rudimentary system for quantification, limited to a theory of quantified propositional functions with two prenex quantifiers [Mitchell, 1883]. In the same year, inspired by Mitchell, Peirce introduced quantifiers as operations on propositional functions over a specific domain and part of the semantics of first-order logic for prenex formulas over this domain [Peirce, 1883]. This direction of research culminated two years later in Peirce's system of first-order logic, which is expressively equivalent to our modern-day first-order logic with functions.

These contributions are later than Frege's, but they are independent from them, born in developing the Boolean kind of logic.

Therefore, quantification theory evidently cannot be the criterion for distinguishing the two big traditions in the history of logic!

References

Anellis, I. (1994). *Van Heijenoort – Logic and its History in the Work and Writings of Jean van Heijenoort.* Ames, Iowa: Modern Logic Publishing.

Beaney, M. (Ed.) (1997). *The Frege Reader.* Oxford: Blackwell.

Béziau, J.-Y. (1998). Review of Anellis (1994). *Modern Logic, 8*, nos. 1/2 (January 1998–April 2000), 105–117.

Boole, G. (1847). *The Mathematical Analysis of Logic. Being an Essay Towards a Calculus of Deductive Reasoning.* Cambridge: Macmillan, Barclay, and Macmillan / London: George Bell; reprinted Oxford: Basil Blackwell (1951).

Boole, G. (1848). The Calculus of Logic. *The Cambridge and Dublin Mathematical Journal*, 183–198.

Boole, G. (1854). *An Investigation of the Laws of Thought, on which are Founded the Mathematical Theories of Logic and Probabilities.* London: Walton & Maberly; reprinted New York: Dover [ca. 1958].

Brady, B. (2000). From Peirce to Skolem. A Neglected Chapter in the History of Logic. *Studies in the History and Philosophy of Mathematics, 4.* Amsterdam etc.: Elsevier.

Frege, G. (1879). *Begriffsschrift, eine der arithmetischen nachgebildete Formelsprache des reinen Denkens.* Halle a. S.: Louis Nebert. Reprint in (Frege 1977); partial translation in Beaney (Ed.) (1997), pp. 47–78.

Frege, G. (1977). *Begriffsschrift und andere Aufsätze.* Dritte Auflage. Mit E. Husserls und H. Scholz' Anmerkungen, ed. by Ignacio Angelelli. Darmstadt: Wissenschaftliche Buchgesellschaft.

Frege, G. (1983). *Nachgelassene Schriften.* Eds. Hans Hermes/Friedrich Kambartel/Friedrich Kaulbach, 2nd rev. ed. Hamburg: Felix Meiner. (= Frege, *Nachgelassene Schriften und Wissenschaftlicher Briefwechsel*, 1.)

Grattan-Guinness, I. (2000). *The Search for Mathematical Roots. 1870–1940. Logics, Set Theories and the Foundations of Mathematics from Cantor through Russell to Gödel.* Princeton/Oxford: Princeton University Press.

Heath, P. (1966). Introduction. In Augustus De Morgan, *On the Syllogism and Other Logical Writings.* Ed. by Peter Heath. London: Routledge & Kegan Paul. (= *Rare Masterpieces of Philosophy and Science*, vii–xxxi.)

Mitchell, O. H. (1883). On a New Algebra of Logic. In Charles S. Peirce (Ed.), *Studies in Logic. By Members of the Johns Hopkins University.* Boston: Little, Brown, and Company; reprinted in *Foundations in Semiotics 1*, 72–106. Amsterdam/Philadelphia: John Benjamins (1983).

Moore, G. H. (1997). The Prehistory of Infinitary Logic: 1885–1955. In Maria Luisa Dalla Chiara et al. (Eds.), *Structures and Norms in Science. Volume two of the Tenth International Congress of Logic, Methodology and Philosophy of Science, Florence, August 1995.* (= *Synthese Library 260*, 105–123.) Dordrecht: Kluwer.

Peckhaus, V. (1991). Ernst Schröder und die "pasigraphischen Systeme" von Peano und Peirce. *Modern Logic*, nos. 2/3 (Winter 1990/91), 174–205.

Peckhaus, V. (1994). Wozu Algebra der Logik? Ernst Schröders Suche nach einer universalen Theorie der Verknüpfungen. *Modern Logic*, 357–381.

Peckhaus, V. (1997). *Logik, Mathesis universalis und allgemeine Wissenschaft. Leibniz und die Wiederentdeckung der formalen Logik im 19. Jahrhundert* (= *Logica Nova*). Berlin: Akademie Verlag.

Peckhaus, V. (2005). Algèbre de la logique, théorie de la quantification et carré des oppositions. In Pierre Joray (Ed.), *La quantification dans la logique moderne*, pp. 53–71. Paris: L'Harmattan.

Peirce, C. S. (1883). Note B: the Logic of Relatives. In *Studies in Logic. By Members of the Johns Hopkins University.* Boston: Little, Brown, and Company; reprinted in Charles S. Peirce (Ed.), *Studies in Logic by Members of the Johns Hopkins University*, pp. 187–203. Amsterdam/Philadelphia: John Benjamins. Critical edition in Christian J. W. Kloesel et al. (Eds.), *Writings of Charles S. Peirce: a Chronological Edition*, vol. 4: *1879–1884*, pp. 453–466. Bloomington: Indiana University Press (1989).

Schröder, E. (1877). *Der Operationskreis des Logikkalkuls.* Leipzig: Teubner; reprinted as "special edition", Darmstadt: Wissenschaftliche Buchgesellschaft (1966).

Schröder, E. (1880). Review of Frege 1879, *Zeitschrift für Mathematik und Physik. Historisch-literarische Abtheilung* **25**, 80–94.

Schröder, E. (1890). *Vorlesungen über die Algebra der Logik (exakte Logik)*, 3 vols.. Leipzig: B. G. Teubner (1890–1905); reprinted Bronx, N. Y.: Chelsea (1966).

Schröder, E. (1891). *Vorlesungen über die Algebra der Logik (exakte Logik)*, vol. 2, pt. 1. Leipzig: B. G. Teubner.

Sluga, H. D. (1980). *Gottlob Frege.* (= *The Arguments of the Philosophers*). London/Boston/Henley: Routledge & Kegan Paul.

van Heijenoort, J. (1967a). *From Frege to Gödel. A Source Book in Mathematical Logic, 1879–1931.* Cambridge, Mass.: Harvard University Press; reprint San Jose etc.: Excel (2000).

van Heijenoort, J. (1967b). Logic as Calculus and Logic as Language, *Synthese 17*, 324–330.

Vilkko, R. (1988). The Reception of Frege's Begriffsschrift, *Historia Mathematica 25*, 412–422.

William of Sherwood, Singular Propositions and the Hexagon of Opposition

YURII KHOMSKII[1]

Abstract

In Aristotelian logic, the predominant view has always been that there are only two kinds of quantities: universal and particular. For this reason, philosophers have struggled with singular propositions (e.g., "Socrates is running"). One modern approach to this problem, as first proposed in 1955 by Tadeusz Czeżowski, is to extend the traditional Square of Opposition to a Hexagon of Opposition.

We note that the medieval author William of Sherwood developed a similar theory of singular propositions, much earlier than Czeżowski, and that it is not impossible that the Hexagon itself could have been present in Sherwood's writings.

1. Introduction

In traditional Aristotelian logic, the predominant view has always been that there are only two kinds of quantities: *universal* and *particular*. In accordance, the four proposition types which figure in the classical theory of syllogisms are the *universal affirmative* ("Every man is running"), *universal negative* ("No man is running"), *particular affirmative* ("Some man is running") and *particular negative* ("Some man is not running"). We will abbreviate these by UA, UN, PA and PN, respectively. When two propositions share both the subject term and the predicate, they are related to one

1 Institute of Logic, Language and Computation, The University of Amsterdam, e-mail: yurii@deds.nl. The author is supported by a Mosaic grant (no: 017.004.066) by the Netherlands Organisation for Scientific Research (NWO).

another in one of the following ways: *contradictory*, *contrary*, *subcontrary* or *subalternate*. This doctrine has frequently been represented by the Square of Opposition, as in Figure 1 below.

Figure 1: The Square of Opposition

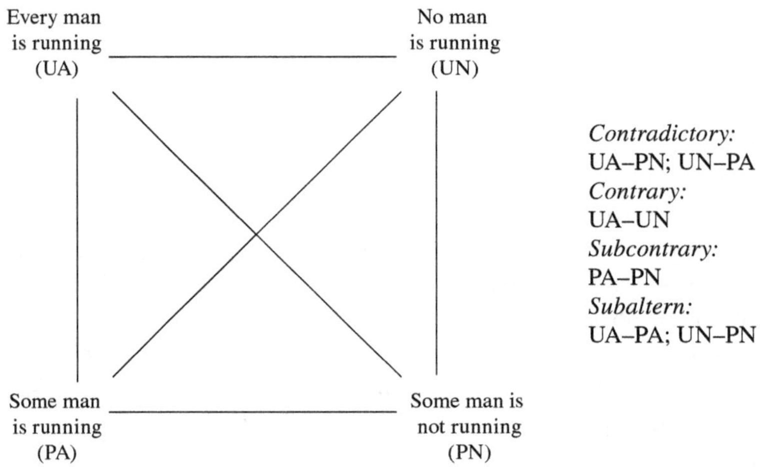

What about *singular* propositions, i.e., propositions of the kind "This man is running" or "Socrates is running"? Handling singular propositions is always difficult in a language based on universal and existential quantification, as witnessed (even in the case of first-order logic) by the intensive discussion on Russell's "The present king of France is bald" example (Russell, 1905). Nevertheless, many authors tried to show that singular proposition were either special kinds of particular, or, more frequently, special kinds of universal propositions (see the appendix for some examples). In 1955 Tadeusz Czeżowski (Czeżowski, 1955) presented an alternative view and showed how singular propositions ought to be analyzed in the Aristotelian context. This, among other things, led him to extend the Square of Opposition to a Hexagon of Opposition, where universal, particular, and singular propositions all play an independent role.

In this paper we note that, before Czeżowski, the medieval logician William of Sherwood developed a similar theory of singular propositions. In particular, the crucial aspect of Czeżowski's "discovery", namely the opposition relations which singular propositions form relative to other categorical propositions, is present in Sherwood's writing. Moreover, we argue that, perhaps, even the Hexagon of Opposition *as a diagram* could have been present in Sherwood's original writings (although it has not been preserved in the surviving manuscript). The paper ends with an appendix with some related historical background.

2. Czeżowski

Towards the end of the 19th century and beginning of the 20th century, a standard view in the theory of Aristotelian logic was that singular propositions are a special kind of universal propositions. The argument was that in a singular proposition, predication is of the *whole* of the subject, just like in a universal one, whereas in a particular proposition it is only of a part of the subject (see, e.g., Keynes, 1906, 102). This argument can be traced back to John Wallis in 1631, and we refer the reader to the appendix for a brief overview of this and subsequent historical developments.

Interestingly, it was known since Aristotle that a singular affirmative and a singular negative proposition are neither mutually contrary, as universal ones, nor subcontrary, as particular ones, but contradictory.[2] Moreover, it was typically believed that singular propositions *cannot* even have contraries:

> Taking the proposition "Socrates is wise", its contradictory is "Socrates is not wise"; and so long as we keep to the same terms, we cannot go beyond this simple denial. The proposition has, therefore, no formal contrary. (Keynes, 1906, section 82, p. 115.)

This, however, was not reason enough to abandon the traditional view that singulars are a special kind of universals.

In 1955, Tadeusz Czeżowski (Czeżowski, 1955) advocated the separation of categorical propositions into *three* classes: universal, particular, and singular. He analyzed the function of singular propositions in syllogisms and the opposition relations they form with other categorical propositions. This led him to extend the traditional Square of Opposition to a Hexagon

2 See, e.g., *De Int.*, VII, 17b.

of Opposition, as illustrated in Figure 2 below (where SA abbreviates *singular affirmative* and SN *singular negative*).

Figure 2: Czeżowski's Hexagon of Opposition

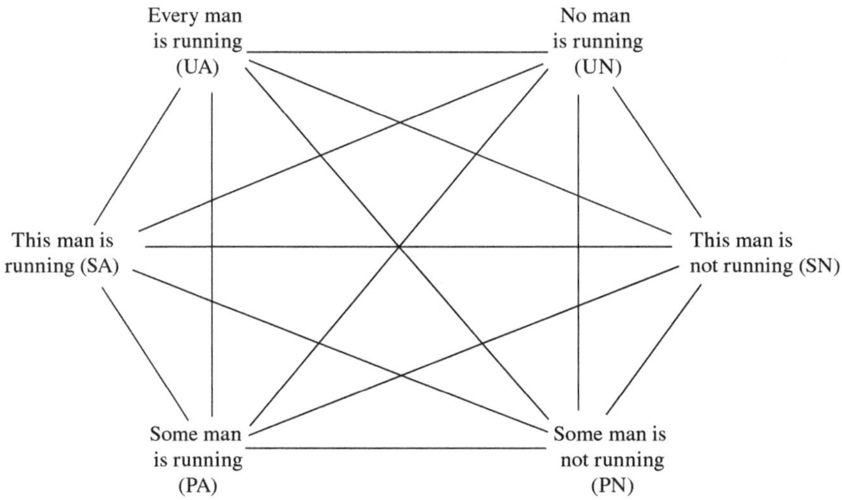

Contradictory:
UA–PN; UN–PA; SA–SN
Contrary:
UA–UN; UA–SN
Subcontrary:
PA–SN; PA–PN
Subaltern:
UA–SA; SA–PA: UA–PA;
UN–SN; SN–PN: UN–PN

The Hexagon of Opposition will be of special interest to us. In particular, we see in it two features of singular propositions worth of notice. Firstly, as we already mentioned, the singular affirmative and singular negative propositions are contradictory. More interestingly, both singular propositions have, besides their contradictories, also their contraries and their subcontraries.

Czeżowski, therefore, explicitly refutes the thesis that singulars do not have contraries,[3] and his overall aim is to correct the misconceptions about singular propositions:

3 Of course, there is one subtlety going on here, namely the issue whether we consider "this man is wise" or "Socrates is wise" to be the generic form of a singular

Classical logic textbooks all concur in the view that singular propositions ought to be regarded as universal The relation between an affirmative singular proposition and a negative singular proposition is the relation of contradiction – and not of contrariety, as between an affirmative and a negative universal proposition. Hence it has been inferred that there is no proposition that might properly be the contrary of a singular proposition. This inference is wrong. In reality, the above-mentioned difference is merely an indication that a distinction ought to be made between singular and universal propositions and that trichotomy into universal (All S is P), singular (This S is P), and particular (Some S is P) propositions should be introduced in place of the customary dichotomy according to quantity, into universal and particular propositions.

Opposition relations among the six propositions thus distinguished will be represented on a hexagon, analogously to the logical square.

Contrary to the belief quoted above, both singular propositions do have their contraries – namely, universal propositions. But at the same time they are placed in a relation of subcontrariety to particular propositions, just as the latter are among themselves. (Czeżowski, 1955, pp. 392–394)

It is obvious from this quotation that Czeżowski considered his hexagon and the new perspective on singular propositions a novel discovery. However, going back some eight-hundred years, we find a medieval logician who, remarkably, came very close to Czeżowski's theory.

proposition – and in the latter case, whether "Socrates" has a term in common with "every man" or "some man". In fact, it seems to be this very issue that lies at the core of the debate between View 1 and View 2. But here we adhere to Czeżowski's view, who consciously does away with this problem: "I make the subject of the singular proposition take the form of 'This S', in order to enable the singular proposition to enter the opposition square, i.e., relations between propositions having the same terms in the subject and predicate respectively. That, however, alters that particular property of singular propositions which had served in classical logic as an argument for placing singular propositions together with universals – namely, that in a singular proposition the subject is predicated in all its extension. If it be 'S' that is to be regarded as subject – and that is how it must be, when a singular proposition is to be opposed to universal and particular propositions – then it must be admitted that is is only one of the 'S's that is predicated on, and not the whole of their extension." (Czeżowski, 1955, p. 392). There has been at least one criticism of Czeżowski on this account (Mackie, 1958) and at least one defence against that criticism (Gumanski, 1960).

3. William of Sherwood

William of Sherwood was an English philosopher and logician, born some time between 1200 and 1210 and died between 1266–1271.[4] Not much is known of his life, and only two works can be definitely ascribed to him: *Introductiones in Logicam*, a logic textbook dealing with Aristotelian logic, and *Syncategoremata*, dealing with the theory of syncategorematic terms. The *Introductiones* have survived in just one manuscript, not written by Sherwood himself, presumably dating from the late thirteenth or early fourteenth century.[5]

There is a wide variety of topics treated in the *Introductiones*, but all we need is in Chapter 1, which is an exposition of Aristotelian syllogistic. There Sherwood discusses the four standard categorical propositions, and, following Aristotle, correctly analyzes the opposition relations – contradictory, contrary, subcontrary and subaltern – that appear between them. Then he mentions *singular propositions*, at first implying that these are really equivalent to particular propositions. Then, however, we read the following:

> [Singular statements] differ [from particular statements] in this respect: if two statements are singular and of different quality they are not subcontraries but, in accordance with the theory, contradictories – e.g., "Socrates is running", "Socrates is not running." Note, moreover, that a universal affirmative and a singular negative, as well as a universal negative and a singular affirmative, are mutually contrary (at least as far as the law goes) because they can be false at the same time and cannot be true at the same time. Suppose that Socrates is running and no one else; in that case these statements are false: "every man is running", "Socrates is not running." Again, suppose that Socrates is not running but everyone except him [is running]; then these statements are false: "no man is running", "Socrates is running." (*Int. Log.*, 2^v, 210 ff.)[6]

4 For a detailed discussion of Sherwood's life, see, e.g., the introduction in (Kretzmann, 1966, pp. 3–20).

5 Bibliothèque nationale MS. Lat. 16, 617 (formerly Sorbonne 1797).

6 "in tantum differunt, quod si utraque sit singularis et diversae qualitatis, non erunt subcontrariae, sed ratione contradictoriae, ut sunt hae: 'Socrates currit', 'Socrates non currit'. Item. Notandum, quod universalis affirmativa et singularis negativa et etiam universalis negativa et singularis affirmativa contrariantur ad minus quantum ad legem, quia possunt simul esse falsae et non simul verae. Posito, quod Socrates currat et nullus alius, hae sunt tunc falsae: 'Omnis homo currit', 'Socrates non currit'. Item. Posito, quod Socrates non currit, sed omnes alii ab ipso, tunc istae sunt falsae: 'nullus homo currit', 'Socrates currit'."

Sherwood clearly notes the two essential properties of singular propositions, present in Czeżowski's writings as well: the contradiction of the singular affirmative and negative, and the contrary relation they form with the universal ones. According to Kretzmann (Kretzmann, 1966, footnote 38 on p. 31) Sherwood is the first known medieval logician to make this particular observation. Of course, this should be seen in the wider context of medieval theories of singular propositions (see, e.g., Ashworth, 2008). Many medieval authors were concerned with the nature of singular terms, in particular with ontological, epistemological and theological issues, but also with logical ones. It is beyond the scope of this paper to discuss all of these, since we want to focus on Sherwood's writing and the relation to Czeżowski's (Czeżowski, 1955). It should however be noted that the obsevations of Sherwood quoted above were not a standard part of medieval theories of singular propositions. For example, it is not present in Peter of Spain's influential *Summulae Logicales* (*Summulae*) (cf. Kretzmann, 1966, footnote 38 on p. 31).

The quoted passage makes it clear that Sherwood's theory comes very close to that of Czeżowski. Although the subcontrary relations between singular and particular propositions are not mentioned, they can easily be inferred from the contrary and the contradictory relations which *are* explicitly mentioned. Thus, Czeżowski's Hexagon of Opposition, or at least the logical essence thereof, was already there in the writings of Sherwood.

In the next sections, we will investigate the possibility of the Hexagon being present in Sherwood's writings not just as a theory, but as an actual diagram as well.

4. Sherwood's Hexagon

Norman Kretzmann, in his annotated translation of the *Introductiones*, shows a Hexagon of Opposition to represent Sherwood's treatment of singular propositions (Kretzmann, 1966, p. 33). He does not mention any possibility of the Hexagon being due to Sherwood, but makes it clear that it is added, for the benefit of the reader, by himself.[7] We would like to take this

7 He even credits a certain Mr. Gerald W. Lilje for the suggestion (footnote 39 on p. 32). Moreover, he does not note the connection to (Czeżowski 1955).

one step further and ask the following question: could Sherwood himself have conceived of the Hexagon of Opposition *as a diagram*? Clearly, if that were even partly true, it would be an interesting discovery for the history of the Square of Opposition and the history of logical diagrams in general.

In the only surviving manuscript of the *Introductiones*, we do not find any Hexagon of Opposition but only two Squares of Opposition – one for categorical propositions and one for modal ones. Nevertheless, in the following we shall speculate that in Sherwood's original text the categorical square might have been a Hexagon or at least something like it, which could have been altered, by the scribe responsible for the surviving manuscript, into a Square. To support this speculation, we shall give corroborating evidence showing why it is indeed likely that the scribe did make changes to the diagrams in the *Introductiones*.

Let us return to the paragraph quoted above, about the opposition relations between singular propositions and other categorical ones. Just after that, we read the following lines:

> This, then, is the division of statements arising from the arrangement or relation of one statement with another – viz., some are contraries, some subcontraries, some subalterns, and some contradictories, *as in the figure below*.' (*Int. Log.*, 2ᵛ, 225 ff.)[8]

In the manuscript, the figure we see is the categorical Square of Opposition. But this is surprising, since in the preceding two paragraphs Sherwood has just elaborately talked about singular propositions. And if we are to relate those paragraphs at least somewhat with the phrase "as in the figure below", we would expect that figure to be some kind of Hexagon of Opposition.

So, since Sherwood's text suggests the need of a Hexagon, could it be that Sherwood did in fact present a Hexagon but that the scribe responsible for our manuscript changed it? For one, the scribe might have misunderstood Sherwood and his Hexagon, or have been under the mistaken impression that the Hexagon was an error on Sherwood's part, which it was his duty to correct. It might also have been that the Hexagon, as presented in the document that the scribe was copying from, really did look confusing, bloated or otherwise unconvincing, and that the scribe felt justified in changing it to the standard version.

8 My emphasis. "Est igitur haec divisio enuntiationis, quae accidit ei in ordinatione ad alterum sive secundum comparationem, scilicet quod quaedam sunt contrariae, quaedam subcontrariae, quaedam subalternae, quaedam contradictoriae, ut in subiecta patet figura."

How realistic is such a scenario? We will now give another piece of evidence which, besides being very interesting in its own right, can serve as an argument for why our claim is not as far-fetched as it may seem.

5. The Modal Square of Opposition

We move ahead a few sections in the *Introductiones*. There, we come to the point where Sherwood discusses modality. Following Aristotle, he defines the four moods: *necessary*, *possible*, *non-necessary* and *impossible*, gives the standard duality principles such as "it is necessary" being equivalent to "it is impossible that not" etc. and analyzes the opposition relations – contrary, subcontrary, contradictory and subaltern – between modal propositions. He then, just as before, refers the reader to a diagram, meant as a visual aid to undertanding and memorizing these relationships. The diagram we see in the manuscript is the standard modal Square of Opposition (Figure 3).

Figure 3: The Modal Square of Opposition

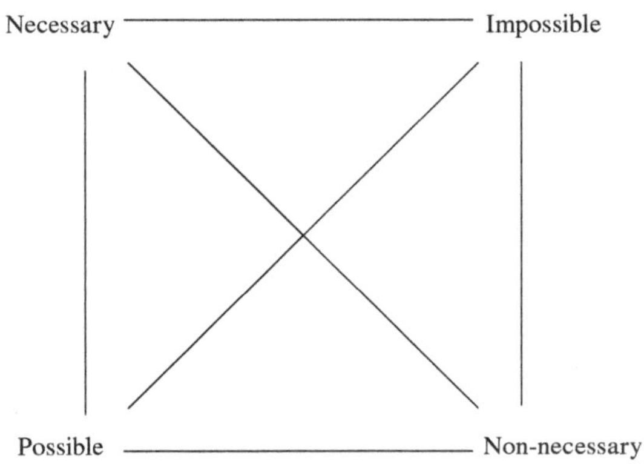

Contradictory:
Necessary – Non-necessary
Possible – Impossible
Contrary:
Necessary – Impossible
Subcontrary:
Possible – Non-necessary
Subaltern:
Necessary – Possible
Impossible – Non-necessary

However, this is what Sherwood has to say about the diagram:

> All these relations also appear in the accompanying figure. The figure could be arranged differently, however, so that the contrary series could be put in the first, or upper, line, and the subcontraries in the lower. But [the arrangement as given] coincides more closely with Aristotle's. (*Int. Log.*, 5^v, 588)[9]

Surprisingly, then, there is a clear discrepancy between Sherwood's description of the diagram and the diagram we actually find in the manuscript. The one that Sherwood describes as an *alternative* and *better* version, is actually the one we find. What, then, was Sherwood's original diagram?

Going back to Aristotle's *On Interpretation*, we indeed find that, although Aristotle did not have a Square of Opposition as such, he listed the four modalities in a square table, arranged as follows: (*De Int.*, XIII, 22a)[10]

Possible	Impossible
Non-necessary	Necessary

Presumably, then, Sherwood's original modal Square had the four modalities arranged precisely the same way, while adding all the logical relations. A reconstruction of such a square would look something like Figure 4 below, not a pretty sight indeed.

Sherwood's clever suggestion, of course, was to rearrange the modalities in order to bring the modal Square of Opposition in line with the standard, categorical Square of Opposition. So presumably, the scribe responsible for our manuscript decided to follow up on Sherwood's suggestion and actually replace the original modal Square with the alternative version. But the result, as we clearly see, is that Sherwood's original diagram has not been preserved.

9 "Et haec omnia patent in figura. Posset tamen figura aliter ordinari, ut ordines contrarii ponerentur in prima linea, quae est superior, et subcontrarii in inferiori. Sed iste magis competit modo Aristotelis."

10 The phrase "θεωρείσθω δὲ ἐκ τῆς ὑπογραφῆς λέγομεν" suggests that the tabular arrangement stems from Aristotle himself. On the other hand, in order to corroborate the argument given in this paper, it would be important to check the diagrammatic representation of the modal Square of Oppositions in manuscripts of *De Interpretatione* in Sherwood's time.

Figure 4: Reconstruction of Sherwood's Modal Square

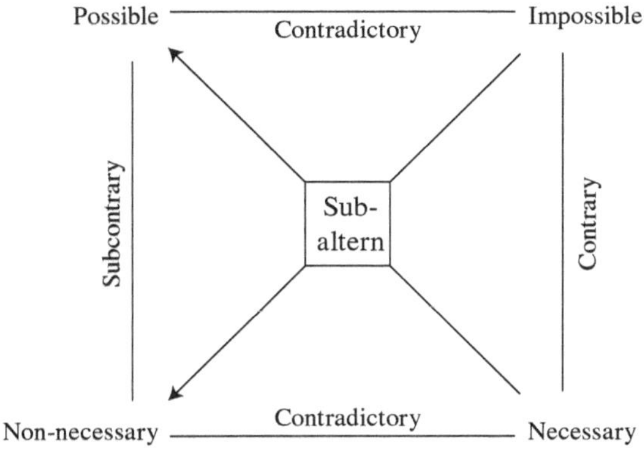

Contradictory:
Necessary – Non-necessary
Possible – Impossible
Contrary:
Necessary – Impossible
Subcontrary:
Possible – Non-necessary
Subaltern:
Necessary – Possible
Impossible – Non-necessary

Now we have a rather clear indication that the diagrams in our manuscript of the *Introductiones* are not necessarily those of Sherwood himself, and thus we can also look at the Hexagon of Opposition in a different light. And since we are talking about the same document and the same kind of diagram, it is really not impossible that the same thing that happened to the modal Square might have happened to the Hexagon.

6. Conclusions

It is not my convinced opinion that Sherwood was indeed the author of the Hexagon. There are still many arguments against it, the strongest one perhaps being Sherwood's own formulation a few paragraphs back, which creates the impression that singular propositions are not really that important. Consequently we could see the paragraph *"Est igitur haec divisio enuntiationis"* not as relating to what immediately precedes it, but as summarizing the whole discussion of categorical propositions in general, in which singular propositions play no important role.

Be that as it may, the idea of the Hexagon occurring in Sherwood's writings is very intriguing, and worthwhile to entertain as a possibility. And our arguments do show why that is not an unrealistic possibility. What would, of course, shed more light on this issue, is if a new manuscript of the *Introductiones* were uncovered.

On the other hand, we definitely know that Sherwood's account does contain the *logical essence* of the Hexagon, even if not the *diagram*, in the description of the opposition relations between universal, particular, and singular propositions.

7. Appendix

In order to put both Czeżowski's and Sherwood's observations in a historical perspective, it is interesting to look at some other accounts of singular propositions, dating from the 17th century onwards.[11] As mentioned, it was customary to rank singular propositions either with particulars or with universals. To be precise, we find two predominant views:

- **View 1:** In a singular proposition, predication is of one individual, which is even less than in a particular. Therefore a singular proposition is certainly not a universal, but a very special kind of particular proposition.
- **View 2:** In a singular proposition, predication is of the whole of the subject, just like in a universal. Therefore a singular proposition is (a special kind of) universal.

The second view was first made explicit by John Wallis in 1631:[12] "A singular proposition, in a syllogistic disposition, always has universal force." (*Inst.*, Appendix).[13] Apparently, this view was quite popular and was, for example, taken up by Arnauld and Nicole into what became known as the *Port Royal* logic (1662):

11 This account of the history of singular propositions is largely influenced by the research conducted by Jaap Maat (private communication).

12 There have been previous attempts to implicitly reduce singular propositions to universal ones, for example in Lambert of Auxerre's *Logica*: "'Sortes currit, ergo homo currit', empthimema est quod sic potest reduci ad sillogismum: 'omne quod est Sortes currit, Sortes est homo, ergo homo currit.'" (*Logica*, De Locis, p. 140)

13 "Propositio Singularis, in dispositione Syllogistica, semper habet vim Universalis."

> Although a singular proposition differs from a universal in not having a common subject, it should nevertheless be classified with them rather than with particulars; because its subject, precisely by being singular, is necessarily taken through its entire extension, which is the essence of a universal proposition and which distinguishes it from the particular. (*Port Royal*, Part II, Ch. 3.)[14]

On the other hand, View 1 seems to have been held by a number of authors, as illustrated by the following informative account by Euler:

> Certain authors insist, that a singular proposition must be ranked in the class of particulars; it being considered, that a particular proposition speaks only of some beings comprehended in the notion, whereas a universal speaks of all. Now, say these authors, when we speak of only a singular being, this is still less than when we speak of some: and, consequently, a singular proposition must be considered as very particular.
>
> However well founded this reasoning may appear, it cannot be admitted. The essence of a particular proposition consists in this, that it does not speak of all the beings, comprehended in the notion of the subject, whereas an universal proposition speaks of all, without exception. Thus, when it is said: *Some citizens of Berlin are rich*, the subject of this proposition is the notion of *all the citizens* of Berlin; but this subject is not taken in all its extent, its signification is expressly restricted to *some*: and, by this, particular propositions are essentially distinguished from universal
>
> It is clearly evident, from this remark, *that a singular proposition must be considered as universal*; as, in speaking of an individual, say Virgil, it, in no respect, restricts the notion of the subject which is Virgil himself, but rather admits it in all its extent. (*Lettres*, CVII)[15]

14 "Mais quoique cette proposition singuliere soit différente de l'universelle en ce que son sujet n'est pas commun, elle s'y doit néanmoins plutôt rapporter qu'à la particuliere; parce que son sujet, par cela même qu'il est singulier, est nécessairement pris dans toute son étendue, ce qui fait l'essence d'une proposition universelle, et qui la distingue de la particuliere."

15 "Quelques auteurs ont prétendu qu'une proposition singulière doit être rangée dans la classe des particulières; attendu qu'une proposition particulière ne parle que de quelques êtres compris dans la notion, pendant qu'une proposition universelle parle de tout. Or, disent ces auteurs, quand on ne parle que d'un être singulier, c'est encore moins que si l'on parle de quelques-uns: et par conséquent une proposition singulière doit être regardée comme très-particulière. Quelque fondée que puisse paraître cette raison, elle ne saurait être admise. L'essentiel d'une proposition particulière consiste en ce qu'elle ne parle pas de tous les êtres compris dans la notion du sujet; pendant qu'une proposition universelle parle de tous sans exception. Ainsi, quand on dit: *Quelques habitants de Berlin sont riches,* le sujet de cette proposition est la notion de *touse les habitants* de Berlin; mais on ne prend pas ce sujet dans toute son étendue, sa signification est expressément restreinte à *quelques-uns:* et c'est par là que les propositions particulières sont essentiellement distinguée des universelles Il est très-évident, après cette remarque, *qu'une proposition singulière doit être regardée comme universelle;* puisqu'en parlant

Leibniz was another prominent figure who wrote at least twice on singular propositions. In this account he seems to adhere strictly to View 2, giving essentially the same argument as Euler and his predecessors:

> It should be noted that (as far as the form is concerned) singular sentences are put with the universals. For, although it is true that there was only one Apostle Peter, one can nevertheless say that whoever has been the Apostle Peter has denied his master. (*Nouv. Ess.*, IV. XVII. 8.)[16]

But later Leibniz had to changed his mind, and wrote the following contrasting passage:

> How is it that opposition is valid in the case of singular propositions – e.g. "The Apostle Peter is a soldier" and "The Apostle Peter is not a soldier" – since elsewhere a universal affirmative and a particular negative are opposed? Should we say that a singular proposition is equivalent to a particular and to a universal proposition? Yes, we should. (*Diff.*)[17]

The debate between View 1 and View 2 continued into the late nineteenth and early twentieth century, with little innovation. For example, in the following exchange we see R. F. Clarke (1889) saying:

> *London is a large city* must necessarily be a more restricted proposition than *Some cities are large cities*; and if the latter should be reckoned under particulars, much more the former. (Clarke, 1889, p. 274)

To this, John N. Keynes (1906) replies:

> This view fails to recognise that what is really characteristic of the particular proposition is not its *restricted* character – since the particular is not inconsistent with the universal – but its *indefinite* character. (Keynes, 1906, footnote 1 on p. 102)
>
> Singular propositions may be regarded as forming a sub-class of universals, since in every singular proposition the affirmation or denial is of the *whole* of the subject. (Keynes, 1906, p. 102)

d'un individu, comme de Virgile, elle ne restreint en aucune manieère la notion du sujet, qui est Virgile même, mais elle l'admet plutôt dans toute son étendue."

16 "Il est bon pourtant de remarquer qu'on comprend (quant à la forme) les propositions singulières sous les universelles. Car quoiqu'il soit vrai qu'il n'y a qu'un seul saint Pierre Apôtre, on peut pourtant dire que quiconque a été saint Pierre l'Apôtre a renié son maître."

17 "Qui fit quod in singularibus procedit oppositio: Petrus Apostolus est miles, et Petrus Apostolus non est miles, cum tamen opponatur alias universalis affirmativa et particularis negativa? An dicemus, singulare aequivalere particulari et universali? Recte."

8. Acknowledgements

The author would like to thank Jaap Maat for his contribution regarding the history of singular propositions, Benedikt Löwe, Sara Uckelman and Christoph Kann for comments on earlier versions of this paper and valuable input, and all the participants of the International Congress on the Square of Opposition, June 2007, Montreux, Switzerland for their fruitful discussions.

Primary Sources

De Int. Aristotle, *De Interpretatione*, in: Cook and Tredennick 1938.

Port Royal Antoine Arnauld, Pierre Nicole, *La Logique ou l'Art de Penser* (1662), *in:* Marin 1970.

Logica Lambert of Auxerre, *Logica, in:* Alessio, 1971.

Lettres Leonhard Euler, *Lettres à Une Princesse d'Allemagne* (1761), *in:* Dubois and Drapier, 1839, *trans.* Brewster and Griscom, 1854.

Summulae Petrus Hispanus, *Summulae Logicales, in:* de Rijk, 1972.

Nouv. Ess. Gottfried Wilhelm von Leibniz, *Nouveaux Essais sur l'entendement humain* (1704), *in:* Brunschwig, 1966, *trans.* Remnant and Bennett, 1996.

Diff. Gottfried Wilhelm von Leibniz, *Difficultates Quaedam Logicae, in:* Gerhardt 1965, pp. 211–217, *trans.* Parkinson, 1966, pp. 115–121.

Int. Log. William of Sherwood, *Introductiones in Logicam, in:* Brands and Kann, 1995, *trans. with annotation in:* Kretzmann, 1966.

Inst. John Wallis, *Institutio Logicae*, Oxford, 1687.

References

Alessio, Franco (Ed.). (1971). *Logica (Summa Lamberti)*. Florence.
Ashworth, E. Jennifer (2008). Medieval Theories of Singular Terms. In Edward N. Zalta (Ed.), *The Stanford Encyclopedia of Philosophy* (Fall 2008 Edition); http://plato.stanford.edu/archives/fall2008/entries/singular-terms-medieval/.
Brands, Hartmut, & Christoph Kann (Eds.). (Trans. 1995). *William of Sherwood: Introductiones in logicam*. Hamburg: Felix Meiner Verlag.
Brewster, D., & J. Griscom (Eds.). (Trans. 1854). *Letters of Euler on different subjects in natural philosophy, addressed to a German princess*. New York: Harper.
Brunschwig, Jacques (Ed.). (1966). *Nouveaux essais sur l'ententement humain*. Paris: Garnier-flammarion.
Clarke, Richard Frederick (1889). *Logic*. London: Longmans & Green.
Cooke, Harold P., & Hugh Tredennick (Eds.). (Trans. 1938). *The Categories, On Interpretation and Prior Analyics*. UK: The Loeb Classical Library.
Czeżowski, Tadeusz (1955). On Certain Peculiarities of Singular Propositions. *Mind*, *64*, 392–295.
Dubois, Drapier et. al. (Eds.). (1839). *Œuvres complètes en français de L. Euler: Lettres a une princesse d'Allemagne, Tome I*. Bruxelles.
Gerhardt, C. I. (Ed.) (1965). *Die Philosophischen Schriften von Gottfried Wilhelm Leibniz*. Berlin.
Gumanski, Leon (1960). Singular Propositions, and "This" as a Quantifier, *Mind 69*, 534–543.
Keynes, John Neville (1906). *Studies and Exercises in Formal Logic*. London.
Kretzmann, Norman (Ed.). (Trans. 1966). *William of Sherwood's Introductions to Logic*. Translation with introduction and notes. Minneapolis: University of Minnesota Press.
Mackie, J. L. (1958). "This" as a Singular Quantifier, *Mind 67*, 522–526.
Marin, L., (Ed.). (Trans. 1970). *Antoine Arnauld, Pierre Nicole, La Logique ou l'Art de Penser*. Paris.
Parkinson, G. (Ed.). (Trans. 1966). *Leibniz: Logical Papers*. Oxford: Clarendon Press.

Remnant, P., & J. Bennett, J. (Eds.). (Trans. 1996). *G. W. Leibniz, New Essays on Human Understanding*. Cambridge: Cambridge University Press.

de Rijk, Lambertus Marie de Rijk (Ed.). (1972). *Tractatus, called afterwards Summule logicales*. Assen: van Gorcum.

Russell, Bertrand (1905). On Denoting, *Mind 14*, 479–493.

Existential Commitment and the Cartesian Semantics of the Port-Royal Logic

JOHN N. MARTIN[1]

Abstract

The paper investigates the truth-conditions put forth in the Port-Royal Logic for categorical propositions in terms of "extension". The new Cartesian semantics was motivated by the rejection of medieval logic's causal theory of reference because of its commitment to the transmission of formal properties from material objects to the mind. Arnauld and Nicole formulate a new referential theory of signification that retains large parts of the medieval semantics of mental language but adopts a dualist metaphysics committed to causal occasionalism. The new account is founded on the use of objective being, a concept developed in medieval philosophy but rejected as problematic by major medieval logicians committed to Aristotelian semantics. Considered as a term in mental language, the objective being of a subject – the idea's comprehension – contains modes that describe the subject and these determine the possible objects outside the mind that the idea signifies. Signification, a relation between mental terms and external things, in turn is used to define extension, which in the Cartesian context is a relation among ideas: the extension of a term, which is understood to be a mental "species", consists of its inferiors, namely those species-ideas that signify entities that instantiate the modes in the higher species' comprehension. Truth-conditions are then defined in terms of extension. Objective being of a subject, as the object of consciousness, also correlates with the propositional knowledge that predicates its content modes of a subject. This correlation is used to unpack the medieval notion of false idea – one with a descriptive content false of every actual being. It is explained how the

1 Department of Philosophy, Charles Phelps Taft Reaserch Fellow, Taft Humanities Center, University of Cincinnati. E-mail: john.martin@uc.edu. This reaserch was made possible in part through the support of the Charles Phelps Taft Humanities Center, University of Cincinnati.

truth-conditions of the categorical propositions, which are stated in terms of extension, and the analysis of false idea entail (contra the interpretation of Jean-Claude Pariente) that the terms of a true affirmative categorical carry existential import.

Keywords: Arnauld, Nicole, Descartes, Pariente, Port-Royal Logic, Categorial Logic, Syllogistic, Existential Import, Reference, Signification, Comprehension, Extension, False Idea, Objective Being

In this paper I will explore the semantic theory of the Port-Royal Logic. My main purpose is conceptual. I will endeavor to explain how the theory manages both to adhere to a correspondence theory of truth and at the same time to define truth solely in terms of ideas denying any causal interaction between concepts and the material world. Central to the explanation are the concepts of a false idea and objective being. As part of the analysis, I will argue, contra to the interpretation of Jean-Claude Pariente[2], that as part of its commitment to correspondence the Cartesian logic remains committed to the claim that affirmative categorical propositions carry existential import and thus conform to the immediate inferences of the traditional Square of Opposition. To facilitate comparison to modern semantics, I will make use as necessary of concepts from metalogic. Since at the same time the paper is interpretive, I will also endeavor to ground its novel or controversial readings in the text itself. To clarify the concepts at issue – especially the novelties imposed by the switch from Aristotelian to Cartesian ontology – it will be useful to compare 17th semantics to the preceding medieval tradition.[3] Since the paper is not about medieval logic, however,

2 See Pariente 1985, pp. 232–244.

3 Buroker 1993 provides a useful summary account of the Port-Royal semantics in modern terms. Also relevant are Dominicy 1984 and Sylvain Auroux 1993, which repeat what I shall be arguing is Pariente's interpretation of affirmatives as lacking existential import. The specific goal of this paper is to explain in modern terms the radical shift in the analysis of reference (signification) and truth imposed by Cartesian dualism, and how this new theory nevertheless incorporates existential import even though truth is defined as a "relation among ideas." See also my paper Martin 2011, to appear in *History and Philosophy of Logic.*

these remarks will draw on the established interpretations of others, with no claim to originality.[4]

To situate 17th century logic, it will be helpful to begin with some remarks on medieval logic. The mature semantics of the 14th century incorporated elements of both a causal and descriptive theory of reference.[5] Logic drew upon a philosophical psychology of concept formation that had been developing since the time of Aristotle. Though opinions differed, Aquinas' account is representative. On his view accidental sensible qualities that are instantiated in an individual outside the mind are causally transferred in stages from the individual to the soul, first to the sensory medium, like air, then to sense organs, like the eyes or ears, and ultimately to the body's central organ of sensation. Though in each stage of the causal transfer an accident *P*, like redness, is instantiated in a substance *S*, like the air or the eye, it is present only in a diminished or "intentional way" in the sense that it is not true of *S* that it is *P*. Though redness is in the air and the eye, they do not become red. Once instantiated in the central sense organ, the agent intellect may then cause the individual originating the perceptual chain to be "experienced" by the agent in all its sensory detail. Aquinas called the individual's sensible accidents experienced as true by the agent a *sensible species*, and the accidents instantiated in the perceiver a *phantasm*. Due to the nature of the causal transfer, the accident in the object of sensation is the same as that transferred to the agent's sense organs though instantiated

4 Because this account is a broad summary of the views of Arnauld and Nicole, I will cite the textual basis for only those interpretations that might be considered controversial or novel. *La Logique ou l'Art de Penser* will be abbreviated as *LAP*, and *Des vraies et des fausses idée* as *VFI*. The standard edition of Arnauld's works, which includes *LAP* and *VFI*, is Elmar Kremer and Denis Moreau, ed., *Oeuvres philosophiques d'Arnauld*, 2003, abbreviated *KM*, and that of Descartes is René Descartes, *Œuvres De Descartes*, ed. C. Adam et P. Tannery, 1897–1909, abbreviated *AT*, both cited by volume and page The standard English translation of *LAP* is Jill Vance Buroker, *Antoine Arnauld and Pierre Nicole. Logic or the Art of Thinking* 1966, abbreviated *B*, and that of *VFI* is Antoine Arnauld, *On true and false ideas*, trans. Stephen Gaukroger 1990, abbreviated *G*.

5 Aquinas' theory of sensation and abstraction is found among other places in *Summa theologica*, p.I.1, q. 79,80,84, and 85. An excellent summary of the relevant medieval psychology of sensation and abstraction in Aquinas and others, and some discussison of its later incorporation into semantic theory see Pasnau 1997. On the implication of the rejection of the the causal account of signification to Descartes' adaptation of objective being as as an alterntive link of concepts to the world see Normore 1986 for an account of medieval semantics relevant to Descartes.

naturally in the object and intentionally in the sense organs. Because it is the same, Aquinas was able to maintain with Aristotle that the accident was the same "form," and because the object and the agent shared the same form, they too.

Once the agent senses the object by instantiating a phantasm, abstraction is possible. This is the action by which the intellect operating on the phantasm as a necessary condition "prescinds" or abstracts some of the individual's form. The abstracted form though true naturally of the object of sensation in instantiated intentionally in the agent, in his "intellect." This form is abstract – it is also called general, common and confused – because it is no longer unique to the sensed individual. Though it is true of the object, it is also naturally true of other actual or possible individuals. Moreover, though the phantasm consists of only sensible accidents of the object sensed, its abstracted form may be of some non-sensible, even essential, form of the sensed object. How it was possible to abstract non-sensible universal form from individual sensible accidents was not explained very well.) Because the abstracted form was actively instantiated in the intellect and done so intentionally, it was regarded as a mental act and was called a concept. Because the form was "understood" by the intellect to be true of the sensed individual though not uniquely, Aquinas called it an *intelligible species*. Because, as in sensation, the form instantiated in the intellect intentionally is true of the sensed object naturally, that form is the same in each, and in that sense the object and the agent's intellect are also "the same." Because a concept was a form that was in general true of more than one individual, Aquinas regarded a concept as a mind-dependent universal.

Two features attributed to a concept in this influential theory are important to the role of concepts in subsequent semantics. These two features duplicate each other in establishing the same concept-world relation, and medieval logicians, some stressing one feature rather than the other, would identify this relation with signification, the medieval reference relation. The first feature is that a concept is that on Aristotle and Aquinas' account a concept is causally linked to objects outside the mind. This linkage may be formulated as rule or natural law: a concept (a mental mode) *M* is causally linked to an object *S* outside the mind if, and only if, *M* could have been abstracted from *S*. The second feature is that a concept is descriptive of the objects it is true of in a way that may also be formulated as a rule or law: *M* describes *S* if, and only if, *M* is a mode naturally true of *S*. According to this rule, the concept grounds the fact that the soul to be the "same" as the sensed because it is true of both though in different ways, one naturally and the other intentionally.[6]

In the 14th century concepts understood in this way were incorporated into a logically sophisticated theory of mental language. Spoken terms, propositions composed of terms, and arguments composed of propositions were understood to be conventionally associated to modes of the intellect – respectively to terms, propositions and arguments understood as mental acts. This conventional pairing was understood to link spoken terms in a many-one relation to mental terms understood as concepts. Mental terms were understood to be syntactic "parts" (subjects or predicates) of mental propositions, and propositions in turn to be parts of mental arguments, all composed in a manner that mirrored the syntax of spoken sentences and arguments. The semantics of mental language was then explained by the psychology of sensation and abstraction described earlier and illustrated by Aquinas' version.

The central semantic concept in these theories is signification, the medieval reference relation. Characterizations varied, but all stressed signification's explanation in terms of a concept's causal-abstractive link to objects outside the mind. They could say that a concept signifies all those possible objects from which it could have been abstracted. This rule's may be formulated in terms of species by those that believed in species and in terms of similarity by those who did not: a concept signifies all possibilia "in the same species as" or "similar to" that individual that caused the sensory phantasm from which it was abstracted. It is precisely this multiple signification that allowed Aquinas to understand a concept as a universal. Though it came to be a matter of dispute whether the distinction between intentional and natural instantiation made any sense, some continued to agree to accept the distinction and to with Aquinas that a concept was "the same" in form both in the intellect and in objects. They could then use this descriptive link to explain signification. They could say a concept that consists of the mode *M* instantiated intentionally in the intellect signifies all possibilia that instantiate *M* naturally.

All agreed that a concept was an accident of the agent, more specifically of his soul or intellect. As such it was counted as a form and was said to

6 See Aquinas, *Summa theologica*, p. I.1, q. 85, a. 2. On Ockham see the discussion of intuitive cognition, which presupposed direct realism, in Karger 1999.

have formal being, *esse formale.* All also agreed in the usage that because a concept signifies objects it was an exemplar or representative of them.

This elaborate theory, however, was unacceptable to Descartes. Both the Aristotelian psychology of concept formation and the semantics based on it, is inconsistent with Cartesian metaphysics. Descartes seems to have denied, as Arnauld and Nicole certainly did, the possibility of Aristotelian property transfer from material to spiritual substance.[7] According to the Cartesians there can be no causal theory of reference using Aristotelian property transfer because no material substance can transmit a form to the soul so as to form a concept. The challenge facing Arnauld and Nicole, then, was to reconceive the semantic link between concepts and things in such a way that reference and description do not depend on Aristotelian causation. What is interesting is that they did so while retaining a large part of the medieval theory of mental language and its semantic apparatus.

Let us turn then to Cartesian semantics. It may be reconstructed as built upon a series of definitions that culminate in the truth-conditions for categorical propositions. The elements are contrived so that the resulting truth-conditions fit a formulation with roots in Aristotle: a universal affirmative is true if the subject and predicate "stand for the same". The "gentlemen of Port-Royal" impart a new reading to this formula by analyzing "sameness" in terms of *extension,* a technical term of their own coinage. It is this new notion that makes their "Cartesian semantics" distinctive because, unlike extension in modern logic, the Cartesian version is a relation among ideas.

The definition of extension is constructed in steps. Adopting a view of medieval nominalists, Arnauld and Nicole understand genera and species to be abstract ideas. Accordingly, the inferiority-superiority relation among species is understood as a relation among ideas. Extension is then defined in terms of this relation: the extension of a term in mental language is defined as consisting of all its inferior species.

7 Suárez, Arnauld and Nicole clearly deny the possibility of Aristotelian transmission of modes from matter to the soul, and espouse a version of occasionalism to the effect that on the occasion of the instantiation of a material mode in the sense organs, God causes there to be an idea in the mind in which that mode occurs as an intentional "content". See *LAP* I,i, *KM* V, 132–33, *B* 29–30; I,ix, *KM* V, 157–78, *B* 49–50; I,xii, *KM* V,168–170, *B* 58–60; *VFI* Chapt. 6, *KM* I,204, *G* 71–71; Chapt. 27, *KM* I,349–50, *G* 208. On Suárez see *De anima* iv 2:13,17,18, and 4,1; *Opera* III,719,721,731; and the discussion in Cronin 1966. Descartes' official view seems to be similar. See *AT* 8b 358:20–359:12, and the discussion in Garber 1993.

This use by Arnauld and Nicole of the French *étendue* and Latin *extensio* to refer to a collection of subordinate species is new to logic. In Latin philosophy *extensio* had always occurred as a technical term in physics used to mean the property that makes matter continuous,[8] and it continued to be used in this sense by Descartes and his followers in their physics and ontology where it describes the essential mode of matter.

Though the explanatory role and technical vocabulary of the second sense is new, the idea itself is not. A term's Cartesian extension in the terms of Aristotle's *Categories* is nothing other than the species that the term is truly "said of." Michael Thompson has recently argued that Aristotle himself understood the quantified subject term in a categorical proposition as ranging over the set of all species subordinate to the term. Thompson observes that when Aristotle asserts the particular affirmative *some animals are viviparous*, he does not give individuals like Helen and Penelope as instances. Rather he cites the species man, horse, and camel. Thompson argues that in its canonical form an I-proposition should be understood as *for some terrestrial life form S, the S is viviparous*. Likewise, he argues that when Aristotle asserts *some animals shed their front teeth, but there is no instance of an animal that loses its molars*, he will not concede its falsity when faced with an actual denture wearer. He does not do so because it is a natural *species* he is "quantifying over," namely the species animal, not individual humans who wear dentures.[9] In a Cartesian context in which a species is understood as an idea, this quantificational range is nothing other than the term's Cartesian extension.

Though the *Logic* appears to be the first work in which the abstract noun *extensio* is used in its semantic sense, there are earlier uses of its cognates to refer to the species inferior to a genus, though these occur not in the statements of propositional truth-conditions, but in commentaries on Porphyry's use in the *Isagoge* of "more" when he says of a genus that it is "more of a collection" than its species.[10]

8 See, for example, Aristotle, *Physics*, V,3–4.

9 See Thompson 1995.

10 Descendentibus igitur ad specialissima necesse est diuidentem per multitudinem ire, ascendentibus uero ad generalissima necesse est colligere multitudinem (collectiuum enim multorum in unam naturam species est, et magis id quod genus est, particularia uero et singularia semper in multitudinem e contrario diuidunt quod unum est; participatione enim speciei plures homines unus, particularibus autem unus et communis

Duns Scotus, for example, uses the verb *extendere* to explain what Porphyry is saying about the relation of a genus to its species. He writes:

> ... I say that [the reason why] a genus is [said to be] a greater [*magis*] universal is not because "greater" [here] speaks about the intension of its adjoined form, but rather because in a certain way it is "more" [*maior*] a universal because it is extended [*extenditur*] to more [*plura*], as a fourfold division is more numerous than a binary. Also as one most specialized species is not called more of a species than another, it is permitted that it may have more contained under it.[11]

Similarly Cajetan uses the adverb *extensive,* which he *contrasts* with *intensive,* to remark that *magis* is being used in this passage "extensively" rather that "intensively."

> To this it is briefly said that being "more [*magis*] a collective of many" can be understood in two ways. In one way intensively, and in this way a species is more of a collective because, being unified, it "more" forms a unit, as the quoted definition [i.e. of *human*] shows. The other way is extensively; and in this way the genus is "more" of a collective, because, as "a many", more [inferiors] fall under its unification than fall under the breadth of the species. Whence species and genus stand to one another as two generals one of whom has an army that is small but of a single mind, the other a large army but divided into factions. Porphyry thus was speaking here about an extensive collection, and therefore said that a genus is "more" of a collective.[12]

Toletus repeats the same commentary again describing the use of *magis* by means of the adverb *extensive.*[13]

plures; diuisiuum enim est semper quod singulare est, collectiuum autem et adunatiuum quod commune est). Isagoge [02] II. *De speciem* 12.

11 Ad aliud dico quod genus non est magis uniuersale, quia "magis" dicit intensionem formae eius cui adiungitur , sed quodammodo maius uniuersale, quia extenditur ad plura, sicut quaternarius est maior numerus binario, non magis. Sicut etiam una species specialissima non dicitur magis species quam alia, licet habeat plura contenta sub se. III, 7–8.23. *Quaestiones in librum Porphyrii Isogoge, B.* 1999.

12 Ad hoc breviter dicitur, quod esse magis collectivum multorum potest intelligi dupliciter. Uno modo intensive; et sic species est magis collectiva, quia magis unit adunata, ut ratio adducta probat. Alio modo extensive; et sic genus est magis collectivum, quia multo plura sub sua adunatione cadunt, quam sub speciei ambitu. Unde species et genus se habent sicut duo duces, quorum alter habet exercitum parvum, sed valde unanimem, alter exercitum magum, sed diversam factionum. Porhyrius autem loquebatur hic de extensiva collectione, et ideo dixit genus est magis collectivum. Caietanus, p. 56., *Commentaria in Porphyrii Isagogen, ad Praedicamenta Aristotelis*1936.

13 Hic notandum est dupliciter aliquid posse dici magis collectiuum; ut notat Cajetanus: priori modo id quod est magis unum, quod dicitur magi collectiuum intensive; altero modo id, quod plura comprehendit, & sic dicitur magis collectiuum extensive: Iuxta hoc intellige minus universale esse magis collectiuum intensive; quia magis sunt unum

The theoretical role of this new concept of extension is to serve as the key idea in the statement of truth-conditions for affirmative propositions. But before we can see how this is done we must first delve more deeply into how mental terms related to the world outside the mind.

Like most logicians of their time, Arnauld and Nicole were realists.[14] On their view, the world consists of substances (material or spiritual) and their modes. Material modes, which all derive from the basic mode of corporeal extension (in the sense found in physics), are quasi-mathematical, like motion and figure. They determine a plenum of extended substance moving in vortices – a natural world rather unlike that of Aristotle's animals and plants classified by the Tree of Porphyry. Among the modes of spiritual substances are ideas, and in their version of the medieval semantics, these serve as terms in mental language. Because they are "forms" of the soul, terms as mental acts possess formal being. Because the Cartesian denied a causal link between the mental form and the material world, however, they could not appeal to the more usual causal account of signification to explain reference to objects outside the mind.

Instead they made use of another device employed in medieval psychology to explain how concepts represent the world. According to this theory in addition to its being as a form of the mind, a concept, as an act of perception or understanding, is simultaneously an intentional object, or in the technical usage of the day it possesses objective being (*esse objectivum*).

Descartes also makes use of objective being, most famously in the ontological argument of *Meditations* III.[15] Arnauld's view, as detailed most completely in his essay *On True and False Ideas,* is that in both perception and

quae in minus universali conveniunt quam que solum in magis universali: at magis universale est magis collectiuum extensive quia sub se plura continet; & sic loquitur Porphyrius. Caput II, p. 53. Toletus, *Cur de Specie post Genus & non de Differentia egerit?* 1985.

14 The authors declare the existence of modes outside the mind at the outset of the *Logic* (*LAP* I.i, *VFI* Chapt 2, *KM* I,204, *G* 71–71), and dismiss the realism-nominalism controversy as one of various scholastic debates that are difficult and of little use, *LAP, Discours* 1, *KM* V, 112–113; *B* 9–10. A commitment to the existence of forms outside the mind was, for example, standard among the Conimbricences. See I.87 in *Commentarii collegii Conimbricensis* 1976 and IV.4.C in Fonseca, *Institutionum dialecticarum libri octo*, 1964.

15 See also *Preface to the Meditations* § 4, and *Responses* 2a, *AT* 7, 161:7–9, 162:8–9, 166:14–16, 166:23–25, and 2ae, *AT* 7 161:10–13, 166:22–25.

thought what one understands (what is "present to" the mind[16]) is the idea as objective being. In the *Logic* the authors do not use the term *esse objectivum* as such, but rather make the same distinction using the somewhat awkward phrase "object as represented by an idea":[17] The more technical *être objectivement*[18] and *réalité objective*[19] occur in *On True and False Ideas*. Ontologically, objective being has no reality apart from the idea itself.

Medieval accounts of objective being varied widely and were developed over a long period of time.[20] One of the earliest clear cases is the view of Duns Scotus, who uses it to explain the object of God's understanding when he knows a creature prior to its creation. What he knows is a concept as an objective being. Likewise Ockham in an early theory, which he later rejects, says that it is a concept as an objective being that is the object of knowledge when we know an abstract idea that is not realized by an object outside the mind. Peter Aureol similarly claims that it is an objective being that is the object of perception when we see an illusion that does not accurately represent what is outside the mind. Some like Suarez identified a possible being with an object of understanding in God's mind understood as a concept in objective being.

What is important in all these accounts for Cartesian semantics is the fact that as an objective being a concept comes with a descriptive content. In medieval accounts it is fair to say that a concept as objective being is understood as an exemplar because it literally possess modes. As such, it can be said to represent something outside the mind because it is literally similar to it by instantiating the very same mode or modes. Accordingly, concepts as objective beings were allotted the role of "object of understanding" and "object of perception" in representational accounts that reject direct perception.[21] By and large medieval philosophers, however, remained committed to Aristotelian psychology and versions of direct perception, and appeals to objective being in representational accounts were the exception. The distinction between the formal and objective being of concepts, however, was

16 *VFI* Chapt 6, *KM* I,204, *G* 71–71.

17 In the *Logic* the authors use: « *Tout ce que nous consevons est représenté à notre esprit ou comme chose, ou comme manière de chose, or comme chose modifiée* », *LAP* I,2, *KM* V,134, *B* 30.

18 *VFI* Chapt 4; *KM* I,193, 198–199, 200–201; *G* 61, 66, 68.

19 *VFI* Chapt 6, *KM* I,204, *G* 71–71.

20 For a general account see Tachau 1988.

21 For a survey of various views see King 2007.

widely enough recognized that it was included in the standard lore about concepts detailed in the encyclopedic logic summaries known to Descartes and Arnauld, and which seem to be their source for the idea.[22]

The Cartesian notion of objective being, however, is somewhat different and in an important way more abstract. Arnauld certainly and probably Descartes rejected representationalism, subscribing instead to a version of direct realism in perception – indeed the falsity of representationalism is one of Arnauld's main claims in his famous dispute with Malebranche. The Cartesians did not believe that ideas possess, in any sense, modes shared with material substances – to do so would violate their dualism. Accordingly ideas for them could not be literally similar to objects outside the mind, and could not be representatives or exemplars defined as such by appeal to a relation of similarity.[23] Arnauld and Nicole nevertheless maintain that associated with an idea is a "content" understood as a mode or modes. In this sense understanding an idea is not the same as it is for Malebranche who holds we "see" an abstract idea with its properties. Rather understanding an idea is simply the understanding of the mode or modes in the idea's content, the way you might understand life insurance

22 See the discussion in Wells 1990, pp. 31–61. Some of the 16th and 17th century sources that discuss objective being are: Toletus, *Commentaria una cum quaestionibus in universam Aristotelis logicam* 1598, q. I, pp. 3. 30 & 32; Fonseca, *Commentarii in xii libros metaphysicarum Aristotelis.* 1599, q ii, §§1; Eustace-of-St.-Paul, *Summa philosophiae quadripartita, de rebus dialecticis, moralibus, physicis et metaphysicis* 1609, p. 1. For a discussion of Descartes' sources see Cronin 1966.

23 The interpretation that Descartes and Arnauld subscribed to direct realism is now widely shared. For Descartes see Alanen 1990) and Brown 2007. On Arnauld see Nadler 1989 and 1992. The case for Descartes' direct realism must contend with the fact that the argument in Meditation III for God's existence appears to depend on a version of causation as the transmission of modes. The interpretation requires that Descartes be committed to the rather baroque causal principle:
the reality of a mode in the content of an idea's objective being must be as great as the reality of the mode in its cause,
rather than to the more straightforward version of modal transmission:
the reality of the mode possessed by an idea as an objective being must be as great as the reality of the mode possessed by its cause.
To the extent that Arnauld accepts Descartes' argument (as he does in *Logic*), he too would have to accept the more complex version of the principle. Nothing in the semantics of the *Logic* actually turns on direct realism. The definition of signfication would work as well if comprehension were defined simply as the set of modes instantiated in an idea as an objective being.

or non-finite axiomatizability – cases in which it is not plausible to think there is some "object" abstract or otherwise that is there to be "looked at".

As Arnauld and Nicole explain, an idea "contains" (*contenir*) and "encloses" (*enfermer)* modes. This "content" is called the term's comprehension. Unlike the modes of possible or actual beings, however, the modes contained in objective being are intentional. Objective beings also vary in abstractness. Though we perceive an individual with the full panoply of sensory properties, when we think of an idea, its content is more limited. Indeed, the *Logic* explains abstraction (*abstraction, précision*) as a process of "selecting out" (*considerer sans faire attention à*) the modal content of perceptual experience[24] to form ideas with a simpler content or comprehension.[25]

It is a term's comprehension, its associated collection of modes, that determine its signification. Consider the case of the term *man*. Its traditional definition, *mortal animate living material substance*, details a list of progressively more general modes that form its essence and determine what the term stands for. It is from standard cases like this that Arnauld and Nicole abstract to nouns and adjectives generally. Each term has a comprehension that consists of modes that inhere in objects that are generally outside the mind. Because these objects are external to the mind, the modes link the idea to the world, and hence serve to define a term's signification.

The full theory posits that there is a grammar to mental language. Mental terms fall into two kinds: substantives (nouns) and adjectives. Nouns in turn are either simple or complex. Simple nouns are either innate or formed by abstraction. Arnauld and Nicole also allow for complex noun-phrases as in medieval logic. A complex noun phrase is understood to be a mental mode constructed from a simple noun and adjective by the grammatical operation called restriction (*determination*): the "complex noun" *SP* is formed by affixing an adjective (or relative clause) *P* to a noun *S*.[26] Adjectives are simple.

Comprehensions vary according to this grammar and according to whether a term is simple or abstract. The comprehension of a simple innate

24 *LAP* I,v,xi; *KM* V,142–43,168–170; *B* 37–38, 58–59; *VFI, G* Chapt. 6, 56–58, 98–100.

25 *LAP* I,v.

26 *LAP* I,viii. For medieval restriction see Buridan, *Treatise on supposition* 4.1.46–47 and 4.63, and *Treatise on consequence* 6.3.1 in Buridan 1985, and Book III, pp. 286, 648, and 835 of *Summulae de dialectica* in Buridan 2001.

idea is fixed by nature and Providence. That of an abstract simple noun consists of the abstracted mode or modes necessarily associated with them through abstraction. The comprehension of a complex noun *SP* is the combination or, as we would say, the union of the comprehensions of *S* and *P*.

The signification of a noun-phrase may now be defined in terms of comprehension: a noun-phrase signifies all objects that possess all the modes in its comprehension.[27] A proper noun signifies a single individual, and as in Aristotle it is possible for nouns to signify individuals in any of the Aristotelian categories, either substances or modes.

As in medieval semantics, the *Logic* refers to adjectives as connotative terms. According to the tradition a connotative term is one that may be paraphrased by two nouns, first by an abstract noun that signifies a kind or species, and second by the name of a mode that inheres in things of that kind. Ockham's example is *white,* which is paraphrased by (has as its "nominal definition") *some thing informed with whiteness.*[28] The connotative term is said in the *Logic* to primarily signify the significata of the kind term and secondarily a mode, and the associated mode. It is the adjective's associated mode that constitutes its comprehension. (In modern logic we would say that the kind term makes explicit the adjective's "type" or significance range.) Both the kind and mode associated with an adjective are fixed by nature as a feature of mental language. The kind is prior to the mode because in standard substance-mode ontology substances are ontologically "prior" to modes that inhere in them. An example given in the *Logic* is *prudent*, which is implicitly associated with the kind term *man* and the modal name *prudence.*[29]

The authors of the *Logic* also refer to the signification of a connotative term as relatively more or less confused. In medieval semantics *confused* is synonymous with *general* or *universal* (*universalis*), and the determinate-confused distinction is used to contrast concepts according to the breadth of the possible objects they signify.[30] Relative to one another, the term *Brunel-*

27 *LAP* I,viii. *KM* V, 153; *B* 46.

28 See "On Connotative and Absolute Terms," *Summa logica* I,10 in Ockham 1974. The *Logic* makes these distinctions in *LAP* I,ii.

29 *LAP* I,ii. *KM* V, 135; *B* 32.

30 See for example the following texts from Aristotle, Abelard, Aquinas, and Buridan: Aristotle, *Physics* I.1 184^{a}22–24; Abelard, *In Isagogen* ed. Geyer, *Logica ingredientibus* L1.01 /21/-/29/ Aquinas, *Summa Theologica* p I.1 ,q. 85, a. 3 *Responsio*; Buridan, Question on Aristotle's *De anima,* line 95 ff., p. 291 ff. Buridan 1989.

lus is more distinct or determinate than the term *donkey,* and *donkey* is more confused than *Brunellus,* because *donkey* signifies a broader range of possible objects than *Brunellus,* which in fact is totally determinate because it signifies a unique individual.

In the *Logic* this distinction is applied to adjectives. An adjective confusedly signifies the significata of its associated kind term. It does so because the signification of the kind term, which is abstract and signifies a broad range of objects, is confused in a prior sense. The adjective signifies distinctly the significatum of its associated modal noun because that noun is proper and signifies a unique mode. Within this general theory of signification there are several additional details that bear directly on the issue of existential import.

First, the *Logic* is very generous in its conception of species. It counts any genuine abstract noun-phrase, including those formed by restriction, as a species, and hence it counts as species many nouns that would not count as such in Aristotle's metaphysics.[31] The *Logic*, nevertheless, continues to apply the traditional terminology of the predicables to this more generous notion of species.[32] A mode is *essential* if it falls in a noun's comprehension; it is a *proprium* if it is non-essential but necessarily true of those entities that satisfy its comprehension; it is *accidental* if it is true but not necessarily true of some entity. The traditional terminology, however, should not obscure the fact that the natural world presumed by the Cartesian semantics is rather non-Aristotelian. It is an extensional plenum informed by quasi-mathematical modes, any possible combination of which qualifies as the comprehension of a species. The species cited in the *Logic* are, in fact, strikingly unlike the biological examples common in Aristotle. The author's examples of material species, i.e. of abstract nouns signifying material objects, include: *body, transparent body, motion, time, even number, odd number, prime number, line, triangle, equilateral triangle, right triangle, quadrilateral, trapezoid, parallelogram, chiliagon, cube, pyramid, cylinder, rational animal, prudent person, knowledgeable person, sun, moon,*

31 *LAP* I,vii. *KM* V, 146–7; *B* 44–41. 17th century logic as a rule subscribes to both mental language and realism, as did many later medieval logicians after Ockham and Buridan. Concepts, or ideas in the Cartesian context, do not serve as they do in nominalism as mental proxies for universals, but rather are mental terms (modes) that correlate with universals (material modes) that have independent existence outside the mind.

32 *LAP* I,vii.

and *star.*[33] This Cartesian world of mathematized matter is more like that of modern physics than the common sense world of macro-sized animals and plants described by Aristotle's essentialism. Within its framework the *Logic* nevertheless succeeds in representing the diminished importance of the Aristotelian essence-accident distinction. It does so not by rejecting essentialism as false in the manner of Quine, but by diluting it to triviality. Species do not differ from "arbitrary sets" because every well-formed abstract noun has a comprehension and, in this sense, has an essence. Species therefore include groupings like *prudent person,* which Aristotle would regard as accidental.

It should be remarked that although Arnauld and Nicole avoid Aristotle's essentialism, their account has oddities of its own, especially when combined with a commitment to knowledge as awareness. If understanding an idea consists of being conscious of its content, then when we cognize an idea, we know its essence, or, to use alternative terminology, we know its real definition. Because this knowledge is a variety of direct awareness, it is *a priori*, even if the idea is itself abstracted from sensation. Moreover, if an idea *S* "contains" a mode *M*, the proposition *S is M* is rightly called analytic. Hence, though Arnauld and Nicole do not use this terminology, they, like other rationalists, hold that we have *a priori* knowledge of real definitions and that these are analytic. Moreover, as we shall see below, the authors hold that if these ideas are experienced "clearly and distinctly", then a proposition predicating the content of the idea is true. This picture is a major departure from medieval semantics in which real definitions are regarded as empirical truths, which are often difficult to discover.

Second, the *Logic* makes very clear that proper names and singular term phrases count as noun-phrases. A singular term may serve as the grammatical subject of a singular categorical proposition. Indeed, the *Logic* espouses the seventeenth century view that a singular proposition is a special case of a universal.[34] Hence a singular term possesses a comprehension and extension. Moreover, an abstract noun, like *pope*, can be restricted, e.g. *the present pope*, so that its comprehension signifies only a unique actual entity.[35]

33 Examples of proper names and singular terms are: *myself, God, Alexander son of Philip, the present Pope, the earth.* Examples of false ideas: *rocks* (that fall from gravity), *fire* (that burns), *wealth* (that makes us happy), *corporeal pain, colored rainbow, bent stick in water.*

34 *LAP* II,iii, *K* V,199, *B* 84; *LAP* III,ix, *K* V,278, *B* 160.

35 *LAP* I,viii, *KM* V,151–152, *B* 44–45.

The extension of a singular term, to be sure, is a kind of degenerate case because it consists of only the idea itself, which signifies the unique individual that satisfies its comprehension. Thus, unlike Aristotle, who denied that a singular term has an essence or could be defined,[36] the *Logic* generalizes comprehension and essence to all noun-phrases including singular terms. Singular terms therefore count as species in a generalized Cartesian sense.

The final step in setting out the theory's core semantic ideas is the formal definition of extension. It is defined by signification, the reference relation. Species S_1 is said to be inferior to species S_2 if all the modes in the comprehension of S_1 are true of all objects that satisfy all the modes in the comprehension of S_2. The extension of S is then defined as the set of species inferior to it. Thus, a term's extension includes any species such that all the modes in the term's comprehension are true of all entities that satisfy the species' comprehension. Nouns may either signify substances or other modes, e.g. *earth* or *heat* (aka *hotness*).

It may be remarked that there is a systematic relation between the Cartesian and the modern notions of extension. It follows from the definition of the Cartesian sense that the extension of any noun literally includes all true singular terms inferior to it, i.e. it includes any singular term with a comprehension that is true of any actual entity that satisfies the modes in the noun's comprehension. Thus, there is a sense in which the *Logic*'s Cartesian extension, which consists of ideas, incorporates the referential notion of extension found in modern logic: there is a 1–1 mapping that pairs an object in a term's modern extension with that singleton species in its Cartesian extension that has that entity as its sole signficatum. It then follows directly from the definitions that sameness of Cartesian extension must be a relation of ideas, but because extension is defined in terms of signification, it also follows that sameness of Cartesian extension entails sameness of extension in the modern sense. As we shall see shortly, it is this pairing that insures a genuine correspondence theory of truth.

Given the concept of extension in its Cartesian sense, we may now state the truth conditions for affirmative categoricals as they are explained in the *Logic*, Book I, vxii. The conditions for both the universal and particular affirmative are formulated as identity statements:

> ...the nature of affirmation is to unite and identify...[37]

36 *Metaphysics* VII.

37 *LAP* II,xvii, *KM* V,247, *B* 129.

This identity is one of extension:

> ...it is the nature of affirmation to put the attribute in everything expressed in the subject according to its extension in the proposition.[38]

The relevant extension of the predicate (*attribut*) is determined by the subject:

> Thus because affirmation puts the idea of the attribute in the subject, it is properly speaking the subject that determines the extension of the attribute in the affirmative proposition. The identity it indicates takes the attribute as restricted to the extension equal to that of the subject...[39]

Here "determinate" is understood in its medieval sense of "less general". The operation that limits the extension of the predicate is restriction. As defined earlier, restriction is a grammatical operation that joins an adjective to a noun to form a noun-phrase that has as its comprehension the union of the comprehensions of the noun and adjective. As the authors describe restriction for an A-proposition, it is the subject *S* that restricts the predicate *P*. Hence in this application it is the subject that is understood in the role of adjective, which as a connotative term would signify secondarily the modes in its comprehension. Because the comprehension of the complex noun *SP* is the union of those of *S* and *P*, its extension will be a restricted subset of that of *S* and *P*.

In addition to the identity of extension, we must add an implied secondary condition. This is necessary in order to retain the standard inferences of the Square of Opposition in which A-propositions entail I-propositions, both understood as possessing existential import. To be added is the condition, which was standard in the logic of the period, that the subject term of true affirmatives must be non-empty. This is a substantive requirement, and its inclusion as part of the interpretation of Port-Royal Logic is somewhat controversial. For the purposes of exposition let us postpone for the moment the interpretive argument showing why it was required by Arnauld and Nicole. With the understanding that we will return to the issue of non-empty terms below, we may state the truth-conditions for the universal affirmative as follows:

Every S is P is true iff the non-empty extension of *S* is identical to that of *SP*.

38 *Ibid.*
39 *LAP* II,xvii, *KM* V,248, *B* 130.

If *Ext* is the operation that assigns to each term its extension, the conditions may be expressed in a more modern notation as:

$$\textit{Every S is P} \text{ is true iff, } Ext(S) \neq \varnothing \text{ and } Ext(S) = Ext(SP).$$

Because of the 1 to 1 correspondence noted earlier in the paper between the modern and Cartesian notions of extension, the universal affirmative is true if and only if *Ext*(S) is included in that of *Ext*(P), as in the equivalence in modern set theory between $A = A \cap B$ and $A \subseteq B$. It also follows that the *Logic*'s truth-conditions are essentially equivalent to George Boole's representation of an A-proposition in terms of sets as $a = ba$.

From a modern perspective there is an odd consequence to this analysis of universal affirmatives as conceptual identities between ideas. Since as terms in mental language the subjects and predicate of a true proposition are literally identical, it follows that for any given subject, all true propositions predicated of that subject in mental language – regardless of any defference in the verbal form of their predicates in speech – posses in mental language literally the very same predicate, namely the subject-idea.[40] Hence the ideas identified with the subject in *every man is rational, every man breathes, every man is risible, every man is a descendent of Eve* are the same, and hence the propositions formed from these ideas must also be the same. It seems that not only do *Scott is the author of Waverley* and *Scott is Scott* express the same proposition, but so do any true propositions with the same subject.

How then would the *Logic* explain differences in "information content"?" The only explanation that seems available might be that on different occasions of thought the same mental act may come about by different acts of restriction and in this sense on different occasions have distinct generative histories – grammatical trees in modern terms. Thus if you assert to me *every S is P*, I learn that you think that if I form the idea with the content of S and that with the content of both S and P, then those two will extend to the same species in the sense that a species S' signifies only entities that S signifies if and only if S' signifies only entities that the restriction of P by S signifies. God presumably would be an exception since his ideas are

40 The *Logic* does not allow for ambiguity in mental language – that their might be two mental terms for the same "idea." Since ontologically mental words simply are ideas, it follows trivially to each mental word there is a unique idea.

fixed and ungenerated. For him all true propositions about *S* would have the same content, namely *S*.

The *Logic* is somewhat less precise in its statement of the truth-conditions for the particular affirmative. This much is clear. We have been told in the passage quoted above that like the universal case the particular affirmative asserts the identity of two ideas. As the authors explain, in this case too the two ideas are formed by restriction:

> So in particular affirmative propositions, for example, when we say "some people are just": the subject and the attribute are both particular, since the subject "people" is particular by the mark of particularity added to it. The attribute "just" is also particular, because its extension is restricted by that of the subject, and so it signifies merely the justice found in some people. [41]

This passage tells us that the predicate *just* is restricted by the subject *people*. Unlike the universal case in which the predicate would be restricted by the whole of the subject, in this case it is restricted so that it signifies only some of the entities signified by the subject. At one point the authors explain that restriction "can be done in two ways" [*se peut faire en deux manières*]:

> Now the extension of a general idea can be restricted or narrowed in two ways.
>
> The first is by joining another distinct or determinate idea to it, as when I join the idea of having a right angle to the general idea of a triangle. Then I narrow this idea to a single species of a triangle, namely the right triangle.
>
> The other is by joining to it merely an indistinct and indeterminate idea of a part, as when I say "some triangle." In that case the common term is said to become particular because it now extends only to a part of the subjects to which it formerly extended, without, however, the part to which it is narrowed being determined.[42]

It is the second way of restriction that is employed in the truth-conditions for an I-proposition. Here the predicate is restricted by an "indistinct or indeterminate idea" formed from the subject – the *some triangle* of this example or the *some people* of the example above.

Jean-Claude Pariente has interpreted this text as positing a second *sui generis* variety of restriction used in the truth-conditions for I-propositions that is different from the restriction operation that is used to form complex ideas generally, which is also employed in the truth-conditions of A-propositions. This second restriction, he suggests, operates on an idea – in this case the subject – to form a new "indeterminate or indistinct" idea. The predicate

41 *LAP* II,xviii, *KM* V,249–250, *B* 131.
42 *LAP* II,vi, *KM* V,145, *B* 40.

is in turn restricted by this indeterminate idea, and the proposition is true if the indeterminate idea is identical with the restriction of the predicate.[43]

The authors of the *Logic*, however, do not define or give examples of either a second restriction operation or a new category of "indeterminate" ideas. These phrases occur only in the passages quoted. I would like to suggest a simpler interpretation that does not require new distinctions.

Aristotle sometimes explains why an I-proposition is true by ecthesis, the "setting out" of a species that is a subspecies of both the subject and predicate.[44] If two terms can be restricted so as to form a common species, then their extensions are identical, and the corresponding I-proposition is true. Arnauld and Nicole suggest that an I-proposition be understood in just this way. As they put it, the extension of the predicate is determined or restricted by "a part" of the subject:

> If the subject is particular, the attribute is conceived
> only in a part of the extension of the subject.[45]

The authors indicate what they mean here by the phrase "conceived of only in a part of the extension." They use the same phrase in an adjacent text to describe the restriction of the predicate in the truth-conditions of the universal affirmative.[46] As they use the expression to explain the A-proposition, "to be conceived in a part of the extension" of *a* term *P* means simply that a new idea *SP* is formed from *P* by restricting it in terms of a second idea *S*, with the result that the extension of *SP* is included in that of *P*. Partial-conception in the truth-conditions for an I-proposition should be understood in the same way. As in the case of a true A-proposition, two ideas are identical. In the case of a true I-proposition, however, the first idea is the partially conceived subject term, and the second is a restriction of a predicate by this partially conceived subject.

Which ideas are appropriate for this restriction? Normally it would be some species in the extension of the subject. In a given case a suitable term may be implicit from the context[47], similar to the way that a connotative term has implicitly associated with it a kind term that specifies the range of

43 Pp. 247–238 in Pariente 1985.
44 On ecthesis see Martin, Ecthesis and Existence in the Syllogistic in Martin 2004.
45 Third axiom, *LAP* II,xvii, *KM* V,249, *B* 129.
46 See the remarks about the universal case in the fourth axiom, *ibid.*
47 Terance Parsons has suggest this to me in correspondence.

application. However, all that is required for a statement of the truth-conditions is that there be some term that meets the relevant conditions.

Why do the authors call the restricted subject in this case "indefinite or indistinct"? The way to understand "indefinite" here, I suggest, is as a higher-order property. It is not that the idea in terms of which the predicate is restricted is itself indefinite. The extension of both the subject and its restricting subspecies may or may not be broad. Their extensional scope is irrelevant so long as they are non-empty. Nor need we appeal to any restriction operation other than the single operation that has already been mentioned, namely that used in the formation of noun-phrases generally and in the interpretation of A-propositions. Rather "indefinite" is to be read as a true second-order description of the meta-name for the class of species relevant to the subject's restriction; i.e. *indefinite* is a meta-adjective that is true when it modifies the meta-name *suitable idea in terms of which the restriction is preformed.* It is that second intention that is indefinite, or in the medieval sense "confused". It is so because any number of ideas could serve that purpose of restricting the subject so long as it is a subspecies[48] of the subject term. In the examples quoted, then, *some people* is short for the metalinguistic expression "some term in the extension of *people*", and "restriction by *some people*" means "restriction by some term in the extension of *people*"

Before stating the truth-conditions formally, we must also note again that the terms mentioned in the truth-conditions must be non-empty. Not only is this requirement necessary to insure that I-propositions with false ideas as subjects are false, it is also required to coordinate syllogistic inferences with A-propositions. As previously stated, A-propositions have a similar clause requiring non-empty terms, and the *Logic* validates subalternation.[49]

48 If there are shared entities that satisfy jointly the comprehensions of the subject and predicate, and if every singular term is a species, then there will exist appropriate subspecies of the subject for a restriction that will make the appropriate identity true, namely the names for the shared entities.

49 *LAP* II,iv. *KM* V,201; *B* 36. It would be an odd syllogistic system indeed if A-propositions and singular propositions carried existential import, I-propositions did not, and subalternation held. Ecthesis, the implication from *some S is P* to *every T is S* and *every T is P,* for some term *T,* would fail even though *T* is essentially the same as the *Logic*'s "indefinite" restriction of *S*. Though the *Logic*'s treatment of the syllogistic is relatively cursory, there is no textual evidence that the authors intend to depart from the standard set of inferences. Indeed one reason the treatment is cursory is that they regarded the standard account as true. See *LAP Discours* I, *KM* V,111, *B* 10*; Discours* I, *KM* V,121, *B*.19.

With these understandings we may state the truth-conditions for an I-proposition in general form:

> *Some S is P* is true iff the restriction of *S* by some term has the same non-empty extension as the restriction of *SP* by that term.

In the idiom of modern metatheory this would be expressed in terms of the existential quantifier:

Some S is P is true iff, for some *T*, $Ext(TS) \neq \varnothing$ and $Ext(TS) = Ext(TSP)$.

Because of the correspondence between the modern and Cartesian notions of extension, this analysis is similar to George Boole's representation of an I-proposition as *ax=bax* in which the variable *x* serves the role of restricting the set-theoretic extensions of both the subject and predicate.[50]

To complete the statement of truth-conditions, it is necessary to state those for negative propositions. For our purposes here it is sufficient to remark that the *Logic* stipulates that the universal negative is the contradictory of the particular affirmative, and the particular negative of the universal affirmative.[51]

Let us now turn directly to the issue of whether the truth-conditions for true affirmative propositions should be formulated so as to require that the subject terms be non-empty. The key interpretive argument we advanced here is that the requirement of non-empty terms is entailed by the *Logic*'s doctrine of false idea and the correspondence theory of truth it presupposes. Both false ideas and truth as correspondence are required by Descartes' epistemology. Of central importance to understanding Cartesian epistemology is the relation that comprehension bears to propositional knowledge. It is this link that ultimately entails the requirement of existential import for affirmative propositions.

Explaining the relevant epistemology requires a discussion of the operations of the soul. Unfortunately, by modern standards Descartes and his followers do not do a very clear job of explaining the relations among the various mental acts of conceptualization, judgment, and reason. These

50 These formulations are used in, for example, Boole 1948 and 1951. Boole assumes that a relevant *x* can be found.

51 See *LAP* II,4.

relations, however, especially that between, on the one hand, clear and distinct conceptualization and, on the other, warranted judgment, are central to their epistemology. This much is clear: the act of being mentally "aware of" the content of an idea – of "understanding" the idea in an abstract sense as an objective being – entails propositional knowledge. The relation can be formulated as a rule: if an agent conceives an idea *S* with modal content *P* clearly and distinctly then he is warranted in asserting the proposition *S is P*. In the *Logic*'s words:

> everything that is contained in (*contenu dans*) the true idea of a thing (i.e. in the clear perception that we have of it) can be truthfully asserted (*affirmé*).[52]

By modern standards what is unclear are the details of act individuation among the related mental acts of conception, proposition formation, judgment (i.e. assertion and denial), and reasoning. Medieval logicians, in contrast, clearly maintain that these acts are distinct, and have views on the details of their relations. Ockham, for example, holds to a part-whole theory in which concepts can be parts of one another, and these in turn parts of propositions. Buridan, in contrast, denies that propositions have parts but holds that acts of conceptualization are necessary conditions for and ontologically prior to acts of judgment.[53]

Though Descartes and the Port-Royal logicians are silent on the criteria for individuating mental acts in general, when they are careful, they too distinguish among conceptualization, judgment, and reasoning. What is unclear is the difference between, on the one hand, the act of being conscious clearly and distinctly of the idea S with content *P* and the act of knowingly asserting the proposition *every S is P*, on the other. The issue is complicated by the *Logic*'s doctrine that a true universal affirmative is to be understood as an identity assertion. On this view the different predicates said of the subject *S* cannot be used to differentiate the various propositions in which *S* is the subject because it is literally the case that every proposition true of *S* simply identifies *S* with the same idea, namely *S* itself. To distinguish among the clear and distinct conceptualizations of the subject and predicate, and the propositional act that identifies them, it would help to have some account of how the various acts involved in their grammatical construction differ from one another. On this topic, however, the *Logic* says little.

52 *VFI* Chapt. 6: *KM* I,209; *G* 76.

53 See Panaccio 2004, and the introduction by King to Buridan 1985.

When the authors are careful, they write as if there are two distinct mental acts: clear and distinct perception of an idea *S* as *P* and affirming with epistemic warrant the proposition *every S is P*. Indeed the authors begin the *Logic* by distinguishing among four mental acts.[54] They draw a tripartite distinction, common in medieval logic, among conception (to conceive, *conçevoir*), judgment (to judge, *juger*, i.e. to either affirm or deny), and reasoning (to reason or deduce, *raisonner*). To these they add a fourth distinction, which was common in 15th and 16th century logic, the methodological organization of knowledge (to order, *ordonner*). In drawing these distinctions, the authors explain that the proper object of conception is an idea, that of a judgment is a proposition (i.e. the judging of one idea that it is or is not another idea), and that of reasoning is a series of related judgments in which one judgment is formed or concluded (*former, concluir)* to be true as a result of judging others to be true.

On the other hand, the authors often write less precisely using terms for one sort of act that strictly speaking should apply to others. They use *to conceive, to contain,* and *to signify,* which are properly appropriate for conception, in contexts that describe propositional assertion and even steps of logical reasoning.[55] Conversely, they use terms appropriate to judgment and reasoning to describe conceptualization.[56] Indeed, the fluid relations

54 *LAP* I Introduction; *KM* V 125; *B* 23.

55 Examples :

... quand je dis que *tout homme est animal ;* je veux dire & je signifie que tout ce qui est homme est aussi animal ; & ainsi je conçois l'animal dans tous les hommes (*LAP* II,xvii ; *KM* V 247; *B* 129).

... il est vrai que l'on n'exprime pas toujours les deux prémisses, parce que souvent une seule suffit pour fair concevoir deux à l'esprit. (*LAP* III,ii; *KM* V 256; *B* 136).

Sachant donc par ce que nous avons dit dans las seconde partie, ce que c'est que l'étendue & la compréhension de termes, par où l'on peut juger quand une proposition en contient ou n'en contient pas une autre; on peut juger de la bonté ou défaut de tout syllogisme, sans considérer s'il est simple ou composé, complexe ou incomplexe, sans prendre-garde aux figures ni aux modes, par ce seul principe général : Que l'une des deux propositions doit contenir la conclusion, & a'autre fair voir qu'elle la contient. (*LAP* III, xi; *KM* V 284; *B* 175).

The remainder of the III,xi continues to use *contenir* to refer to a relation among propositions.

56 For example:

Les propositions particulièrs son enfermées dan les générales de même nature, & non les générales dan les particuilièrs, I. dans A. & O. dans E. & non A. dans I. ni E. dans O. (*LAP* III, iii; *KM* V 258; *B* 139).

among conceptualizations, judgments and acts of reasoning – Jill Buroker has remarked that for Descartes conceptualizations and propositional acts "shade into" one another[57] – are related to Descartes' understanding of logical inference, which is rather different from that of modern logic and seems to be shared by Arnauld and Nicole. In their view a step in a deduction is a transition from a state of understanding one idea clearly and distinctly to a second state of understanding another idea clearly and distinctly. Being clear and distinct, moreover, each step of the reasoning process consists of a state of knowledge that is independently warranted in its own right[58]

For our purposes here it will suffice to note that the relations among these acts is very close, and that a clear and distinct idea of *S* as *P* entails knowledge that *S* is *P*. The contrapositive is also important: if I assert *S is P* but am mistaken – if I fall short of knowledge – then I do not have a clear and distinct idea of *S* as *P*. What is important to our semantic investigation here is that this sort of knowledge and error has implications for the existential import of ideas.

Consider knowledge first. If an idea is perceived clearly and distinctly, the propositional act is said to qualify as certain or "scientific" knowledge. Quoting Descartes with approval, Arnauld writes

> ...he understands the word 'idea' in the proposition that 'everything that I perceive clearly as being in the idea of a thing can correctly be asserted of that thing [*tout que je vois clairement être enfermé dans l'idée d'une chose, eut avec vérité être affirmé de cette chose*],' which he claims, with good reason, to be the foundation of all the natural sciences. If, examining the idea that I have of a triangle (by reflecting on the perception that I have of it), I find that the equality of its three angles to two right angles is contained in [*est enfermé dans*] this idea or perception, I can correctly assert that every triangle has three angles equal to two right angles.[59]

The case of error is more complex, and it is this that has implications for the semantics of existential import. In error, as the *Logic* explains, the proposition-idea entailment typically works in its contrapositive form: propositional error leads to the formation of defective ideas. The *Logic*'s epistemology uses this relation to explain why people commonly believe that sensory qualities affect the mind. The explanation proffered has a general form. We

57 P. 457 in Buroker 1993.

58 For this interpretation see Gaukroger 1989. For a similar understanding by Arnauld see *VFI* Chapt. 5: *KM* I,199, *G* 67; and Chapt. 6, *KM* I,204, *G* 71–72; *KM* I,206, *G* 73; *KM* I,207–208, *G* 74–75; *KM* I,209, *G* 76.

59 *VFI* Chapt. 6: *KM* I,206; *G* 73.

form the habit of judging the proposition *S is P*. This habit in turn leads us to form the complex idea *SP* by combining the subject and predicate. If the proposition is false – for example *fire causes pain* – the corresponding idea, e.g. *corporeal pain,* is false or even confused.

Though in modern semantics truth and falsity are understood as properties of sentences or propositions rather than of terms or ideas, the notion of a true or false idea has a long history in traditional logic, and requires some explanation here.

In medieval semantics truth and falsity are properties that inhere in mental acts like sensations or concepts, but only because they inhere in a prior way in propositions. Aristotle, for example, refers to sensations as true or false[60], but he does so only because in a more basic sense acts of composition or assertion, and of division or denial are true or false:

> For falsehood always involves a synthesis; for even if you assert that what is white is not white you have included not white in a synthesis. It is possible also to call all these cases division as well as combination. However that may be, there is not only the true or false assertion that Cleon is white but also the true or false assertion that he was or will be white. In each and every case that which unifies is mind.[61]
>
> Imagination is different from assertion [*phaseōs*] and denial [*apophaseōs*]; for what is true or false involves a synthesis of concepts [*noēmata*]. In what will the primary concepts [*noēmata*] differ from images [*phantasmata*]?[62]

In his view thoughts, even non-referring terms like *goat-stag*, are not true or false until they are used in propositions:

> Nouns and verbs, provided nothing is added, are like thoughts without combination or separation; 'man' and 'white', as isolated terms, are not yet either true or false. In proof of this, consider the word 'goat-stag.' It has significance, but there is no truth or falsity about it, unless 'is' or 'is not' is added, either in the present or in some other tense.[63]

The two-stage application of truth and falsity first to propositions and then concepts became standard in medieval semantics. In *De veritate*, for example, Aquinas explains the more basic role of propositional truth this way.

> Just as it is said that the true is found in a more prior way in the intellect than in things, so also is it to be found more in the intellect's act of putting together and dividing than in its act of forming the quiddity of things.

60 See for example *De anima,* iii:3, 428^a11.

61 *De anima* III,6, 430^a25-$^b6$.

62 *De anima* III,8, 432^a8–14. See also *De interpretatione* I, 16a10–19.

63 *Parts of Animals* I,16^a13–19.

At the same time, however, he acknowledges the secondary usage in which a definition or a "composite thought" (*ratio compositionis*) is true or false. He cites as an example of a false definition *insensible animal*. He explains that it is false because the corresponding proposition *animal is insensible* that affirms one term of the other is false.[64]

The *Logic*'s account of error in terms of false ideas derives from Descartes, who in *Meditation* III includes false ideas as a sub-variety of factitious ideas, which is what he calls a complex idea of our own composition. In his usage a false idea is one that fails to be "like or to conform to things that are external to us" or that fails to be "an idea of a real object".[65] He wonders about the composition of the ideas *heat* and *cold*, whether *heat* is a composite from *negation* (*privation*) and *cold*, or cold from *privation* and *heat*. When construed as containing in addition the idea *caused by the external world*, both ideas are false.

Arnauld and Nicole appeal to false ideas to account for vulgar beliefs generally. In their account defective ideas fall into two types. The first and more general is called simply a false idea. This is a complex idea with a comprehension that consists of modes that are jointly true of no actual object.[66] Some false ideas are possible in the sense that they contain modes

64 See *De veritate* q. 1, a.3 co.; 51615. See also 54811.

65 *Meditation* III, 7 and 19.

66 The view that there could be definitions or, to use the terminology of the *Logic*, comprehensions that are unsatisfied dates from Aristotle. He says in the actual *Posterior analytics*, Book 7 (92a34–92b19):
How then by definition shall we prove substance or essential nature? We cannot show it as a fresh fact necessarily following from the assumption of premises admitted to be facts – the method of demonstration: we may not proceed as by induction to establish a universal on the evidence of groups of particulars which offer no exception, because induction proves not what the essential nature of a thing is but that it has or has not some attribute. Therefore, since presumably one cannot prove essential nature by an appeal to sense perception or by pointing with the finger, what other method remains? To put it another way: how shall we by definition prove essential nature? He who knows what human – or any other – nature is, must know also that man exists; for no one knows the nature of what does not exist – one can know the meaning of the phrase or name 'goat-stag' but not what the essential nature of a goat-stag is. But further, if definition can prove what is the essential nature of a thing, can it also prove that it exists? And how will it prove them both by the same process, since definition exhibits one single thing and demonstration another single thing, and what human nature is and the fact that man exists are not the same thing? Then too we hold that it is by demonstration that the being of everything must be proved – unless indeed to be were its essence; and, since being is not a genus, it is not the essence of anything. Hence the being of anything

that could be instantiated, for example *gold* and *mountain*; the *Logic* numbers objective beings of this sort – again using medieval terminology – among *beings of reason.*[67]

A second and more grievously defective type of idea has a comprehension consisting of modes that cannot possibly be instantiated. In the *Logic*'s technical vocabulary such an idea is called *confused.* Its modes may contain a logical contradiction or simply a metaphysical impossibility. Vulgar beliefs that attribute to the same subject a material mode like *has a material cause* and a mental mode like *is suffering pain* generate confused ideas of this sort because they describe an impossible content. It is impossible for the propositions *S has a material cause* and *S is suffering pain* to be jointly true. Examples of confused ideas that result from habitual false judgment include *corporeal pain, heat caused by fire, gravity*, and *happiness caused by material wealth.*[68]

The motivation for calling such ideas "confused" comes from a rather specialized and uniquely Cartesian application of the standard distinct-confused distinction from medieval semantics. According to this distinction a concept with broader range of signification is more confused and less distinct than one with a narrower range. When applied to defective ideas, however, this distinction is used in combination with another.

Medieval semantics and in like manner the Port-Royal Logic distinguish in addition between first- and second-order concepts. In its normal use the spoken word, for example, *donkey* is associated with the concept *donkey*

as fact is matter for demonstration; and this is the actual procedure of the sciences, for the geometer assumes the meaning of the word triangle, but that it is possessed of some attribute he proves. What is it, then, that we shall prove in defining essential nature? Triangle? In that case a man will know by definition what a thing's nature is without knowing whether it exists. But that is impossible. G. R. G. Mure, trans.

67 Que si les objets rerésentés par ces idées, soit de substances, soit des modes, sont en effet tels qu'ils nous sont représentés, on les appelle véritables : que s'ils ne sont pas tels elles sont fausses en la mannière qu,elle les peuvent être; & c'est ce qu'on appelle dans l'école êtres de raison, qui consistent ordinairement dans l'assemblage que l'esprit fait de deux idées réelles en soit, même qui ne sont pas jointes dan la verité pour en former une même idée, comme celle qu'on se peut former d'une montagne d'or, est un être de raison, parce qu'elle est composée des deux idées de montagne & d'or, qu'elle représente comme unies, quoiqu'elle ne le soient point véritablement. *LAP* II,ii, *KM* V,136, *B* 32.

68 For passages in which the formation of such ideas are described see: *LAP Discour* I,. *KM* V, 110, *B* 9–10; I,ix. *KM* V, 157–78; *B* 49–50; I,xi. *KM* V, 168–170; *B* 58–60.

which signifies things outside the mind, as in the proposition *Brunellus is a donkey,* but any word also has a second-order use in which it stands for an idea that signifies mental entities like other ideas, as in the sentence *donkey is a species.* A concept that signifies things is said to be a first intention and one that signifies ideas that in turn signify things is a second intention.[69]

In medieval semantics the determinate-confused distinction is normally applied to primary intentions. Relative to one another the concept *Brunellus* is more distinct or determinate than *donkey* and *donkey* is more confused that *Brunellus* because *Brunellus* signifies a narrower range of possible objects than *donkey.* If a term is completely determinate, like the proper noun *Brunellus,* it signifies a unique individual.

When the Cartesians apply the determinate-confused distinction to false ideas, the distinction is applied to ideas in second intention. A term in second intention that signifies ideas is confused if it signifies multiple ideas, and it is totally determinate or distinct if it signifies a unique idea. (Strictly, it is not correct to call a confused second-order idea ambiguous because such an idea does not signify one idea in one context and a different idea in a second context. Rather, it is confused because given its unambiguous signification, it signifies multiple ideas.)

This rather careful distinction is applied by Descartes, and by Arnauld and Nicole, to the false ideas of vulgar opinion. The complexes *rational animal* and *golden mountain* are false, but in the Cartesian use they are not called confused. Though false they are nevertheless well-formed complex ideas. Each describes a possible though non-actual object because the comprehensions of its parts are mutually compatible. The new complex is a well-formed abstract noun. Because this idea is unique, its second-order name[70] has determinate signification: "the idea *rational animal*" and "the idea *golden mountain*" each signifies a unique idea.

The idea *corporeal pain*, on the other hand, is a defective complex. The abstract nouns *body* and *pain* do not succeed in forming an abstract combination because the comprehensions of its component ideas *body* and *pain* fail jointly to describe a possible object. Ontologically the two sets of modes are necessarily disjoint. Though we might try to combine the

69 *LAP* I,ii. *KM* V,136; *B* 32.

70 In the *Logic* the idea of an idea is called the idea in second intention, which is its standard name in medieval logic, *LAP* II,ii, *K* V,136, *B* 32. Cf. Book 6.4, p. 428 in Buridan, *Summulae de Dialectica* 2001.

ideas, the effort is futile. At best, we can form the second-order name "the idea *corporeal body*", but this would necessarily be a confused idea in the technical sense. Its signification ranges over two different ideas. Though it is a well-formed name in second intention – it is an idea signifying ideas – it fails to be determinate. This second intention, therefore, signifies confusedly.[71]

It remains to explain which among the ideas that describe possible objects are true and which false. As a general rule there is a correlation between objective being and possibility: there is an idea *S* with modal content *P* (in at least God's mind) iff the proposition *every S is P* is possibly true.[72] However, from what has been said so far, it is perfectly possible that the content of our idea fails to signify anything actual. It is possible that this idea fails to refer.

The theoretical commitment that insures the existential import of an idea with objective being is the *Logic*'s particular version of occasionalism.[73] On the authors' view, an undeceiving God would not allow the mind to possess an innate idea of a being, or to have a perceptual experience of one, if that being did not exist in the actual world.[74] It follows that the content of simple ideas, both those that are innate and those that are abstracted from perceptual experience, is actually realized in existing things.[75]

Like Descartes, the Port-Royal logicians hold that it is possible to distinguish in mental experience between those ideas that are distinct but false and those that are distinct and true: the criterion is clarity. Here *clarté* seems to be like Aquinas' *claritas,* a subjective property of ideas that enables

71 Antonius Rubius, a Spanish logician who worked in Mexico, anticipates this Cartesian usage by calling a term *confused* because it ambiguously names more than one *esse objectivum*. See Ashworth 1966.

72 *VFI* Chapt 5–6, *KM* I,200,204, *G* 67,71–72. Though some in the medieval tradition identified *esse objectivum* and *esse possible*, e.g. Scotus and Suárez, and possibly Descartes, Arnauld himself viewed the notion of a possible non-actual substance as incoherent. See Letter Arnauld to Leibniz, May 13, 1686, *KM VI,* pp. 31–32. Accordingly the proper correlation as understood by Arnauld is one between a genuine *esse objectivum* and a proposition that is possibly true. On the earlier tradition see Suárez, *DM* XXXI.2.1 (Vivès XXVI.229) and *DM* XXXI.2.10 (Vivès XXVI.232), and the discussion in Normore 1986.

73 For references to occasionalism by the authors of the *Logic* see note 6 above.

74 *VFI* Chapt 5, *KM* I,202, *G* 69–70;Chapt 28, *KM* I,351,353–354, *G* 209,313–214.

75 This is the Cartesian version of Aristotle's view that "perception is always true." See, for example, *De anima* 427^b6–22.

conscious experience to occur, just as light enables vision.[76] Just as we are able to see an object only when it is illuminated by light, so we are able to be conscious of the content of an idea only when it possesses clarity. On this view, God has set up the world so that only distinct ideas that are true are experienced as clear.[77] This particular epistemic or psychological view has implications beyond semantics. What is important here, however, are the implications of the doctrine of clear and distinct ideas for existential import.

According to the theory, clear and distinct ideas are true and warrant knowledge. Moreover, true ideas carry existential import and false ideas do not. This result follows trivially from the definition of a false idea as one that has content that is not instantiated in an actual object. The definition also entails the more interesting result that a true affirmative categorical proposition cannot be composed of false ideas. It follows, for example, that *some golden mountain is gold* is true only if some subject is jointly golden and a mountain, a conclusion that would contradict the assumption that *golden mountain* is a false idea.

With this background we may address directly the issue of existential import. It follows directly from the truth-conditions for the universal affirmative and the earlier definition of false idea, that the terms of a true affirmative proposition carry existential import. The argument goes as follows. By definition, a false idea is one such that if it is affirmed universally of a subject, the proposition is false. But if a false term were the subject of any universal affirmative, its extension and that of the predicate restricted by

76 The *Logic* (I, ix) characterizes a clear idea as one that "*nous frappe vivement*," like pain. For Aquinas clarity, which finds its ultimate form in the beatific vision, is analogous to light in vision. It is something that when added to an essence brings it to consciousness. When Paul was knocked to the ground in the blinding light, he was, according to Thomas, experiencing God's essence with clarity. He writes,

But *claritas*, which is the principle of the vision of the divine and the actualization of glory, is glory's endowment. If therefore the mind of Paul saw God through an essence, and was illuminated by the light that was the principle of his vision, he was simply glorified. [*Sed claritas quae est principium divinae visionis, quae est actus gloriae, est dos gloriae. Si igitur mens Pauli Deum vidit per essentiam, et illustrata fuit luce quae est principium huius visionis, fuit simpliciter glorificata.*] *De veritate*, q. 13 a. 2 arg. 4.

Hence, clarity is a necessary condition for being conscious of the content of objective being. It comes in degrees and presumably, as Aquinas holds, is a gift of God. See *De veritate*, q. 13 a. 2 co., *Responsio*.

77 See, for example, *LAP* I,ix. *KM* V,157; *B* 14.

the subject would both be empty and hence identical. The assertion would hence be true. That is, any universal affirmative with a false term as subject would be true. Hence, for any term, that term is the predicate of some true universal affirmative that affirms, according to its truth-conditions, a self-identity. Hence there is no false term. It follows that if a universal affirmative is true when the extension of the subject is identical to that of the restricted predicate with no further condition, and if the content of a false idea is false of every subject, then there are no false ideas. The same result may be expressed as a contrapositive: a universal affirmative with an empty subject term is false even though the extensions of the subject and restricted predicate are identical.

The argument may be formulated as a reductio. Suppose both that *every S is P* is true and that *S* is a false idea. It follows that *S* is true of no actual object. But the truth or falsity of an idea is a function of the truth or falsity of propositions. What proposition could it be in this case? Because the term is false of every object, any universal affirmative with *S* as its predicate must be false: for any term *T*, *every T is S* is false. Then by the definition of restriction, it follows that *SP* is false of every term. By the truth conditions for the universal affirmative, then, it follows for every *T* that *every T is SP* is false. Hence *every S is SP* and *every S is P* are false. Hence the original assumption that *every S is P* is true and *S* is a false idea has led to a contradiction.

False ideas, however, have an important place in the *Logic*'s Cartesian epistemology. Accordingly, to accommodate them within the theory, it is necessary to attribute existential import to true affirmatives.

It is instructive to note the unacceptable consequences of the opposite interpretation. If the truth-conditions for the universal affirmative were defined so as to be true when the extensions of the subject and predicate are empty and hence the same, the following counter-intuitive propositions would be true:

every square circle is an infinite being,
every dragon is a god,
every dragon is a square,
every dragon is a round square,
some dragon is a god,
some square circle is a square,
some square circle is a cube,
something is a round square.

It would also follow that any traditional contrary pair, like *round* and *square*, or *odd* and *even* could both be true of the same subject. It would also follow that *nothing is a square circle* is false.

As Jennifer Ashworth as shown, logical theory from the Middle Ages to the 17th century as a standard practice required that a true universal affirmative have a non-empty subject.[78] If the *Logic* had rejected this assumption, it would have been contradicting centuries of doctrine. However, nowhere does the *Logic* explicitly reject this practice, nor does it cite as true any example of a universal affirmative with an empty subject.[79]

Jean-Claude Pariente has advanced the contrary interpretation, namely that the truth-conditions for categorical propositions do not carry existential import.[80] His reasoning is that the truth-conditions do not carry existential import because they are formulated in terms of relations on extensions, which are relations among ideas, not things in the world, and that in principle ideas may be empty. What he fails to observe is that if A-propositions with empty terms are true, then there can be no false ideas. Since the *Logic* is committed to false ideas, its semantics must also be assuming, in common with the logic of the period, that the terms of a true affirmative are non-empty.

It should also be remarked that particular affirmatives with a false idea as subject term, like *some square circle is a cube,* are counter-intuitive indeed. Even first-order logic, which accepts as true universal conditionals with false antecedents, treats these existential cases as false. Indeed, it is hard to conceive of a rationale for a logic that would accept the truth of particular affirmatives with empty terms. But given that the *Logic* is committed to subalternation, it would follow that if the *Logic* were also committed to the truth of universal affirmatives with empty terms, these particular affirmatives with empty terms would also be true. Again, the *Logic* cites no examples of true particular affirmatives with empty terms. Nor need it be interpreted as doing so. As I have shown above, its logic of ideas

78 Subalternation was a fixed feature of syllogistic logic. On the existential presupposition of universal affirmatives see Ashworth 1973.

79 Pariente 1985, p. 243 remarks correctly that two affirmative propositions with empty subjects could both express different propositions, but he fails to remark on their truth-values. In the conventional logic of the period, as well as the Port-Royal Logic, though non-synonymous both would be false.

80 See the references in note 1.

is perfectly coherent while at the same time maintaining traditional commitments to existential import.

In conclusion, I hope I have shown in some detail that not only is the existential import of affirmative propositions consistent with the Cartesian semantics; the theory of false ideas requires it. Its epistemology and semantics are formulated in terms of ideas, to be sure, but the Cartesians were dualists, not idealists. Their commitment to occasionalism and the truth of clear and distinct ideas ground signification in actual objects, and as a consequence, an affirmative proposition is true if two conditions are met: its terms must be non-empty and it must correspond to the world.

Works Cited

—— (1976). *Commentarii collegii Conimbricensis e societate Jesu in universam dialecticam Aristotelis.* Hildesheim: Georg Olms Verlag [reprint of 1607].

Abelard, Peter (1994). From the "Glosses on Porphyry" in his *Logica ingrendientibus.* In Paul Vincent Spade (Ed.), *Five texts on the mediaeval problem of universals.* Indianapolis: Hackett.

Arnauld, Antoine (1990). Stephen Gaukroger (Trans.), *On true and false ideas.* Manchester: Manchester University Press.

Alanen, L. (1990). Cartesian Ideas and Intentionality. In M. K. L. Haaparanta, I Niiniluto (Eds.), *Language, knowledge, and intentionality: perspectives on the philosophy of Jaakko Hintikka* (pp. 344–369). Helsinki: Philosophical Society of Finland.

Arnauld, Antoine and Pierre Nicole (1996). Jill Vance Buroker (Trans.), *Logic or the art of thinking.* Cambridge: Cambridge University Press.

Ashworth, E. J (1996). Antonius Rubius on objective being and analogy: One of the routes from early fourteenth-century discussions to Descartes' third meditation. In *Meeting of the minds: the relations between medieval and classical modern European philosophy. Acts of the international colloquium held at Boston College, June 14–16, 1996 organized by the société internationale pour l'étude de la philosophie médiévale.*

Ashworth, E. J. (1973). Existential sssumptions in late medieval logic. *American Philosophical Quarterly, 10*, 141–47.

Auroux, Sylvain (1993). *La logique des ideés*. Montréal, Paris: Bellamin, Vrin.

Boole, George (1951). *An investigation of the laws of thought, on which are founded the mathematical theories of logic and probabilities*. N.Y.: Dover.

Boole, George (1948). *The Mathematical Analysis of Logic* [1847]. Oxford: Blackwell.

Brown, D. (2007). Objective being in descartes: That which we know or that by which we know? In H. Lagerlund (Ed.), *Representation and objects of thought in medieval philosophy* (pp. 135–153). Aldershot: Ashgate.

Buridan, John (1985). Peter King (Trans.), *Jean Buridan's logic: The treatise on supposition, the treatise on consequence*. Dordrecht: Reidel.

Buridan, John (1988). Question on Aristotle's *De anima* (third redaction). In John Alexander Zupko (Ed.), *John Buridan's philosophy of mind: An ddition and translation of book iii of his 'Questions on Aristotle's De anima (third redaction), with commentary and critical and interpretative essays* (pp. 288–312). PhD. Dissertation, Cornell University. Ann Arbor, MI: University Microfilm.

Buridan, John (2001). Gyula Klima (Trans.), *Summulae de dialectica*. New Haven: Yale University Press.

Buroker, Jill Vance (1993). The Port-Royal semantics of terms. *Synthese, 96:3*, 455–475.

Caietanus, Thomas de Vio (1936). P. Isnardus, M. Marega O.P. (Eds.), *Commentaria in Porphyrii Isagogen, ad Praedicamenta Aristotelis* scripta philosophica. Roma: Insitutum Angelicum.

Cronin, T. J. (1966). *Objective being in Descartes and Suárez*. Vol. 154, *Analecta Gregoriana, 154*. Rome: Gregorian University Press.

Descartes, René (1897–1909). C. Adam et P. Tannery (Eds.). *Œuvres de Descartes*. 11 vols. Paris: Vrin.

Dominicy, Marc (1984). *La naissance de la grammaire moderne*. Bruxelles: Pierre Mardaga.

Duns Scotus, John (1987). Peter King (Trans.), Ordinatio II d. 3 p. 1 q. 1. Ms.

Duns Scotus, John (1999). *Quaestiones in librum Porphyrii Isogoge*, R. Andrews, O. Bychkov, S. Ebbesen, G. Etzkorn, Fr. G. Gál, Fr. R. Green, R., Plevano, A. Traver, R Wood, T. Noone (Eds.), *B. Iohannis Duns Scoti opera philosophica*. St. Bonaventure: Franciscan Institute.

Eustace-of-St.-Paul (1609). *Summa philosophiae quadripartita, de rebus dialecticis, moralibus, physicis et metaphysicis*. Paris.

Fonseca, Pedro da (1964). *Institutionum dialecticarum libri octo*. Vols. 1 and 2. Coimbra: Universidade de Coimbra.

Fonseca S.J., Petrus (1599). *Commentarii in xii libros metaphysicarum Aristotelis*. Frankfurt.

Garber, Daniel (1993). Descartes and occasionalism. In Steven M. Nadler (Ed.), *Causation in early modern philosophy* (pp. 9–26). University Park, PA: Pennsylvania State University Press.

Gaukroger, Stephen (1989). *Cartesian logic*. Oxford: Clarendon Press.

Karger, Elizabeth (1999). Ockham's misunderstood theory of intuitive cognition and abstractive cognition. In Paul Vincent Spade (Ed.), *The cambridge companion to Ockham* (pp. 204–26). Cambridge: Cambridge University Press.

King, Peter (1985). *Jean Buridan's logic, The treatise on supposition, The treatise on consequences*. Dordrecht: Reidel.

King, Peter (2007). Rethinking representation in the middle ages: A vademecum to medieval theories of mental representation. In H. Langerlund (Ed.), *Representation and_objects of thought in medieval philosophy* (pp. 81–100). Aldershot: Ashgate.

Kremer, Elmar and Denis Moreau (Eds.). (2003). *Œuvres philosophique d'Arnauld*. 6 vols. Bristol: Theommes Press.

Marenbon, John (1997). *The philosophy of Peter Abelard*. New York: Cambridge University Press.

Martin, John N. (2004). Ecthesis and existence in the syllogistic. In *Themes in neoplatonic and aristotelian Logic* (pp. 19–24). Aldershot: Ashgate.

Martin, John N. (2011). Existential Import in Cartesian Semantics, *History and Philosophy of Logic, 32:3*, 211–239.

Nadler, S. M. (1989). *Arnauld and the cartesian philosophy of ideas*. Manchester: Manchester University Press.

Nadler, S. M. (1992). *Malebranche and ideas*. New York, Oxford.

Nadler, S. M. (Ed.) (1993). *Causation in early modern philosophy : Cartesianism, occasionalism, and pre-established harmony*. University Park, PA: Pennsylvania State Univeristy Press.

Normore, Calvin (1986). Meaning and objective being: Descartes and his sources. In Amélie Oksenberg Rorty (Ed.), *Essays on Descartes' Meditations* (pp. 223–242). Berkeley: University of California Press.

Ockham, Willam of (1974). Michael J. Loux (Ed. & Trans.), *Ockham's theory of terms: Part I of the summa logicae*. Notre Dame, IN: University of Notre Dame.

Ockham, Willam of (1981). *Reportatio, question XIII. Opera philosophica et theologica,* Book II. St Boniventure, N.Y.: Franciscan Institute.

Panaccio, Claude (2004). *Ockham on concepts.* Aldershot: Ashgate.

Pariente, Jean-Claude (1985). *L'Analyse du language à Port-Royal.* Paris: C.N.R.S., Éditions de Minuit.

Robert Pasnau (1997), *Theories of cognition in the later middle ages.* Cambridge: Cambridge University Press.

Raconis, M. de (1648). *Totius philosophia, hoc est logicae, moralis, physicae et metaphysicae, brevis et accurata, facilisque et clara methodo disposita tractatio.* Paris.

Spade, Paul Vincent (2007). *Thoughts, words and things; An introduction to late mediaeval logic and semantic theory.* Ms. http://pvspade.com/Logic/docs/Thoughts,%20Words%20and%20Things1_2.pdf.

Suárez, Francisco (1856–1878). *Opera omnia.* Paris: Ludovicum Vivès.

Tachau, K. (1988). *Vision and certitude in the age of Ockham.* Leiden: Brill.

Thompson, Michael. The representation of life. In Gavin Laurence, Rosalind Hursthouse, Warren Quinn (Eds.), *Philippa Foot and moral theory.* Oxford: Clarendon Press.

Toletus S.J., F. (1596). *Commentaria una cum quaestionibus in universam Aristotelis logicam.* Cologne.

Toletus S.J., F. (1985). *Cur de specie post genus & non de differentia egerit? Commentaria in universam Aristotelis logicam.* Wilhelm Risse (Ed.), *Opera omnia philosophica,* Vol. I [Köln 1615/16]. Hildesheim: Georg Olms.

Wells, Norman J. (1990). Objective reality of ideas in Descartes, Caterus, and Suárez, *Journal of the history of philosophy 28:1,* 31–61.

The Medieval Modal Octagon and the S5 Lewis Modal System

JUAN M. CAMPOS-BENÍTEZ[1]

Abstract

The modal Medieval Octagon for both quantification and modality shows several interesting relations. We take modal conversion for AE and II sentences below (where the first vowel stands for quantifiers and the second for modes. For example: II: some German may be a philosopher, some philosopher may be a German). The informal reading poses no problems, but when formalized they cannot be proved unless we make some assumptions and use the S5 Lewis Modal System.

Keywords: Medieval Modal Octagon, modal conversion, S5 Modal System

1. The Equivalences and the Squares

There is a modal square combining modality and quantification[2]. It presupposes both the equivalences between modes and quantifiers. We take both modes and quantifiers as primitives. The equivalences are[3]

1 Benemérita Universidad Autónoma de Puebla, México.

2 The *doctrine* of the Octagon may be found in fourteenth century logical works, Albert of Saxony's *Perutilis logica* (1988) and Jean Buridan's *Summulae de dialectica*. The reader may take a look on an early figure of the Octagon at http://www.st-andrews.ac.uk/~slr/medieval.html (visited on January 23th, 2009). Buridan seems to be the first one to use the figure of the Octagon.

3 The symbols are $\Diamond$ and $\Box$ for possibility and necessity; $(\exists x)$ and (x) for particular and universal quantifyers; $\sim$, $\wedge$, $\vee$, $\supset$, $\equiv$ for negation, conjunction, disjunction and equivalence; $\rightarrow$ for strict implication and ◄--► for logical independence respectively.

$(x)Fx$	$\equiv$	$\sim(\exists x)\sim Fx$	$\Box p$	$\equiv$	$\sim\Diamond\sim p$
$(x)\sim Fx$	$\equiv$	$\sim(\exists x)Fx$	$\Box\sim p$	$\equiv$	$\sim\Diamond p$
$\sim(x)Fx$	$\equiv$	$(\exists x)\sim Fx$	$\sim\Box p$	$\equiv$	$\Diamond\sim p$
$\sim(x)\sim Fx$	$\equiv$	$(\exists x)Fx$	$\sim\Box\sim p$	$\equiv$	$\Diamond p$

The equivalences are expressed in the square. Let us take "man" as the subject and "argue" as the predicate to get several sentences combining quantifiers, according to the traditional square,

Figure 1

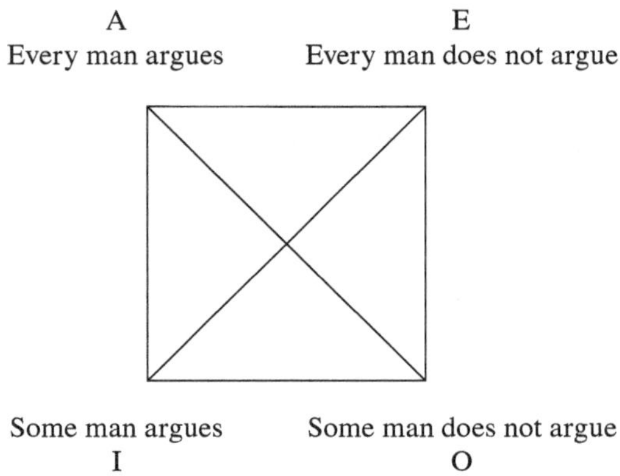

Sentence E may sound a little unusual but I want to stress the universal quantifier "every" and then the negation inside the sentence. It is equivalent to "no man argues". There is a modal square paralleling the quantified square where the extremes are "necessarily", "impossibly", "possibly" and "possibly…not". They are also called universal and particular extremes, as they are in the former square[4]. Every extreme admits equivalences, so it may admit two sentences, but we will just take one form.

4 In an *Opusculum* attributed to Aquinas for instance, it is noticed the similarity: *Attendendum est autem quod necessarium habet similitudinem cum signo universali affirmativo (…) impossibile cum signo universal negativo (…) contingens vero et possibile similitudinem habent cum signo particulari.* De propositonibus modalibus, Aquinas (1976, p. 420).

Figure 2

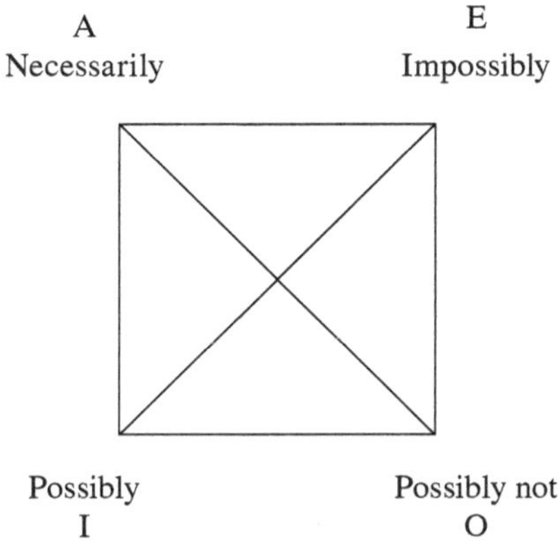

Now, every sentence of the former square admits a mode between subject and predicate. If we take the modes of necessity and possibility in their adverbial form we obtain two modal sentences for each one of the sentences of the square.

2. The Modal Quantified Squares

The square deals with contrary, contradictory, subcontrary and subalternate sentences according to the quantifyer and to the mode. Some sentences are unrelated to each other (they were called *disparatae*, independent sentences). The square shows all these relations regarding

a) both quantifiers and modes (simpliciter)
b) only quantifiers (by reason of the subject which is quantified)
c) only modes (by reason of the mode)

We have two squares, one for quantification, another for modality, and we have to combine them. Both have universal sentences (every, necessarily) and particular sentences (some, possibly). Each quantified sentence admits two modes, so we have eight sentences. Let me use two capital letters, the first for the quantifier and the second for the mode. The quantifier ranges over the subject of the sentence ("man") and I will use "subject" instead of "quantifier" so (AA) means a universal affirmative sentence according to the subject and according to the mode, (AE) means a universal affirmative sentence according to the subject but universal negative according to the mode, (AI) expresses a universal sentence according to the subject but particular respect to the mode, (AO) universal according the subject and negative particular respect the mode. They are[5]

AA Every man necessarily argues
AE Every man impossibly argues
AI Every man possibly argues
AO Every man possibly does not argue

Which are combined in this unusual square since each extreme is universal according to the subject

Figure 3

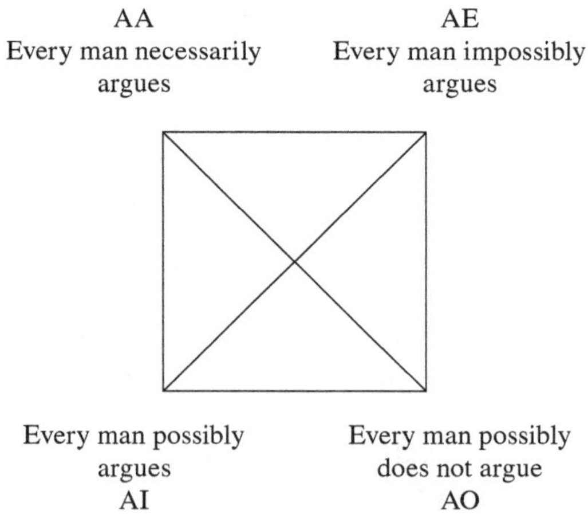

5 I changed the examples, the Latin text in Buridan reads: *Omnem hominem necesse est currere, Omnem hominem impossibile est currere, Omnem hominem possibile est*

(AA) and (AE) are absolutely contrary sentences, regarding both the subject and the mode.

(AI) and (AE) are contraries by reason of the subject (since "impossibly" is equivalent to "necessarily not"); (AA) and (AO) are also contraries by reason of the subject. (AA) (AI) and (AE) (AO) are subalterns by reason of the mode, given that the universal implies the particular. (AI) and (AO) are independent sentences, which means that they are not contrary neither subcontrary nor contradictory or subaltern; they have no relation at all (besides being independent of each other, a symmetrical relationship).[6]

The particular sentences by reason of the subject are[7]

IA	Some man necessarily argues
IE	Some man impossibly argues
II	Some man possibly does argue
IO	Some man possibly does not argue

currere, Omnem hominem possibile est non currere. Cfr. *Summulae de Dialectica*, Book 1, ch. 8. All latin quotations from the online version; he gives nine equivalent forma for each one. Albert of Saxony uses variables: *omne B possibile est esse A, nullum B potest esse A, omne B necesse est esse A.* Cfr. his *Perutilis logica*, (1988, p. 322).

6 Gyula Klima presents this relation as a negation of contradictories. He also shows further applications of the Octagon: it also fits sentences with oblique terms (genitive) and sentence where the predicate is explicitely quantified, cfr. his translation of The *Summulae*, Klyma,*(*2001, p. 45).

7 Again from Buridan: *Quendam hominem necesse est currere, Quendam hominem impossibile est currere, Quendam hominem possibile est currere, Quendam hominem possibile est non currere.* Ibidem.

Which form the following also unusual square since all extremes are particulars

Figure 4

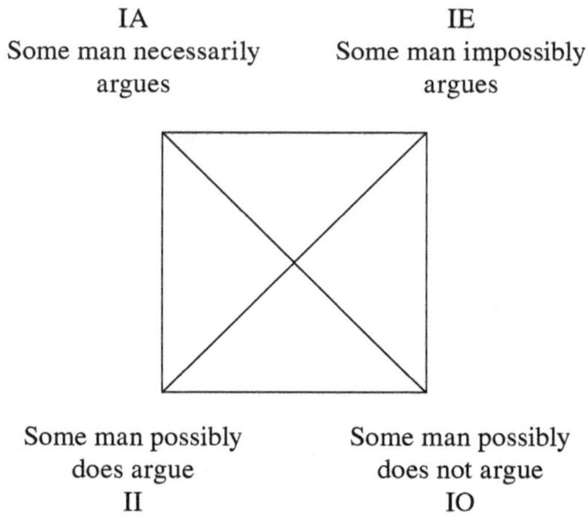

There are several squares that can be formed from the two squares above but let me just join the two squares above to get the picture, the modal Octagon

Figure 5

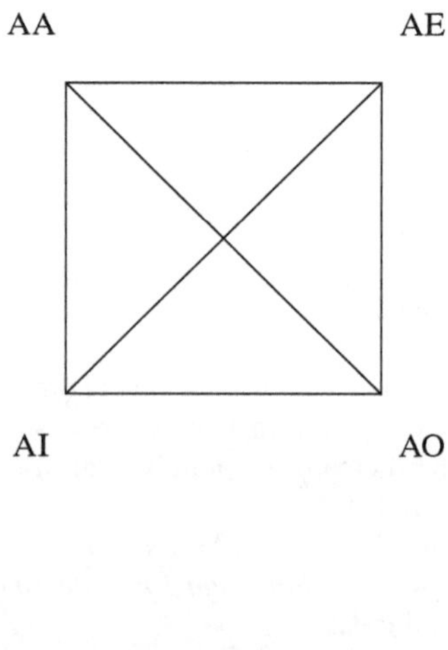

Figure 6

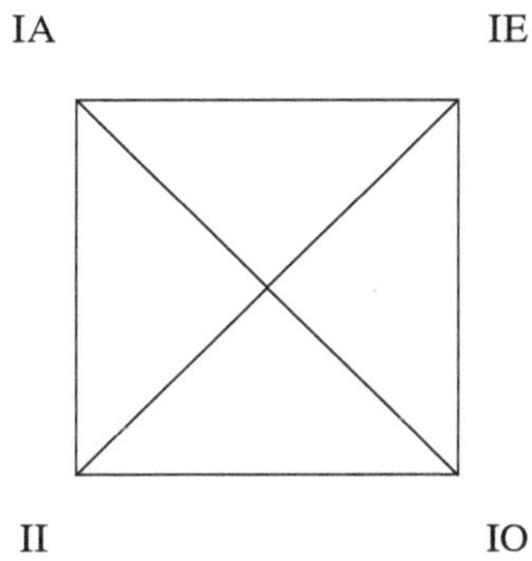

There is a blank between these two squares since (AI) and (IA) have no relation at all: they are neither subordinates nor subcontraries. They cannot be contradictories given their same quantification. The same holds for (AO) and (IE) but also for (AI) and (AO) and (IA) and (IE), so we may form another unusual square for independent sentences, like this

Figure 7

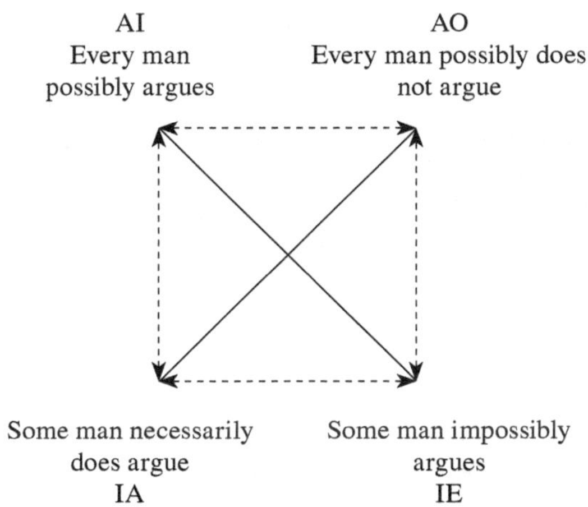

where the double arrows expresses the independency but the diagonals point to contradictories. We could as well duplicate the number of the squares but it is time to get the whole picture.

3. The Modal Octagon

This is the square sketchelly expressed: all the relationships of the traditional square hold, plus the new one. Independent sentences were called "disparatae".

Using the predicate calculus for the square we have these formulas

AA:	$(x)[Mx \supset \Box Ax]$	AE:	$(x)[Mx \supset \sim\Diamond Ax]$
AI:	$(x)[Mx \supset \Diamond Ax]$	AO:	$(x)[Mx \supset \Diamond\sim Ax]$
IA:	$(\exists x)[Mx \,\&\, \Box Ax]$	IE:	$(\exists x)[Mx \,\&\, \sim\Diamond Ax]$
II:	$(\exists x)[Mx \,\&\, \Diamond Ax]$	IO:	$(\exists x)[Mx \,\&\, \Diamond\sim Ax]$

The relations among these eight sentences are expressed in the Octagon. Since they are governed by a quantifier they follow the usual non-modal rules, plus existential import for affirmative universals. However, their proofs in a natural deduction system presuppose the modal equivalences, necessity elimination, possibility introduction rules. They may be proved in the weakest modal system (K).[8] They are also the kind of sentences that appear in modal Syllogistic though it is not easy to deal with it.[9] The sentences of the Octagon are all divided modal sentences. Composite modals are those sentences where the modal operator governs the whole sentence.

8 Cfr. Girle, (2000, p. 31).

9 Modal syllogistic is rather complex. It combines different modal sentences along wiht non modal or assertoric sentences. The result is quite complex. Sometines a special notation is needed to meet its subtleties. The reader may consult Paul Thom (2003, p. 171) who presents the Octagon in a new arrangement where the relations of subalternation and contradiction are emphasized.

Other relations also hold: contra positives, for instance, are equally proved in K, but modal conversion shows special problems. [10]

4. Modal Conversion

4.1 The II Formula

I am interested in two modal quantified sentences and their conversions. They are (AE) "Every man impossibly argues" and (II) "Some man possibly does argue", negative universal and affirmative particular respectevely. It is useful to change the examples to see that they are intuitively valid.

II: "Some philosopher may be a German".

which is equivalent by simple conversion to

II: "Some German may be a philosopher"

Put into symbols

$$(\exists x)[Px \,\&\, \Diamond Gx] \equiv (\exists x)[Gx \,\&\, \Diamond Px]$$

By natural deduction rules or tableaux techniques we can check if the formula is valid. It is not. The problem is not the quantifyer, the problem is the mode or more precisely its absence. Using "a" as a name for an arbitrary individual we may instantiate the formula to see the problem. So a

10 There are modal conversion for sentences different from the ones treated here. Ockham, for instance, takes sentences where both the subject and the predicate are modalized ("Everyting which is possibly B is possibly A", which converse is "Something which is possibly A is possibly A"), and even sentences with singular terms and no quantification ("*creans potest esse Deus, igitur Deus potest esse creans*"), quoted by Lagerlund, (2000, p. 115. Conversion was applied to time as well: " 'This boy can be and old man' is converted into 'A future old man can be this present boy' ", cfr. Knuutila, (1993, p. 111). Medieval Logic took conversion seriously and posed different problems in different contexts. I just want to face one problem taken from the modal Octagon.

is a philosopher and he could be a German; a is German and he could be a philosopher.

$$[Pa \,\&\, \Diamond Ga] \equiv [Ga \,\&\, \Diamond Pa]$$

Taking one side of the equivalence $[Pa \,\&\, \Diamond Fa] \supset [Fa \,\&\, \Diamond Pa]$, we can easily infer one conjunct but not the other and vice versa with the other side of the equivalence. The arrows show what can be inferred and the vertical lines what cannot

Figure 8

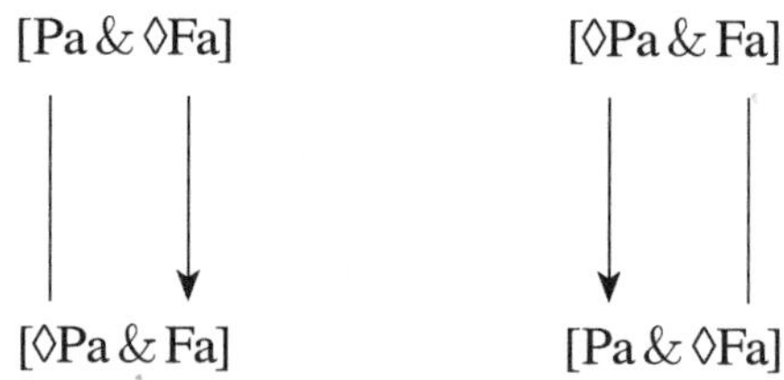

The actual implies the possible; from an assertoric sentence to a modal one holds the inference provided that the modal operator is the operator of possibility. The opposite direction of the inference is not valid: from the possible to the actual there is no inference.

The theorem is not valid.

As it is, it shows a relation from the actual word to the actual world, from an assertoric sentence to another assertoric sentence. Each side of the equivalence is assertoric even though there is a mode *inside* each one. What if we take them to show a relationship from one possible world to another? Let us put each part of the formula in a circle and see what happens

Figure 9

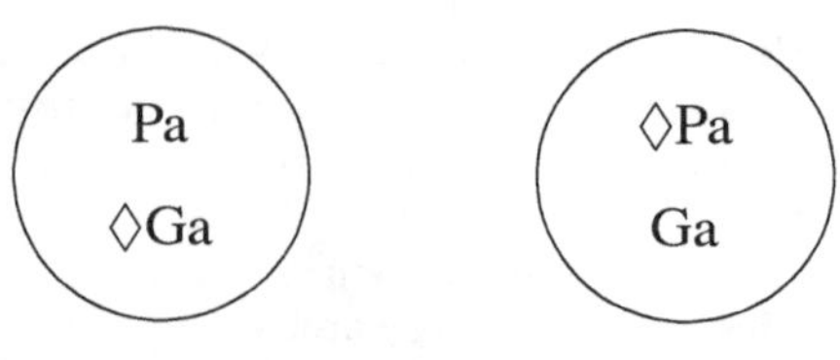

We can see that what it is true in one circle is possible into the other. It suggests that the modal conversion must be a relation between possible worlds, and that they are accesible to each other. But the formula, as it is, is assertoric, it shows no relation between possible worlds. If we take each circle as a possible world and express the accessibility relation with an arrow we get this picture

Figure 10

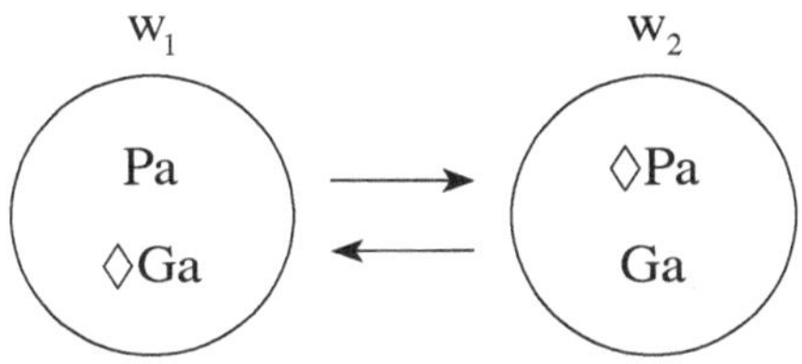

Since the arrows point to one way and to the other way around they are symmetrical, so that w_1Aw_2 and w_2Aw_1, which is characteristic of the S5 Lewis Modal System. The formula is to be rewritten as

$$\Diamond[Pa \,\&\, \Diamond Ga] \equiv \Diamond[Ga \,\&\, \Diamond Pa]$$

or put in a sentential formulation

$$\Diamond[p \,\&\, \Diamond q] \equiv \Diamond[\Diamond p \,\&\, q]$$

And finally the quantified form

$$\Diamond(\exists x)[Px \,\&\, \Diamond Gx] \equiv \Diamond(\exists x)[\Diamond Px \,\&\, Gx]$$

The formulas are S5-valid according to the natural deduction system rules and modal tableux. I present one side of the proof according to natural deduction rules[11]. The other side is symmetrical. See appendix for the proof.

11 See Konyndick, (1986, ch. 2).

4.2 The AE Formula

The AE formula, "Every man impossibly argues" also offers some problems. An E-type sentence may be simply converted. If we say "no man is canine" we obtain "no canine is a man". The modal versions of these are: "no man possibly is canine", "every man necessarily is not a canine", "every man impossibly is a canine" and their converses. The set of humans and the sets of canines never touch nor overlap each other, being this necessarily so. Put in symbols we obtain

$$(x)(Mx \supset \sim\Diamond Cx) \equiv (x)(Cx \supset \sim\Diamond Mx),$$

which is an invalid formula. The problem is not the quantifier. To make it valid does not help adding a possibility operator, as it did in the previous theorem (as a matter of fact it may be modalized using strict implication). Where exactly is the problem? If we try to find a proof for this theorem we prove first one side

$$(x)(Mx \supset \sim\Diamond Cx) \supset (x)(Cx \supset \sim\Diamond Mx)$$

Put it without quantifier we have

$$(Ma \supset \sim\Diamond Ca) \supset (Ca \supset \sim\Diamond Ma)$$

by $\supset$ introduction twice applied we get this set of premises $\{(Ma \supset \sim\Diamond Ca), Ca\}$ and the conclusion $(\sim\Diamond Ma)$. The problem is to pass from the premises to the conclusion. It cannot be since 4 is assertoric and 5 is a necessary sentence. It is not valid to infer from an assertoric sentence to a necessary one.

1.	$(Ma \supset \sim\Diamond Ca)$	assumption
2.	Ca	assumption
3.	$\Diamond Ca$	2 $\Diamond$ introd.
4.	$\sim Ma$	1, 3 modus tollens
5	$\sim\Diamond Ma$	¿?

The same applies to the other side of the theorem.

The answer to this problem, to make valid the inference from "no man is possibly a canine" to "no canine is possibly a man", which is intuitively valid, seems to lie in a hidden necessity. We can say: if x is a human cannot be a canine, and vice versa. Conversion for E-sentences seems to be necessary since an E expresses disjointed sets, but that is not captured by the usual formulation as above. We had to add the operator for possibility to make valid a theorem for II-type sentences, which are particulars. Now we face AE-type sentences, which are universals. It makes sense to resort to the necessity operator. Being this so, I express the formula using strict conditionals

$$(Ma \rightarrow \sim\Diamond Ca) \supset (Ca \rightarrow \sim\Diamond Ma)$$

We go now to our former argument to justify the conclusion by

1.	$(Ma \rightarrow \sim\Diamond Ca)$	assumption
2.	Ca	assumption
3.	$\Diamond Ca$	2 $\Diamond$ introd.
4.	$\sim\Diamond Ma$	1, 3 strict modus tollens

Of course there is no official strict modus tollens rule but the rationale is a proof in sentential logic; the tableaux is closed (see Appendix). It is valid in S5 Lewis Modal System.

Now we can express our formulas both at sentential and quantificational level. In the sentential level the variables p and q trade places, p is antecedent and modalized q is consecuent in one side of the equivalence and vice versa.

$$(p \rightarrow \sim\Diamond q) \equiv (q \rightarrow \sim\Diamond p)$$

The valid AE formula thus is

$$(x)(Mx \rightarrow \sim\Diamond Cx) \equiv (x)(Cx \rightarrow \sim\Diamond Mx)$$

A further problem would be the place for the necessity operator when defining the strict implication as a necessary conditional. It could be before or after the quantifier

$$(x)\Box(Mx \supset \sim\Diamond Cx)$$
$$\Box(x)(Mx \supset \sim\Diamond Cx)$$

And I found they are equivalents in S5 Lewis Modal System.

5. Some Final Words

I tried to justify some modal theorems from the Octagon of Opposition which seem not to be valid according to the Predicate Modal Calculus of our days. I am not sure about adding a modal operator to a formula when it is needed. It seems to be an *ad hoc* strategy but otherwise we cannot endorse some inferences which appear intuitevely right.

Certainly it sounds an oddity to talk about modern Modal systems in a Medieval context. It does not mean that the Medievals knew about the Lewis Systems but it does mean that if we want a full understanding of their logics we need something as complete as S5. Another problem is to "find" some "traces" of S5 in an Octagon which is at least seven centuries old!

Appendix

I. The Modal Theorem of Conversion for Particular Sentences

1.	$\Diamond(\exists x)[Px \,\&\, \Diamond Fx]$	assumption
2.	$\Box{\sim}(\exists x)[Fx \,\&\, \Diamond Px]$	assumption
3.	$\Box$ x $Px \,\&\, \Diamond Fx$	assumption
4.	Px	3 & elim.
5.	$\Diamond Px$	4 $\Diamond$ introd.
6.	$\Diamond Fx$	3 & elim.
7.	$\Box$ Fx	h
8.	**$\Diamond Px$**	**5 S5-iteration**
9.	$Fx \,\&\, \Diamond Px$	7, 8 & introd.
10.	$(\exists x)[Fx \,\&\, \Diamond Px]$	9 $\exists$x introd.
11.	$\Diamond(\exists x)[Fx \,\&\, \Diamond Px]$	10 $\Diamond$ introd.
12.	$\Diamond(\exists x)[Fx \,\&\, \Diamond Px]$	6, 7–11 $\Diamond$ elim.
13.	$\Box{\sim}(\exists x)[Fx \,\&\, \Diamond Px]$	2 S4-iteration
14.	${\sim}\Box{\sim}(\exists x)[Fx \,\&\, \Diamond Px]$	12 equivalence
15.	${\sim}[Px \,\&\, \Diamond Fx]$	3–14 ~ introd.
16	$(x){\sim}[Px \,\&\, \Diamond Fx]$	3–15 (x) introd.
17.	${\sim}(\exists x)[Px \,\&\, \Diamond Fx]$	16 equivalence
18.	$\Box{\sim}(\exists x)[Px \,\&\, \Diamond Fx]$	3–17 $\Box$ introd.
19.	${\sim}\Diamond(\exists x)[Px \,\&\, \Diamond Fx]$	18 equivalence
20.	$\Diamond(\exists x)[Px \,\&\, \Diamond Fx]$	1 iteration
21.	${\sim}\Box{\sim}(\exists x)[Fx \,\&\, \Diamond Px]$	2–20 ~ introd.
22.	$\Diamond(\exists x)[Fx \,\&\, \Diamond Px]$	21 equivalence

II. Strict "Modus Tollens"

1.	$p \rightarrow \sim\Diamond q$	assumption
2.	$\Diamond q$	assumption
3.	$\Box$ p	assumption
4.	$p \supset \sim\Diamond q$	1 ST- iteration
5.	$\sim\Diamond q$	4,5 modus ponens
6.	$\Diamond q$	2 **S5-iteration**
7.	$\sim p$	3–6 ~ introd.
8.	$\Box\sim p$	3–7 □ introd.
9.	$\sim\Diamond p$	8 equivalence

III. Tableaux

We have premises and the negation of the conclusion, all in $world_1$:

1.	$(Ma \rightarrow \sim\Diamond Ca)$		w_1	assumption
2.	Ca		w_1	assumption
3.	$\Diamond Ma$		w_1	negation of the conclusion
4.	Ma		w_2	(3), Accesibility relation m_1 A m_2
5.	$\sim Ma$	$\sim\Diamond Ca$	w_2	from (1) Accesibility relation w_1 A w_2
6.	X	$\Box\sim Ca$	w_2	(5) modal equivalence
7.	4	$\sim Ca$	w_1	(6) Accesibility (S5 symmetry[12]) relation w_2 A w_1
		X		
		2		

The tableaux closes its branches, the formula is S5-valid.

12 Cfr. Graham Priest, (2001, p. 38).

Buridan's *Magna Figura* ("Big Diagram"), as it appears in Johannes Buridanus: *Compendium Totius Logicae*, Venice, 1499, reprinted Minerva: Frankfurt/Main, 1965 (courtesy of Gyula Klima).

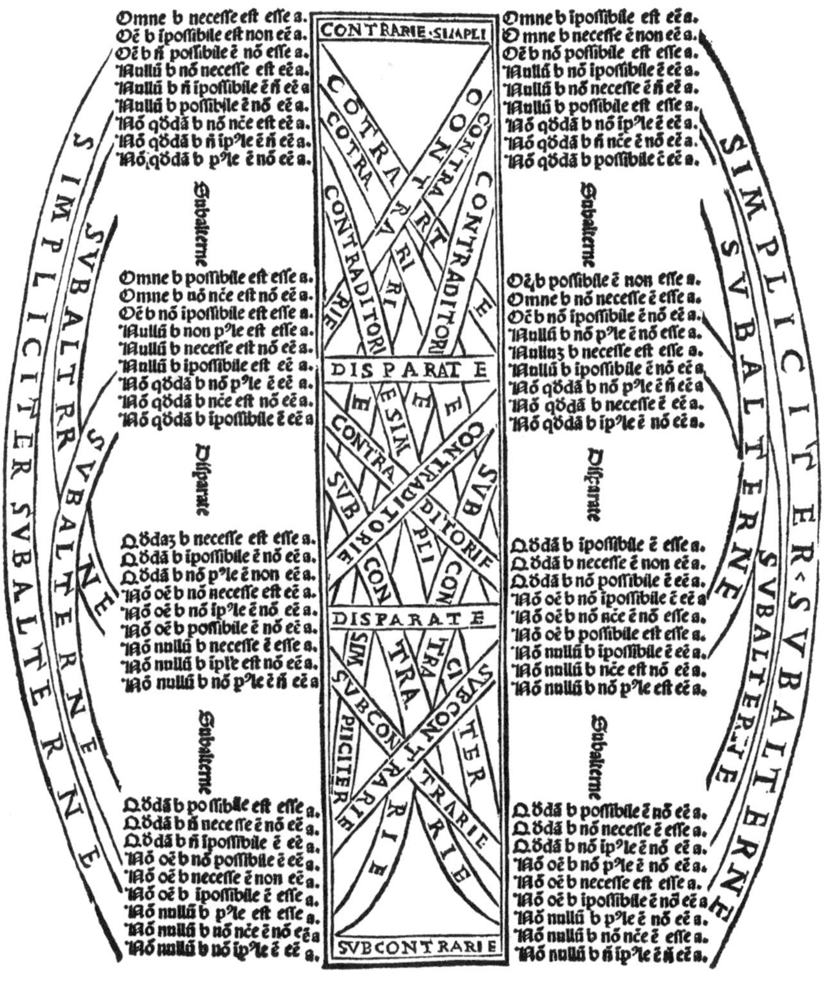

Bibliography

Aquino, Thoma de (1976). De propositionibus modalibus. In *Opera Omnia iussu Leonis XIII P. M. edita*, vol. XLIII. Roma: Editori di San Tommaso.

Buridan Jean, *Summulae de dialectica*, avalaible at http://individual.utoronto.ca/p.king/resources/buridan/Summulae_de_dialectica.txt

Buridan Jean (2001). *Summulae de dialectica*. An annotated translation, with a philosophical Introduction by Gyula Klima, New Haven & London: Yale University Press.

Girle, Rod (2000). *Modal logics and Philosophy*. Montreal: McGill-Queen's University Press.

Knuutila, Simo (1993). *Modalities in Medieval Philosophy,* London and New York: Routledge.

Konyndick Kenneth (1986). *Introductory Modal Logic*. Notre Dame: University of Notre Dame Press.

Lagerlund, Henrik (2000). *Modal Syllogistics in the Middle Ages*. Boston: Brill.

Priest, Graham (2001). *An Introduction to Non-Classical Logic*. Cambrige: Cambrige University Press.

Saxony, Alberto de (1988). *Perutilis logica*, Critical edition of the Latin text and Spanish translation by Angel Muñoz. Mexico: UNAM.

Thom, Paul (2003). *Medieval Modal Systems. Problems and Concepts*. United Kingdom: Ashgate Publishing, Ltd.

Part 2: Diagrams

From the "Logical Square" to the "Logical Poly-Simplexes"

A Quick Survey of what Happened in Between

ALESSIO MORETTI

Abstract

In this paper four things are sketched. First, the philosophical reasons which led different authors in different times to discover, adopt, modify, discard or discuss the logical square (or square of opposition). Second, the history of the recent, unexpected mathematical discoveries which have lead from the notion of logical square to that of logical poly-simplex of dimension m. Third, the main conceptual lines of the flourishing new branch of mathematics thus obtained – that is "NOT" (for *N*-Opposition Theory), alias the "geometry of oppositions" – which nowadays hosts and legitimates, as one of its founding formal elements, the logical square. Fourth, a series of possible epistemological consequences of the preceding, both for logic and for the humanities.

Keywords: logical square, logical hexagon, logical bi-simplexes, logical poly-simplexes, *N*-Opposition Theory (NOT), the geometry of oppositions, Aristotelian p^q-semantics, Aristotelian p^q-lattices, neo-structuralism

Introduction

The logical square is an abstract object of the highest importance for Western culture. It was famous, as a central key in the educational system of the Middle Ages. As such, it also has influenced, even much later, a large part of the Western thought. But the square has also become useless; it has been abandoned by the analytical philosophers and logicians of the 20th century. The main reason for this seems to have been the charge, against it, of implying "undue existential import". Nowadays this charge is seemingly being defeated. But truly speaking, the square remains mysterious: it is a piece of "geometry" in a world – the one of Aristotelian logic and philosophy – where no geometry whatsoever is neither expected nor wished. And, mysteriously, the square is *interdisciplinary*: in the 20th century, if not by the analytical philosophers, it has been favoured by almost all the main structuralist thinkers, who adapted it in several ways for use in their respective pioneering research domains (the main branches of the humanities). Indeed, the square is "lonely", for its aforementioned geometrical and visual flavour is a totally isolated – and suspect – behaviour inside logic and no other comparable object was known for long in the logical realm. Then the square has suddenly turned out to be "prophetic": for it has recently opened to "*n*-opposition theory" (NOT), the general, rich geometry of opposition, which turns out to be a new branch of mathematics (similar to, but different from, say, graph theory or knot theory). The previous cryptic remarks become clearer when one examines how the path of science has led from the logical square, an "unconscious" (and lonely) logical bi-simplex of dimension 1, to the "conscious" (and numerous) logical poly-simplexes of dimension *m*, and beyond.

1. The "Riddle" of the Logical Square: Its Glory and then Its Decay

The "square of opposition", or "logical square", seems to have been, *stricto sensu*, proposed first by the Latin Apuleius. But the underlying logic of what is commonly called "Aristotle's square" appears much sooner, in Aristotle's work, in the context of a crisis of geometry inside Plato's Academy and of a subsequent new foundation of philosophy by "logic" and of logic by Aristotle's "principle of non-contradiction" (PNC)[1]. Contradiction, in turn, is a kind of "opposition" and the square of opposition embodies the theory of that[2]. Opposition theory is motivated by a fight against the relativism of the Sophists. Aristotle's seminal reflections originated while examining the concept of "falsity" contra the Sophists, by combinatorial considerations on the use of linguistic negation leading to an abstract quaternary structure (Aristotle calls "υπογραφη" a sketched drawing of it: *De Int.*, §13). For the use of the two kinds of negations (in Greek "ουκ" and "μη") of a sentence (and this holds for any human language, he says) gives four and only four possible outcomes (cf. Figure 1).

Aristotle's definitions touch only the "anti-gamma" scheme. The full square of opposition appears later in Apuleius' texts. The meaning of the square of opposition is pedagogical: it is a mnemo-technical means for mastering the main properties of logic and this is the main use it had in the "universities" during the Middle Ages. For doing this, it deals with visual left-right symmetry, top-down asymmetry and with combinatorial definitions. It encodes graphically fundamental logical properties like the PNC and its dual, the principle of excluded middle (PEM). It must be noted that the logical square has different applications: the main ones are "quantificational" (the square puts order into the notions of "all", "none", "some", …) and "modal" (it puts order in the notions of "necessary", "impossible", "possible", …, cf. Figure 2).

1 Cf. Moretti, A., "The Geometry of Oppositions and the Opposition of Logic to It", in: Bianchi, I., & Savardi, U. (eds.), *The Perception and Cognition of the Contraries*, McGraw-Hill, Milano, 2009.

2 Cf. Moretti, A., «La complexité formelle de la "contradiction" dans la géométrie des oppositions», in: Guy, B. (ed.), *Ateliers sur la contradiction*, Presses des Mines, Paris, 2010.

To trouble all these virtues, at least two major drawbacks of the logical square have been pointed out almost since its beginning. The first one concerns its being somehow conflicting with another (implicit) similar model, rather triangular (a "triangle of contrariety"), emerging from another philosophical reading of "possible" and of "some" (these points are raised respectively by Aristotle himself and, much later, by Vasil'ev, cf. Figure 3).

The second main problem with the square is the charge of being flawed by "undue existential import" (UEI), a nasty property seemingly allowing, by deduction (through the subalternation arrows), the undesired existential introduction of, say, real unicorns, golden mountains and the like. This last charge seems to be the main cause for its lasting disfavour among the analytical philosophers and logicians: for in order to keep the square free from this flaw one would have to restrict its quantification to non-empty domains (i.e. to avoid speaking about any kind whatsoever of *fictional* entities) and the analytical philosophers do not want to restrict anyhow the power of abstract mathematical logic. So the square was abandoned.

2. The Many Useful and Strange Uses of the Square of Opposition

Despite the aforementioned flaws, the square has further, nice logical and mathematical applications. The classical main formal uses of the square of opposition in mathematics and logic may concern not only quantification and modality, but also, for ex., the order relations of mathematics and the binary connectives of propositional logic (cf. Figure 4).

Remark that a second "formal square", concerning the formalisation of the "denying (mental or pragmatic) *operations*" (rather than *relations*) was discovered around 1950 (independently) by the unnoticed structuralist logicians Piaget and Gottschalk. The former has proposed a theory of the mental basic operations necessary for a child to develop the ability to reason logically, the latter has proposed a generalised theory of formal "dualities" in mathematics (cf. Figure 5).

If the 20th century has witnessed some sort of "decay" of the square of opposition among the logicians and the philosophers of logic (at least the ones calling themselves "analytical"), it has known, on the contrary, an implicit success among the so-called "structuralist thinkers", who were very much concerned with oppositional holism. The first of them seems to have been the anthropologist and sociologist C. Lévi-Strauss. His own uses have been followed by similar structuralist uses of the square in the humanities: as for instance by T. Parsons (in sociology), Greimas (in semiotics and narratology), Lacan (in psychoanalysis), Badiou (in philosophy), Meyer (in argumentation theory), etc (cf. Figure 6).

In these as in other similar cases, leading intellectual figures have sought inspiration in the square's deepness for founding or creating the oppositional core of their own discipline[3].

3. The Rhapsodic Attempts in Going Further, beyond the Square

We saw ways of *applying* the logical square. But what *is* the logical square, fundamentally? The aforementioned theory of Gottschalk sees it as a particular case of a more abstract object, the already mentioned "quaternality square". Gottschalk demonstrated that the quantificational and the modal readings of the logical square fulfil the more abstract requirements of his own theory, of which they are therefore a particular case. So Gottschalk and Piaget seem to propose, by their PG-square, some kind of "primitive" of the logical square[4]. The logical square can also be seen as a particular case of

3 On the relations of the theory of opposition to structuralism, cf. Moretti, A., "The Geometry of Oppositions and the Opposition of Logic to It", op cit. Among other contemporary studies on the square, seeing it as something "alive", let us mention a very interesting approach, where Pieter Seuren, in order to use it in models of natural language, distinguishes between subtle versions of the historical square (the ones of Aristotle, Apuleius, Boethius and Abelard) by means of a formal technique due to B. van Fraassen and in order to cope with the problem of UEI.

4 On Gottschalk, cf. Moretti, A., *The Geometry of Logical Opposition*, PhD Thesis, Neuchâtel, 2009, ch. 5; Smessaert, H., "The Classical Aristotelian hexagon versus the Modern Duality hexagon" (forthcoming); Moretti, A., & Schang F., "A semantics for assertoric sentences" (forthcoming); Schang, F., "Questions and answers about oppositions" (in this volume).

the so-called "generalised quantifiers theory". In a paper thereupon, Mark Brown endorses the point of view of contemporary logic and thence, by restricting progressively the scope of the quantification, he gets the following overall typology of possible qualitative cases of opposition squares, in which the logical square belongs to type II (type I, the "Boolean square", is the case favoured by the analytical philosophers, where the square becomes useless because it loses its four sides, i.e. 3 of its 4 constitutive opposition relations, cf. Figure 7).

Other proposals have tried to start from the square and to add further pieces. Examples of this are offered by Englebretsen, Wolenski (while commenting a modal system of Hugh MacColl) and Béziau (on the latter, cf. next section of this paper) (cf. Figure 8).

In a similar vein, some people have tried to expand the logical square, so to catch opposition situations with more than four terms. For instance, this has notably been the case with a group of logicians and philosophers investigating deontic logic from the point of view of a moral concept usually untreatable logically speaking, the concept of "supererogation", studied by Joerden, McNamara and Wessels. And another interesting philosophical expansion of the square (a nested one) is the one proposed, for epistemic logic, by Lenzen (cf. Figure 9)[5].

Remark that, so far, all these approaches remain nevertheless unrelated – Kant would say "rhapsodic" – and therefore leave, fundamentally, the square to its lasting mystery.

4. The Logical Hexagon: *The* Square's Conservative Extension

One of the most important changes with respect to the deep understanding of the logical square has been the independent discovery by Sesmat (1951) and Blanché (1953) of an extraordinary "geometrically conservative extension" of it, the "logical hexagon" (cf. Figure 10).

5 On these subjects, cf. Moretti, A., "McNamara's 'deontic dodecagon for DWE' in the light of oppositional geometry" (draft) and Moretti, A., "The Geometry of Standard Deontic Logic", *Logica Universalis*, 3, 1, 2009.

Notice that the problems of the logical square with respect to the triangle of contrariety (cf. Figure 3 *supra*) are solved in the logical hexagon, for the latter contains both the square and the triangle. Notice also that the logical "square" is in fact a "*rectangle*"! (the hexagon AUEOYI contains 3 logical "squares": the classical AEOI, but also the YAUO and the EYIU).

As for its applications, again, most of them (as in Blanché himself and in the narratologist P. Gallais) seem to pertain to structuralist thinkers (cf. Figure 11).

But it must be noticed that the logical hexagon is much less known than the logical square.

One of the recent uses of the logical hexagon concerned a famous dispute about the foundations of "paraconsistent logics"[6]. In this context, while arguing against H. Slater who summoned, against paraconsistency, the authority of the logical square, Béziau discovered – by a structuralist move, that is by taking into account as "modalities" the "null modalities", i.e. "α" and "$\neg\alpha$"– that there are, beside the classical Sesmat-Blanché one, two hidden logical hexagons, one paraconsistent and the other paracomplete – i.e. intuitionist (cf. Figure 12).

Béziau's conjecture was that there is a "solid of opposition", allegedly a stellar dodecahedron, gathering the 3 known logical hexagons. And this new object should take the place of Aristotle's square and be the new "transcendental" of logic (a transcendental constraint testifying the naturalness and thus the legitimacy of paraconsistency). Note that instead of the hexagons Béziau used "stars", i.e. logical hexagons deprived of their arrows (cf. Figure 13).

Notice also that Béziau himself never drew his conjectural 3D "solid of opposition". But such a drawing is useful, for it shows that Béziau's conjecture ("the solid of opposition is a 'stellar dodecahedron'") is not right as such (it deserves a correction). Actually, as it has been demonstrated by us (2004), the true reunion (or "gathering") of Béziau's three hexagons (plus a fourth one unknown to the latter) is rather a "logical cuboctahedron" (cf. Figure 14).

6 Cf. Slater, "Paraconsistent logics?", *Journal of Philosophical Logic*, 24, 1995 ; Béziau, "New Light on the Square of Oppositions and its nameless Corner", *Logical Investigations*, 10, 2003 ; Moretti, "The Critics of Paraconsistency and of Many-Valuedness and the Geometry of Oppositions", *Logic and Logical Philosophy*, Vol. 19 (2010).

But is that all there is after the square of opposition of Aristotle and Apuleius?

5. *N*-Opposition Theory (NOT) and Its Hyper-Solid Geometry

We saw that the logical square has an avatar (a successor), which is the logical hexagon. Now, it turns out that the logical hexagon itself has an avatar: the "logical cube" (Moretti, 2004). "Successors" in which sense? As it happens, the square expresses "2-opposition", the hexagon expresses "3-opposition" and the cube expresses "4-opposition" – i.e., respectively, opposition between 2, 3 or 4 terms (cf. Figure 15).

Do these three opposition structures, ordered in a row indexed over the set {2, 3, 4}, have further avatars? The answer to the question whether there is a geometrical expression for any "n-opposition" (with $n \in$ N, $n \geq 2$) is positive: *there is a complete series of the successors of the logical square.* Technically speaking, n-opposition is based on n-contrariety and n-contrariety turns out to be conceived at best as a mathematical "simplex" joining n points (2 by 2 equidistant), i.e. adopting some kind of minimally metrical geometry (cf. Figure 16).

The reading of this is the following: for any "contrariety simplex", any two points (conventionally blue) of it are mutually contrary. Accordingly, for any number n of contrary points, the simplex of n-contrariety is m-dimensional (with m=n-1).

The second main geometrical ingredient for building the avatars of the square is the notion of "central symmetry": two points of a geometry of opposition are contradictory iff they are centrally symmetric (conventionally, the contradiction of a blue point will be a green point). So contradictory negation is conceived as m-dimensional central symmetry. Conventionally, contradiction is depicted as red: intuitively a "red diagonal" (cf. Figure 17).

Now, by virtue of this, it turns out that any simplex of contrariety generates, by central symmetry, a dual "simplex of subcontrariety". Conventionally, the latter are depicted in green. The blue and green simplexes, duly interlaced (i.e. under central symmetry), form together "logical bi-simplexes" (cf. Figure 18).

The geometry of opposition commands (as can be checked logically) that each of the blue vertices shoots an implication arrow (i.e. a subalternation, conventionally grey) towards each of the green vertices (excepting the one symmetric to it). And *vice versa*, each green vertex receives an arrow from each of the blue vertices (excepting the one symmetric to it). This ends the construction of a complete series of full-fledged avatars of the logical square (cf. Figure 19).

This is also called the series of the αn-structures and it expresses "n-opposition" for any n[7].

The first intuitive use of n-opposition theory consists in expressing oppositions less trivial than the usual black/white or male/female ones (cf. Figure 20).

Remark that Joerden's "polygon" (cf. Figure 9 *supra*) is a logical bi-simplex of dimension 4(!)

But there is more, for there seem to be reasons to believe in the existence of a second series of "opposition solids" (or "opposition polyhedra"), the series of the "gatherings of bi-simplexes", totally distinguished from the series of the bi-simplexes and akin to the logical hexagon (which *gathers* 3 squares) and to the logical cuboctahedron (which *gathers* 4 hexagons – cf. Figure 21).

As it happens, Smessaert and Pellissier have demonstrated that the logical cuboctahedron is just a fragment of the real $\beta 3$-structure (the β-avatar of the hexagon), which in Pellissier's terms turns out to be a "logical tetraicosahedron" (a "logical rhombic dodecahedron" in Smessaert's terms, the two solids are equivalent). This 3D solid contains not 4 (as did the logical cuboctahedron) but 6 logical hexagons (Smessaert discovered the last 2) (cf. Figure 22).

The logical tetraicosahedron has been proven to be the "geometrical" closure of S5 (it exhibits *all* the oppositions of S5. Remark that Wolenski's and Béziau's (cf. Figure 8 *supra*) "logical octogon" is a fragment of this logical tetraicosahedron (such logical octogons generally mean a hidden logical tetraicosahedron). Indeed, it has been proven that there is a whole

7 The theory of the logical bi-simplexes has been proposed first in Moretti, A., "Geometry for Modalities? Yes: Through n-Opposition Theory", in: Béziau, J.-Y., Costa Leite, A., & Facchini, A. (eds.), *Aspects of Universal Logic*, special issue of *Travaux de Logique*, N. 17, University of Neuchâtel, 2004; a deepened presentation of it is also in Pellissier, R., "'Setting' n-opposition", *Logica Universalis*, 2, 2, 2008.

series of so-called "βn-structures"(in Figure 23 we give their symbolic, simplified representation).

Remark that the logical hexagon is both an α- (the $\alpha 3$-) and a β-structure (the $\beta 2$-structure).

There is, finally, a third family (but not a *series*, this time) of opposition structures, the so-called "γ-structures", which are a systematic generalisation of the traditional "modal graphs" used informally and namelessly in modal logic (from Prior onwards)[8]. The γ-structures (conventionally black, deprived of a metric, but expressing negation by central symmetry) give, so to say, the "genetic code" of their own modal system (cf. Figure 24).

So, there are three families of opposition structures: the αn-, the βn- and the γ-structures. Now, one of the main results of NOT, after the discovery of this clarifying conceptual classification, is the so-called "($\gamma \rightarrow \beta n \rightarrow \alpha k$)-translation rule" (for $2 \leq k \leq n+1$) allowing to switch systematically between abstract "modal (or quantificational or conceptual) graphs" and oppositional geometry[9]. But the gains offered by NOT to modal logic are much more general, and in fact NOT has a range of applications much wider than the one of modal logic: NOT is a *conceptual* technology. For instance, it can bring conceptual clarity in geometrical formalisations, as we did with respect to McNamara's aforementioned model, which we proved to be in fact expressible (with a gain of clarity) in the format of NOT (cf. Figure 25)[10].

But, again, is that all there is? Is that all that can be deduced from the logical square?

8 Cf. for instance Chellas, B. F., *Modal Logic. An Introduction*, Cambridge University Press, Cambridge, 1980, p. 149–157; Hughes, G. E. & Cresswell, M. J., *A New Introduction to Modal Logic*, Routledge, London, 1996, p. 56, 60. A Scholastic ancestor of the modal graphs (or γ-structures) seems to have been John Buridan (!).

9 For complete examples of uses of NOT's translation rule, respectively on the system S5 and on the system of standard deontic logic, cf. Pellissier, R., "'Setting' *n*-opposition", *op. cit.*; and Moretti, A., "The Geometry of Standard Deontic Logic", *op. cit.*

10 On the usefulness of NOT for modal logic, cf. Moretti, A., "A geometrical method for classifying and studying quantifiers and modalities", (draft). As for philosophical applications of NOT, cf. Pellissier, R., "Taoist logical hexagon. Philosophical meaning of 3-opposition and weak 2-opposition in cosmology", (draft); and Moretti, A., "McNamara's 'deontic dodecagon for DWE' in the light of oppositional geometry", *op. cit.*

6. NOT Again: The Logical Poly-Simplexes (of Dimension *m*)

We saw that the logical square is just one term among an infinite series of logical bi-simplexes (it is the logical bi-simplex of dimension 1 or "logical bi-segment"). It can be shown that this series, in turn, taken as a whole, is just one term inside an infinite series (of series), namely the series of the "logical poly-simplexes" (of dimension *m*). Let us see why.

One deep philosophical problem with NOT (raised to us in 2004 by A. Costa Leite) is that the bi-simplexes, no matter how big, have only four "opposition kinds", i.e. only four colours (blue, red, green, grey). Because the 4 colours express 2 logical simplexes and the 2 possible kinds of links between them (i.e. negations and implications), NOT seems to be destined to remain "binary", and this clearly seems to depend on the *bi*- prefix of the logical *bi*-simplexes (of dimension *m*). Is it possible to change something to what seems here to be a transcendental geometrical constraint of logic? For, otherwise, the geometry of opposition would remain fundamentally Aristotelian. So, put in other words, the question is: can we have things like "logical *tri*-simplexes", "logical *quadri*-simplexes", and the like? (cf. Figure 26)

In order to answer this unusual challenge, we proposed to go back to Aristotle and try to understand the implicit semantics underlying his construction of the logical basis of the square of opposition. As it happens, Aristotle and his successors (i.e. Apuleius) defined the 3 oppositions (i.e. the 3 over the 4 colours) as follows: 2 things are,

contradictory: iff they cannot be both false and they cannot be both true;
contrary: iff they can be both false, but they cannot be both true;
subcontrary: iff they cannot be both false, but they can be both true.

Then we can add, as a fourth definition (for the fourth term of the logical square), provided we add a further specification (over asymmetry) the following: two (ordered) things are,

subaltern: iff they can be both false and they can be both true, *provided that one can never have the first without the second.*

This kind of move is crucial, for it leads to the discovery, at the meta-level of the logical square, of some kind of profound and powerful (and natural) "ask-answer semantics". More precisely, the implicit semantics of the square can be said to be an "Aristotelian 2^2-semantics" constituted by two questions Q1 and Q2 admitting each an answer chosen in the set {0, 1} (interpreted as "no" and "yes"): the answer to Q1 will be called "A1", the answer to Q2 will be called "A2". This Aristotelian 2^2-semantics leads to four possible outcomes, [0|0], [0|1], [1|0] and [1|1] (in [*x*|*y*] "*x*" stays for A1, "*y*" stays for A2) which can be ordered into an "Aristotelian 2^2-lattice" (square, or, more precisely, lozenge-shaped, cf. Figure 27)

This allows us to perceive a possible way of generalising this semantics by viewing it as a particular instance of a conjectural, more general "Aristotelian p^q-semantics" (furnished with a correlative "Aristotelian p^q-lattice") and, hopefully, to be able to build opposition frameworks radically different from the binary one of Aristotle. As it happens, the logical tri-simplexes become thus indeed reachable: through an "Aristotelian 3^2-semantics", that is through shifting the "*p*" parameter from $p=2$ to $p=3$ (i.e. by changing the number of the possible qualities of answer). This gives therefore 3^2, i.e. 9 possible "[*x*|*y*]" outcomes and they happen to be ordered by an Aristotelian 3^2-lattice[11] (still square-shaped, cf. Figure 28).

This semantics and lattice rule the *general* logical tri-simplexes of dimension *m*. In other words, the tri-simplexes constitute a series (like the previous one of the logical bi-simplexes) and, in order to determine one of the terms of this series, one has to specify the (finite) value of the parameter *m*. For instance, the first two tri-simplexes of the general series are the "logical tri-segment" (the counterpart of the square, or "bi-segment") and the "logical tri-triangle" (the counterpart of the hexagon, or "bi-triangle"). In this case, of these two we can give two kinds of representations, a "decomposed" one and a "whole" one (the decomposed representation of the logical tri-segment and tri-triangle is given in Figure 29).

A more precise and faithful graphical "whole representation" of the tri-segment and of the tri-triangle is possible (cf. Figure 30).

But here stops the possibility of having whole representations of the logical tri-simplexes. For the two next tri-simplexes, those of dimension 3 (i.e. the logical tri-tetrahedron, the counterpart of the logical cube, or

11 Cf. Moretti, A., *The Geometry of Logical Opposition*, op. cit., ch. 18.

bi-tetrahedron) and 4 (this one has no special name), have indeed a decomposed representation (cf. Figure 31).

But they do lack a whole representation (for this one would be geometrically very complex).

The relevant point for us is that when one changes progressively the value of "*p*", one gets, one by one, all the terms of the series of the "logical *p*-simplexes of dimension *m*, an "avatar series" of the series of the avatars of the logical square! (for an intuitive example of this series of series, for *m*=3 this gives the series of the logical *p*-tetrahedra, cf. Figure 32).

Note that, because of their construction, i.e. their having changed the number of possible answers, the logical poly-simplexes are *p*-valued (instead of 2-valued) oppositional structures[12]. But there are even more general avatars of the logical square than the logical *p*-simplexes of dimension *m*. In fact, what about the "*q*" parameter of the Aristotelian p^q-semantics and p^q-lattices?

7. The General Aristotelian p^q-Semantics and p^q-Lattices

The simplest change of the parameter "*q*" leads from the (bi-simplicial) Aristotelian 2^2-semantics to the Aristotelian 2^3-semantics, an oppositional semantics whose possible outcomes (the opposition kinds) are generated by asking not two, but three "compatibility questions", "Q1", "Q2" and "Q3". Because here the possible answers are still of two kinds ("0" or "1", i.e. "no" or "yes"), the number of possible "[*x*|*y*|*z*]" cases is 2^3, that is eight. As it happens, they can be ordered into an Aristotelian 2^3-lattice (a cubic lattice, cf. Figure 33).

Notice that at the present day we still do not clearly know the corresponding oppositional geometry of this semantics and lattice, but strong reasons suggest that there is one. In a similar way, a further increase of

12 For a study of the possible relations of the logical poly-simplexes (and therefore of NOT in general) with many-valued logics, cf. Moretti, A., "The Critics of Paraconsistency and of Many-Valuedness and the Geometry of Oppositions", *op. cit.* In this study we examine the possible relations of NOT to Suszko's criticism to many-valuedness and the defence of the latter offered by Malinowski, Shramko and Wansing.

"q" gives us an Aristotelian 2^4-semantics with its correlative Aristotelian 2^4-lattice (a 4-dimensional lattice isomorphic to an hyper-cubic one) ordering 2^4, i.e. 16 oppositional cases of the (not yet geometrical) form $[x|y|z|w]$. Again, at the present day the inner semantic meaning (and use) of these generalised opposition contexts must still be clarified. But this representation gives us at least a hint as for the kind of shape the general series of the "2^q-lattices" (some kind of dual of the p^2-lattices of the logical p-simplexes) has: the shape of the series of the "(hyper-)hyper-cubes" (cf. Figure 34).

The general p^q-semantics and p^q-lattices combine the features of the p^2-semantics and lattices (i.e. the p-simplicial geometry) and those of the 2^q-semantics and lattices (whose geometry is still unknown). It might be the case that the particular p^q-semantics and p^q-lattices where $p=q$ (like the 2^2-semantics and 2^2-lattice of the logical bi-simplexes) bear, for symmetry reasons, a particular importance (this conjecture must be checked). In any case, if we now look for a synthetic representation of the overall structure of the generalised theory of opposition, we get that the Aristotelian p^q-semantics, with their Aristotelian p^q-lattices, form together an infinite set of "opposition structures" which can be ordered 3-dimensionally into some kind of infinite "*npq*-parallelepiped of opposition", whose three spatial orthogonal dimensions are "n" (the number of opposed terms, or the spatial dimensionality – augmented of one unit – of the involved simplexes), "p" (the number of answers, or the number of simplexes put into mutual relation) and "q" (the number of questions – as said, this parameter is still unclear) (Figure 35).

Two remarks deserve being made here. First, to some it may seem strange (and even suspect) that NOT, the generalised theory of opposition, considers as two distinct matters the fact (1) of having three possible answers ("no", "maybe", "yes") to the Aristotelian "compatibility question" of the 3^2-semantics ("can two things be true together?", etc.) and the fact (2) of keeping just two possible answers ("no", "yes") while having three possible questions (as in the 2^3-semantics): in other terms, what is the real difference between (A) answering "maybe" to the question "can two given things be false together?" and (B) answering "yes" to the question "can two things be undetermined (or 'fizz') together?"? As an intuitive answer to this kind of *mistaken objection* I propose to consider NOT, with its Aristotelian p^q-semantics and p^q-lattices, as some kind of analogon of "substructural logic". In the latter it makes indeed sense to distinguish (strongly) between

"$p \wedge q$" (in the object level) and "p, q" (in the meta-level), although *in some sense* both expressions mean the same thing.

A second remark is that as the logical bi-simplexes proposed by us had as decorating technique Pellissier's sets, it seems that our generalised poly-simplexes have, as decorating technique, a new kind of sheaves discovered by Pellissier for that aim. In other words, if the logical bi-simplexes are a series of α-structures generating a series of β-structures, the logical poly-simplexes seem to be, so to say, a series of A-structures generating a series of B-structures (the latter should become accessible *via* Pellissier's new decorating technique).

8. The Plurality of the Logical Oppositional Geometries

We mentioned the existence of a "PG-square", mysteriously different from Aristotle-Apuleius' one. Now, together with F. Schang we demonstrated that a suited modification of the Aristotelian 2^2-semantics yields a suited Piagetian-Gottschalkian 2^2-semantics yielding that square. This semantics, generalised as Piagetian-Gottschalkian p^q-semantics (and lattices) generates a whole new logical geometry, called the PG-geometry, of which I investigated how the changes of "q" give a "PG-cube" (cf. Figure 36) and then a series of "PG-hypercubes".[13]

Interestingly, in this new geometry it is the "p" parameter, rather than the "q", which is quite hard to explore (in the NOT-geometry things go, as we mentioned, the other way round). Inside the (structuralist) framework, proposed by Béziau, of "universal logic", the abstract structure of such generalised p^q-semantics and lattices is being investigated, as possibly constituting an important (provisory) deep invariant of abstract logic. It is in this context that NOT might incorporate, as a special fragment, Varzi and Warglien's sketch of a "geometry of negation" (2003), where central symmetry seems to mean identity (instead of contradiction!).

13 Moretti, A., "A cube extending Piaget-Gottschalk's formal square" (draft).

9. The Epistemological Meaning of the Avatars of the Square

The avatars of the logical square constitute NOT and NOT brings light to several traditional "minor" issues[14]. But, more fundamentally, NOT seems to bring light on a traditional "major" problem, the one of the "deadly fight" between logic and geometry, a question which concerns the interpretation of Ancient philosophy (especially as reconstructed by the Tübingen-Milano school) as well as contemporary philosophy (as recently "awakened" with scandal by P. Gärdenfors' claims on the *geometrical* and not logical deep nature of concepts)[15]. Indeed, it can be shown that such a fight between logic and geometry, although very seldom explicitly thematised as such, is one of the very backbones of Western philosophy, involving major authors as Plato, Aristotle, Descartes, Spinoza, Kant, Bolzano, Bergson, Deleuze, the so-called analytical philosophy and Badiou. More generally, "opposition" being a very important concept for philosophy (involving deeply the proposals of contemporary thinkers like Popper, Perelman, Deleuze, Derrida, Apel, Severino, ...), it seems natural to think that at least some of the discoveries of NOT (the theory stemming from the logical square) should turn out, somewhere in the future, rather important for several issues at stake in contemporary philosophy.

From the point of view of science, many issues should be made clearer about NOT. One is that, despite the fact that NOT is an instance of "spatial logic" (in the sense of Aiello, Pratt-Hartmann and van Benthem), it is important to stress that NOT embodies a major, previously unseen *change* in logic (it is not "just another spatial logic"). With this respect it is important to grasp that, despite some superficial appearances, NOT is not reducible to graph theory[16]. More generally, NOT seems to show that logic, because of its peculiar geometrical hidden essence associated with "negation", is not strictly reducible (this point is shocking) to algebra (put otherwise: it is clumsy to *systematically* read logic through algebraic glasses –

14 An example concerning Aristotle and Leibniz can be found in Moretti, A., "The Geometry of Standard Deontic Logic", *op. cit.*, as well as in Moretti, A., *The Geometry of Logical Opposition*, *op. cit.*, ch. 17.

15 This question is dealt with in Moretti, A., *The Geometry of Logical Opposition*, *op. cit.*, ch. 1–4; and in Moretti, A., "The Geometry of Oppositions and the Opposition of Logic to It", *op. cit.*

16 Cf. Moretti, A., "*N*-opposition theory and graph theory", (draft).

this is a point taught to NOT by the so-called "universal logic" research framework, but which NOT clarifies, in return, to universal logic)[17]. Notice that it is seemingly this fact that makes the reception of the novelty of NOT difficult to "swallow" for many scholars, for NOT shoves off some very deep dogmas and habits of mainstream logic.

Finally, remark that, on the ground of one paradox linked to the logical square, most analytical philosophers, despite their devotion to logic, have dangerously sacrificed the essential geometrical side of logic (related to negation), just for refusing restricting the quantification domain (as their anti-paradox strategy seemed to request). This move (this refusal) may seem reasonable, except if one considers two facts: (a) that there probably are means, not too expensive, to save the square from the charge of undue existential import (as, *mutatis mutandis*, Routley and others have found logical ways, against Russell's violent dogma of 1905, for rescuing brilliantly Meinong's very brilliant "crazy logical ideas" – in other terms, the analytical recurrent ostracisms against "deviant" approaches are not reliable, for they are first of all *ideological*); (b) the sacrifice of the logical square (and of the geometry of logic) appears *crazy* in the light of the richness of the recent developments of the geometry of opposition. With this respect the suspicion may be raised that the dominating paradigm of the analytical philosophy (partnered by the so-called cognitive sciences) is, truly speaking, an ill-fated one, for, fundamentally speaking, it results in restricting abusively but petulantly the abstract notion of "structure" to that of "symbolic calculus". But this is highly unreasonable, for the latter notion is very narrow with respect to the former. Which means, in other words, that it could make much sense in a not too far future to strive for substituting to the analytical paradigm a new version of the old but glorious "structuralist" one. For structuralism relies on a use and on an intuition of the "formal" which is much richer and much more powerful than the one of the "logical fetishists". Put in still another way, NOT, the geometry of opposition is the emergence of a "logical structuralism", which explains why the logical

17 This issue can be made clearer (we won't make it here though) by comparing the ways of dealing with the spatial ordering of the 16 binary connectives of the propositional calculus, a topic where the paradoxical superiority of NOT (which displays for it a logical tetraicosahedron) over Boolean algebra (which displays an hyper-cube) is surprising but striking. Part of this question, involving reflections by Béziau, Smessaert, Pellissier and Moretti can be found in our dissertation *The Geometry of Logical Opposition*, *op. cit.*

square is so powerful and interdisciplinarily appealing and suggests, by developing it powerfully, that things might be soon changing in the formal approaches to philosophy and the humanities.

Bibliography

Aiello, M., Pratt-Hartmann, I., & van Benthem, J. (Eds.). (2007). *Handbook of Spatial Logics*. Berlin: Springer.

Aristotle (1963). *Categories and De Interpretatione*. Translated with notes by J. L. Ackrill. Oxford: Clarendon Aristotle Series.

Béziau, J.-Y. (2001). From paraconsistent to universal logic. *Sorites, 12.*

Béziau, J.-Y. (2003). New Light on the Square of Oppositions and its Nameless Corner. *Logical Investigations, 10.*

Blanché, R. (1953). Sur l'opposition des concepts. *Theoria, 19.*

Blanché, R. (1957). Sur la structuration du tableau des connectifs interpropositionnels binaires. *The Journal of Symbolic Logic*, *22, 1,* March.

Blanché, R. (1966). *Structures intellectuelles. Essai sur l'organisation systématique des concepts.* Paris: Vrin.

Brown, M. (1984). Generalized Quantifiers and the Square of Opposition. *Notre Dame Journal of Formal Logic*, *25, 4*, October.

Englebretsen, G. (1984). Opposition. *Notre Dame Journal of Formal Logic, 25, 1,* January.

Gallais, P. (1982). *Dialectique du récit médiéval (Chrétien de Troyes et l'hexagone logique).* Amsterdam: Rodopi.

Gärdenfors, P. (2000). *Conceptual Spaces. The Geometry of Thought.* Cambridge, MA: The MIT Press.

Gottschalk, W. H. (1953). The theory of quaternality. *The Journal of Symbolic Logic, 18, 3.*

Luzeaux, D., Sallantin, J., & Dartnell, C. (2008). Logical extensions of Aristotle's square. *Logica Universalis, 2, 1.*

Moretti, A. (2004). Geometry for Modalities? Yes: Though *n*-Opposition Theory. In J.-Y. Béziau, A. Costa Leite and A. Facchini (Eds.), Aspects of Universal Logic. *Travaux de Logique 17.* University of Neuchâtel.

Moretti, A (2009). *The Geometry of Logical Opposition*. PhD Thesis, University of Neuchâtel, Switzerland, March.
Moretti, A. (2009). The Geometry of Standard Deontic Logic. *Logica Universalis, 3, 1.*
Moretti, A. (2009). The Geometry of Opposition and the Opposition of Logic to It. In U. Savardi (Ed.), *The perception and the cognition of the contraries*. Milano: McGraw-Hill.
Moretti, A. (2010). La complexité formelle de la "contradiction" dans la géométrie des oppositions. In B. Guy (Ed.), *Ateliers sur la contradiction*. Paris: Presses des Mines.
Moretti, A. (2010). The Critics of Paraconsistency and of Many-Valuedness and the Geometry of Oppositions. *Logic and Logical Philosophy, 19.*
Moretti, A. (Draft). A cube extending Piaget-Gottschalk's formal square.
Moretti, A. (Draft). McNamara's 'deontic dodecagon for DWE' in the light of oppositional geometry.
Moretti, A. (Draft). A geometrical method for classifying and studying quantifiers and modalities.
Moretti, A. (Draft). *N*-opposition theory and graph theory.
Parsons, T. (2006). The traditional square of opposition. *Stanford Encyclopedia of Philosophy.*
Pellissier, R. (2008). 'Setting' *n*-opposition. *Logica Universalis, 2, 2.*
Pellissier, R. (2011). 2-opposition and the topological hexagon. In this volume.
Pellissier, R. *Taoist logical hexagon. Philosophical meaning of 3-opposition and weak 2-opposition in cosmology* (draft).
Piaget, J. (1972 [1949]). *Traité de logique. Essai de logistique opératoire.* Paris: A. Colin.
Piaget, J. (1970 [1968]). *Structuralism*. New York: Basic Books.
Schang, F. (2011). Questions and answers about oppositions. In this volume.
Sesmat, A. (1951). *La logique II. Les raisonnements, la logistique*. Paris: Hermann.
Seuren, P. (2010). *The Logic of Language*. Oxford University Press, Oxford.
Smessaert, H. (2009). On the 3D visualisation of the logical relations. *Logica Universalis, 3, 2.*
Smessaert, H. *The Classical Aristotelian hexagon versus the Modern Duality hexagon* (forthcoming).
Varzi A., & Warglien M. (2003). The Geometry of Negation. *Journal of Applied Non-Classical Logics, 13, 1.*

Wolenski, J. (1999). MacColl on modalities. *Nordic Journal of Philosophical Logic, 3, 1.*

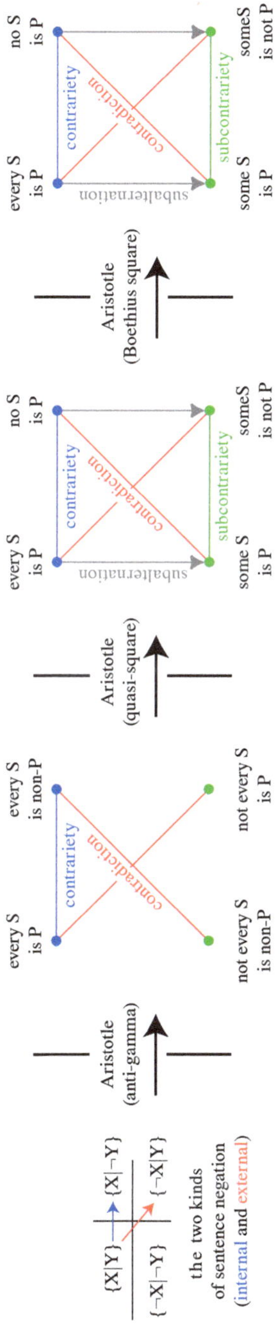

Figure 1: The Progressive Discovery of the Logical Square

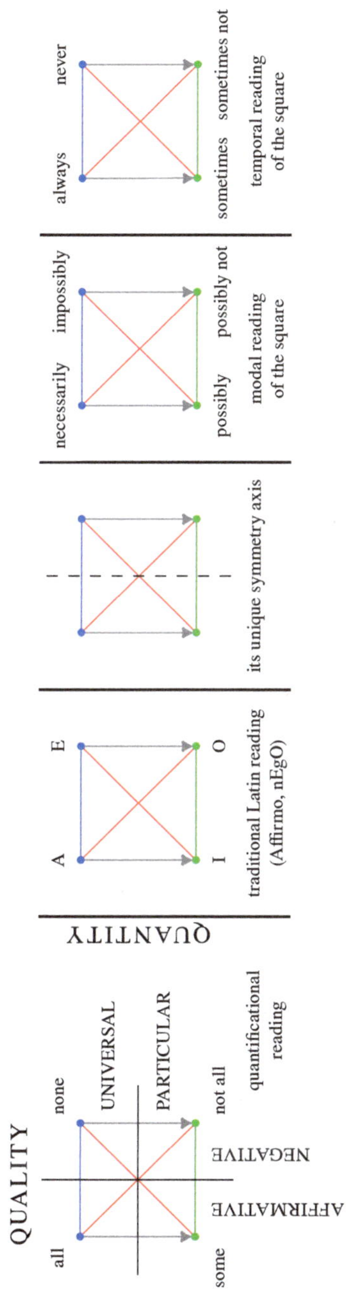

Figure 2: Some Traditional Aristotelian Readings of the Square of Opposition

Figure 3: One of the Two Traditional Problems with the Square of Opposition

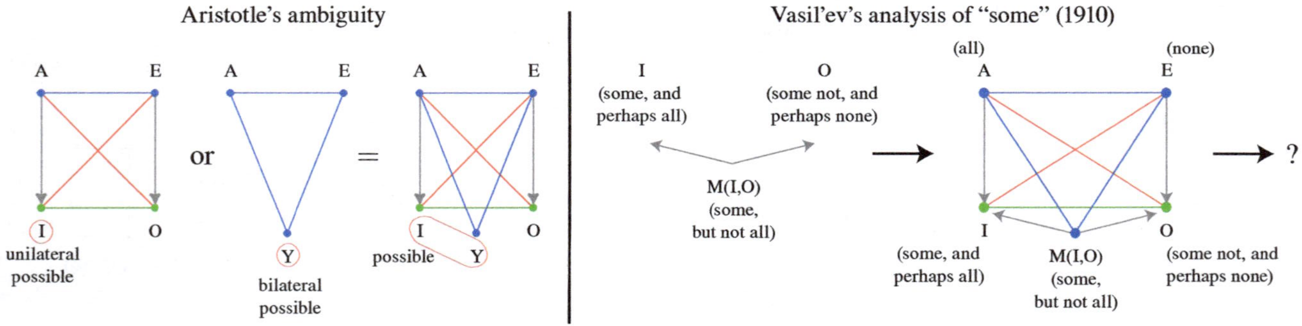

Figure 4: Some Contemporary Uses of the Logical Square

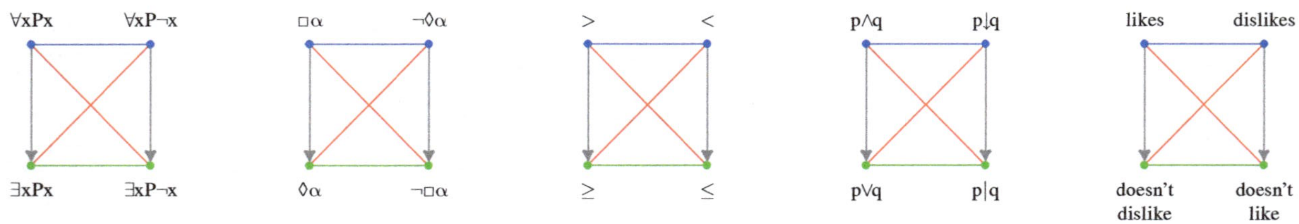

Figure 5: Piaget's (1949) and Gottschalk's (1953) Formal Square (the "P-G Square")

Piaget's INRC group

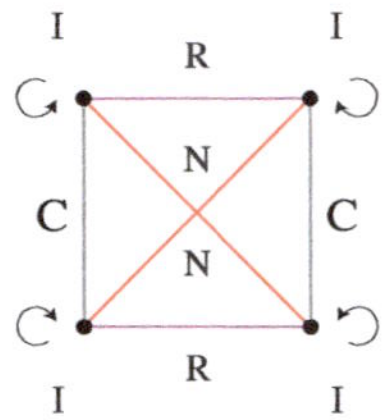

(its operations)

	I	N	R	C
II	N	R	C	
N	N	I	C	R
R	R	C	I	N
C	C	R	N	I

Gottschalk's square of quaternality

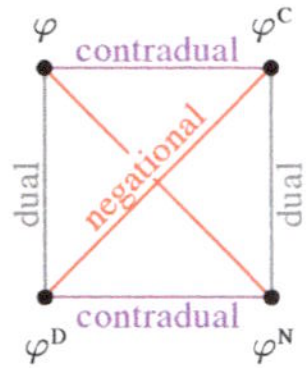

Piaget-Gottschalk Aristotle-Apuleius

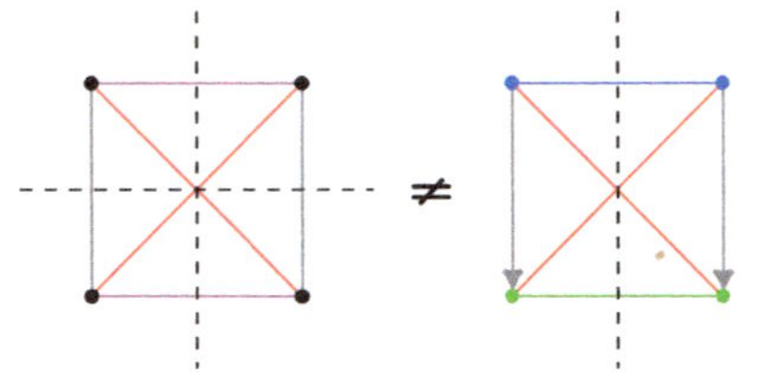

Figure 6: Some Structuralist Models Formally Close to the Square of Opposition

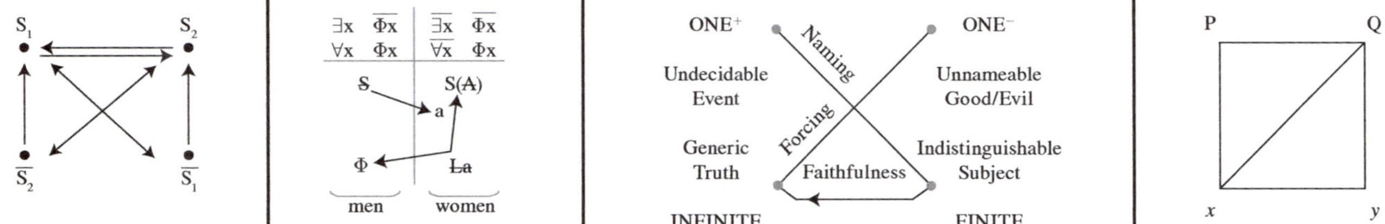

Figure 7: The Four Kinds of Square of Opposition According to Mark Brown (1984)

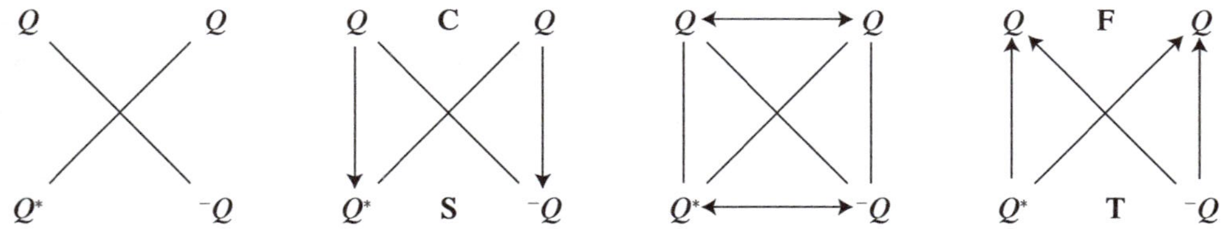

Type I: Boolean Type II: Classical Type III: Semi-degenerate Type IV: Degenerate

Figure 8: Some Recurrent Patterns of Expansion of the Logical Square

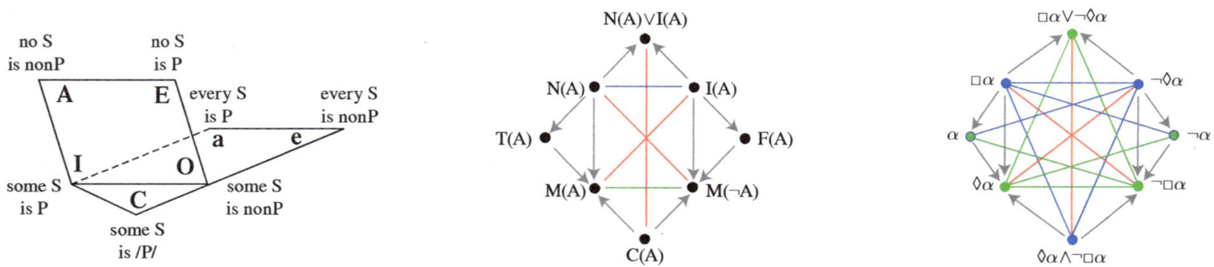

Englebretsen (1984) Wolenski (1999) Béziau (2003)

Figure 9: Some More Complex Expansions of the Logical Square

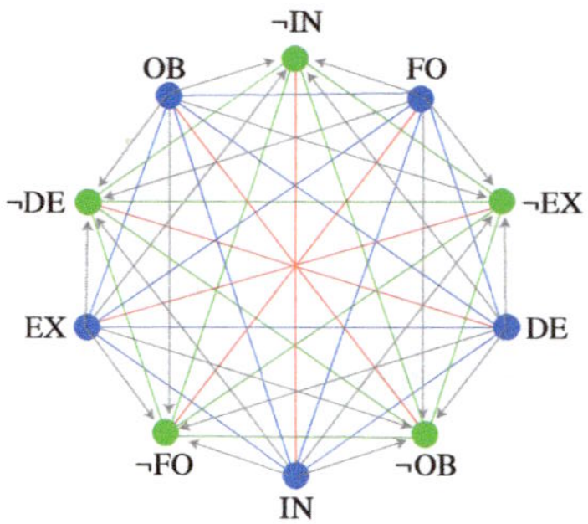

Joerden's "deonzic dragon"
for supererogation (1987)

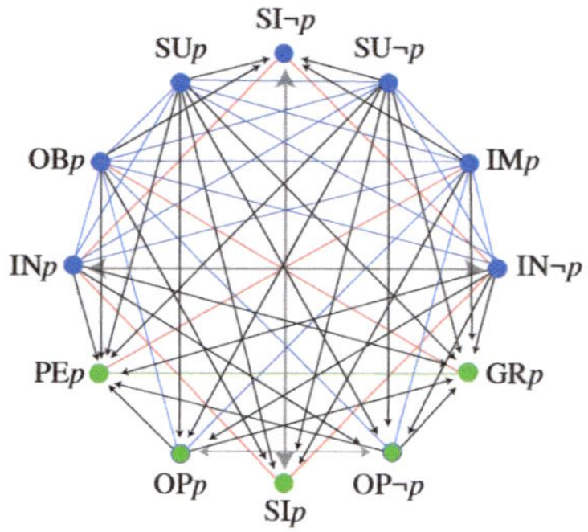

one of McNamara's models
for the deontic system DWE (1996)

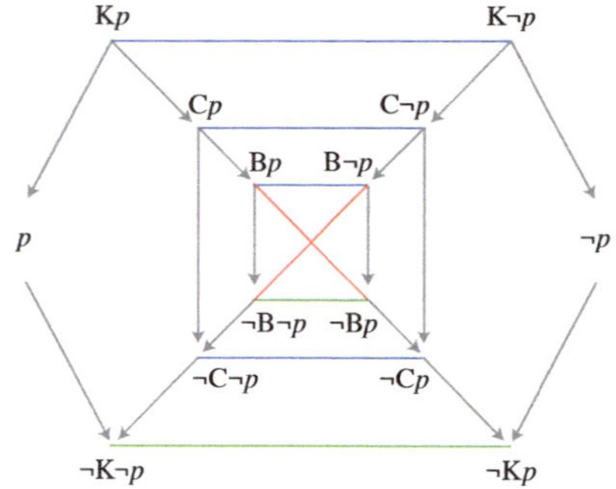

Lenzen's model for epistemic logic
(personal communication, 2007)

Figure 10: The Logical Hexagon (1951), the Square's Conservative Extension

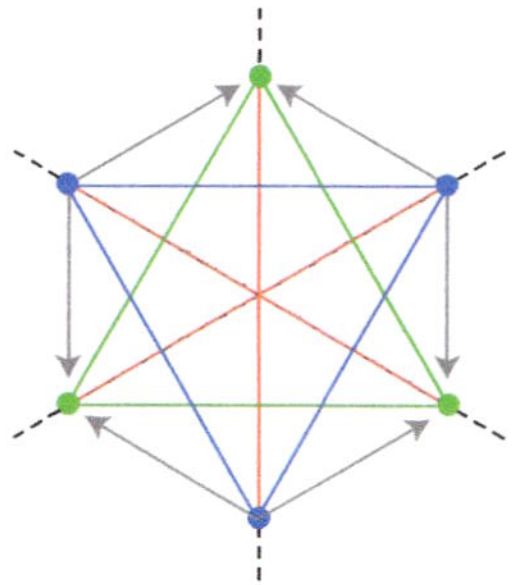

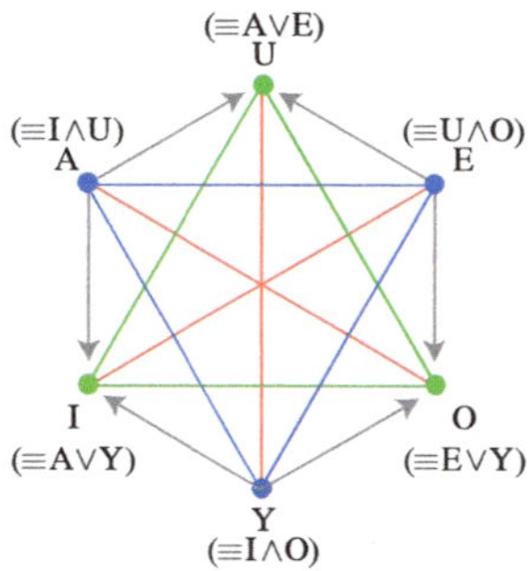

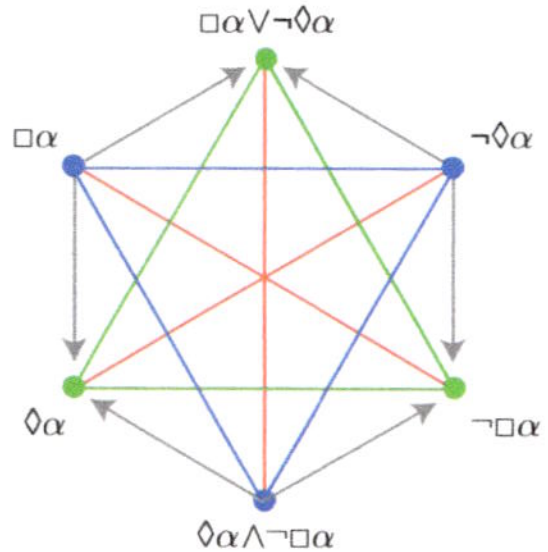

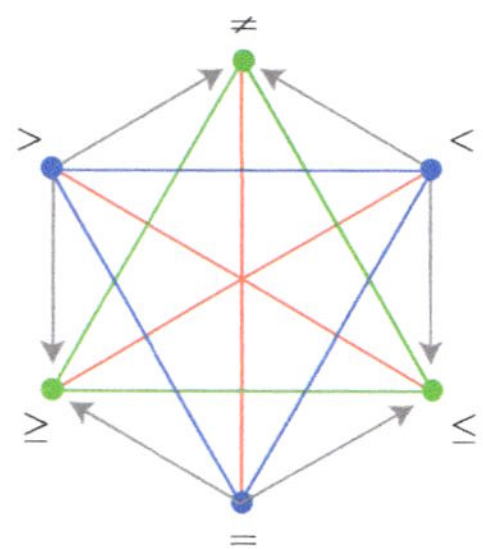

Figure 11: Some Applications of the Logical Hexagon (Logic, Philosophy, Narratology

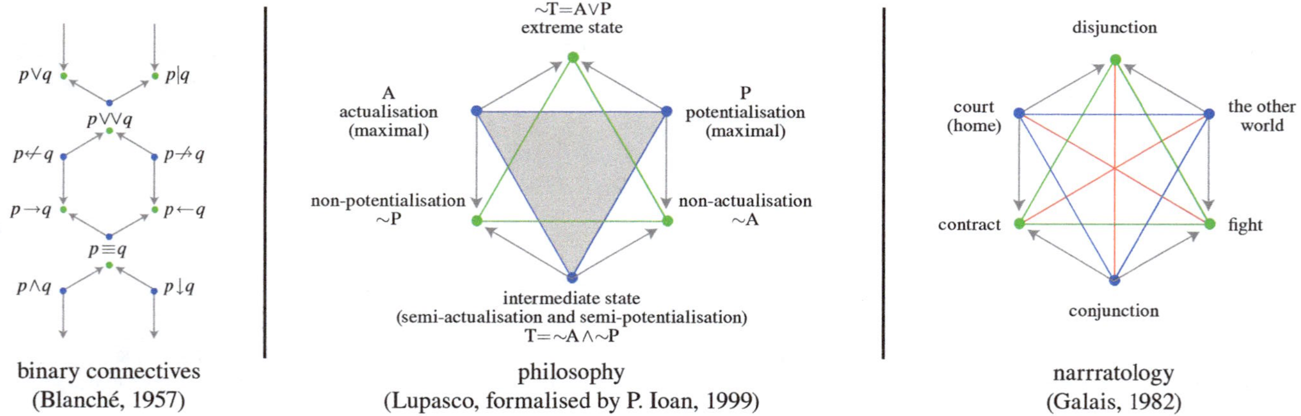

Figure 12: Béziau's Structuralist Discovery of Two New Logical Hexagons (Paraconsistent and Paracomplete)

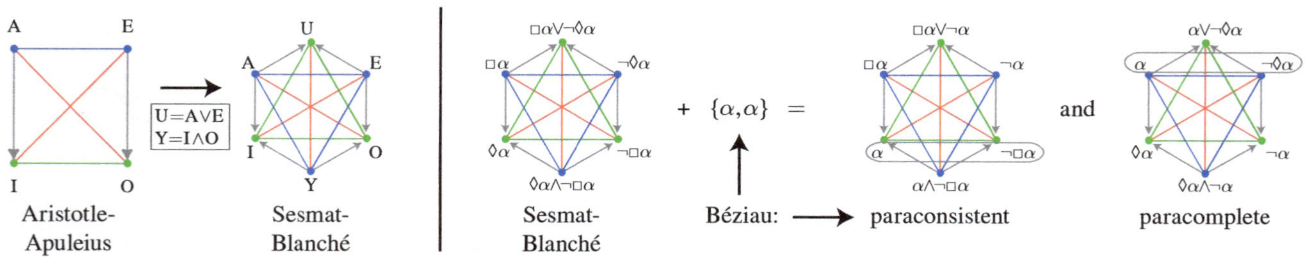

Figure 13: Béziau's Conjecture of a 3D Gathering of the Three Known Logical Hexagons

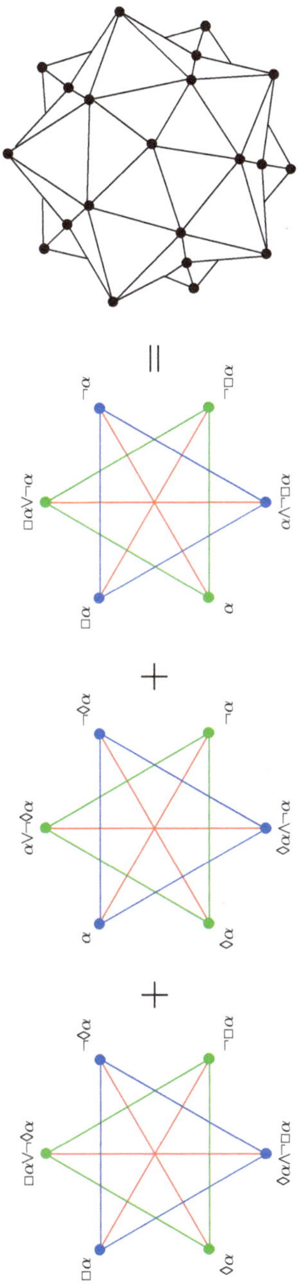

classic star of oppositions (Sesmat and Blanché) + paracomplete star (Béziau) + paraconsistent star (Béziau) = Béziau's "solid of opposition" (the "stellar dodecahedron")

Figure 14: The Logical Cubohectahedron (Gathering, in the 3D Space, 4 Logical Hexagons)

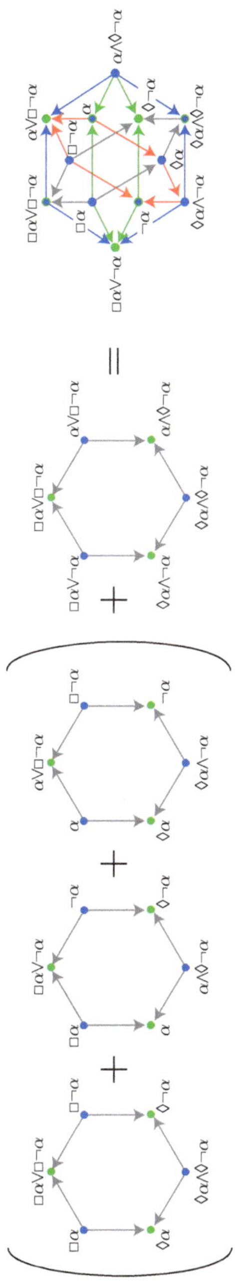

Sesmat-Blanché + Béziau 1 + Béziau 2 + Moretti-Smessaert = Moretti

Figure 15: The Logical Cube (the Successor of the Logical Square and Hexagon) Is a Logical Bi-Tetrahedron

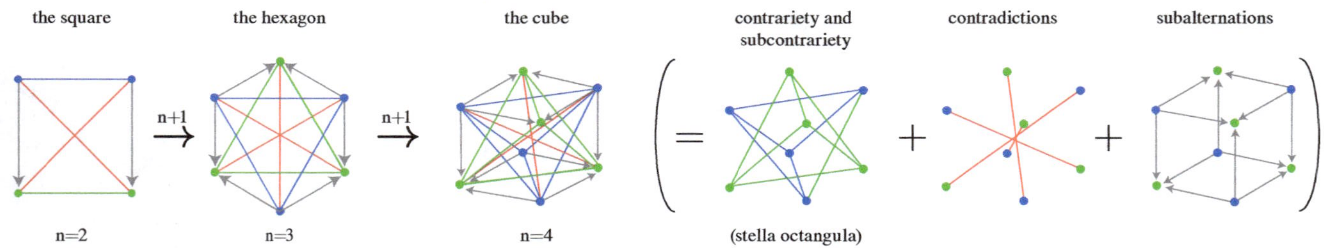

Figure 16: The Series of the Geometrical Simplexes of Dimension m (Expressing Contrariety)

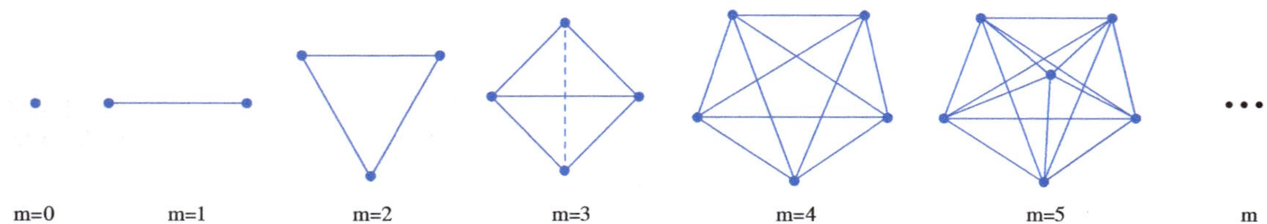

Figure 17: "Contradiction" is Expressed, for any Dimension m, as Central Symmetry

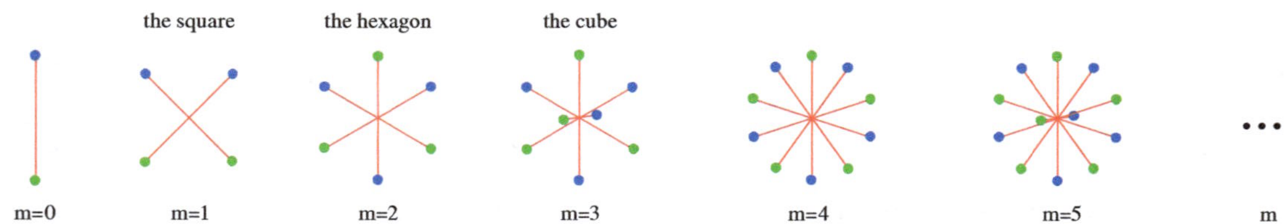

Figure 18: Contrariety Simplexes and Subcontrariety Simplexes Combine into Logical Bi-Simplexes

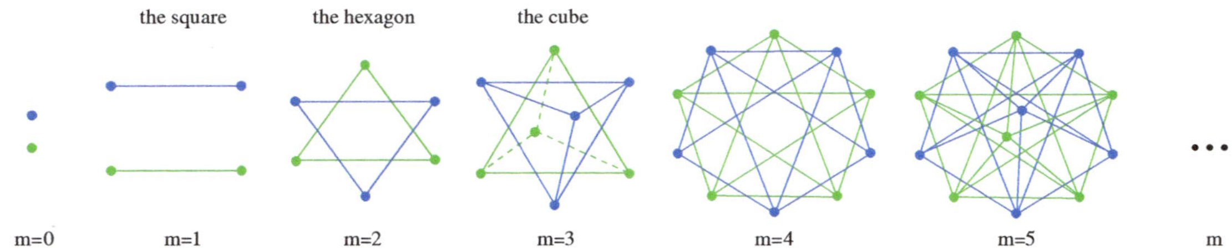

Figure 19: The Series of the αn-Structures (the Avatars of the Logical Square)

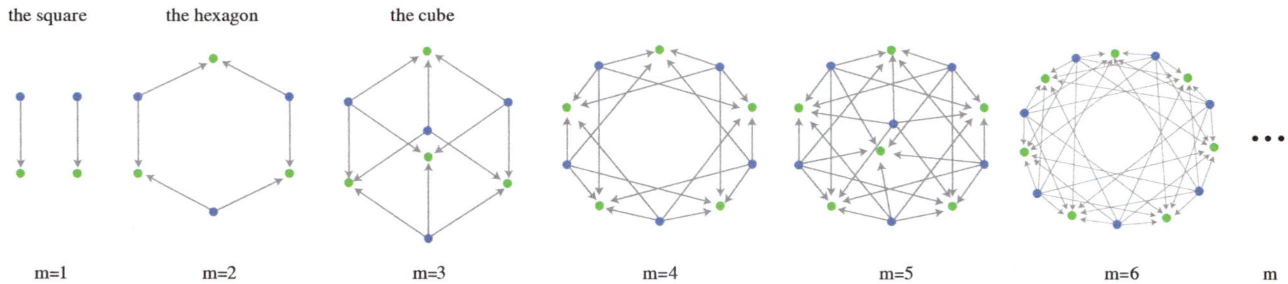

Figure 20: Using NOT: Expressing Gender Oppositions of Growing Complexity

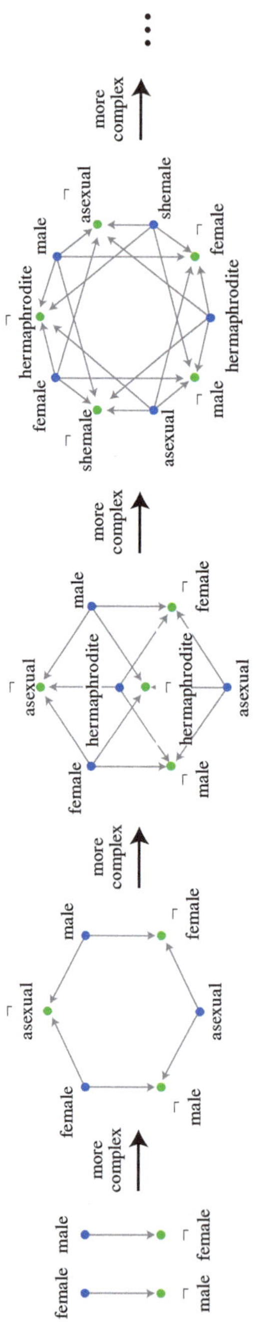

Figure 21: The Conjecture over the Existence of "Gatherings" of Logical Bi-Simplexes: The β*n*-Structures

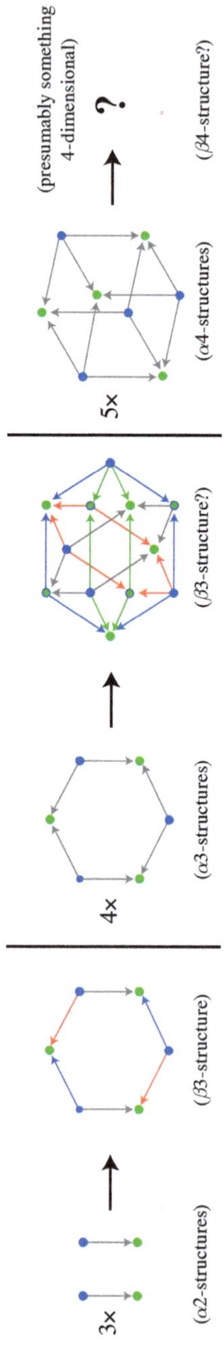

Figure 22: The Exact Determinant of the β3-Structure (Pellissier-Smessaert)

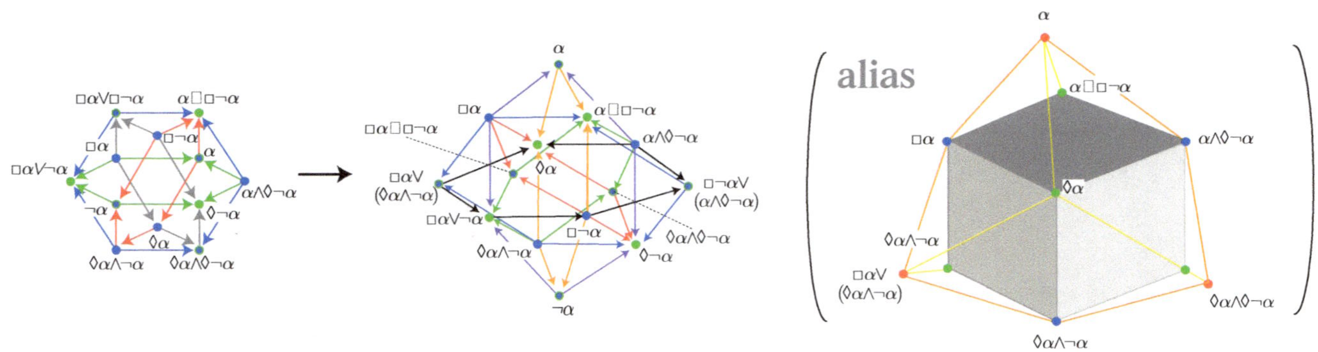

logical cuboctahedron
(Moretti, 2004 [2003])

logical tetraicosahedron
(Pellissier, 2008 [2004])

logical rhombic cuboctahedron
(Smessaert, 2009 [2004])

Figure 23: The Series of the β*n*-Structures (Simplified Representation)

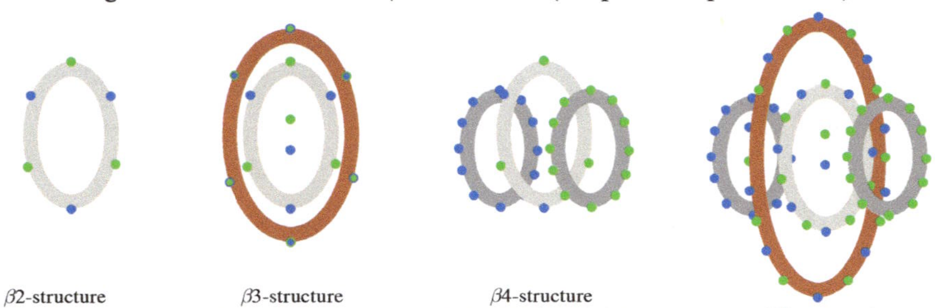

*β*2-structure
(hexagon)

*β*3-structure
(tetraicosahedron)

*β*4-structure
(hyper-tetraicosahedron)

(*β*5-structure)

Figure 24: Some Well-Known γ-Structures (i.e. Modal Graphs)

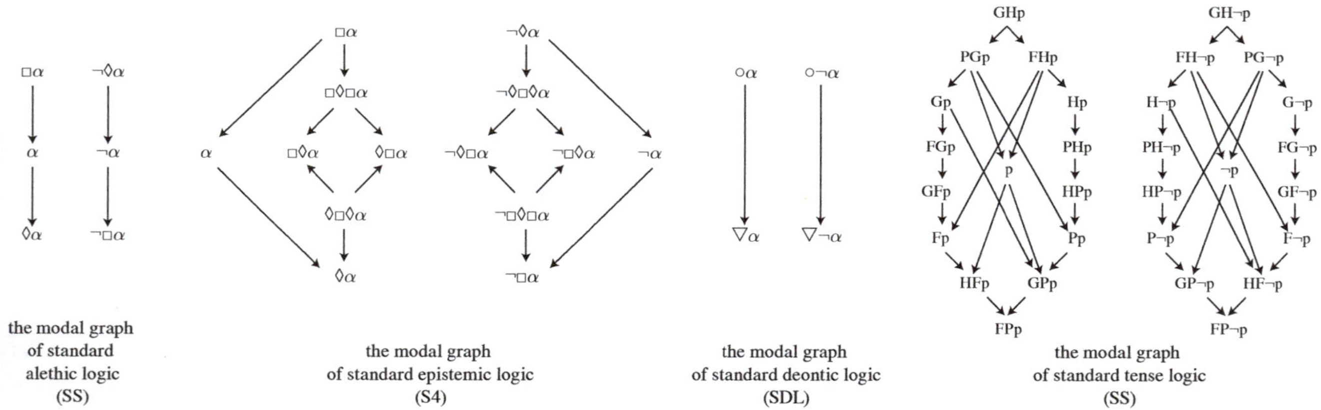

Figure 25: The NOT-Analysis of McNamara's "Deontic Dodecagon": It is Made of Three Logical Bi-Simplexes

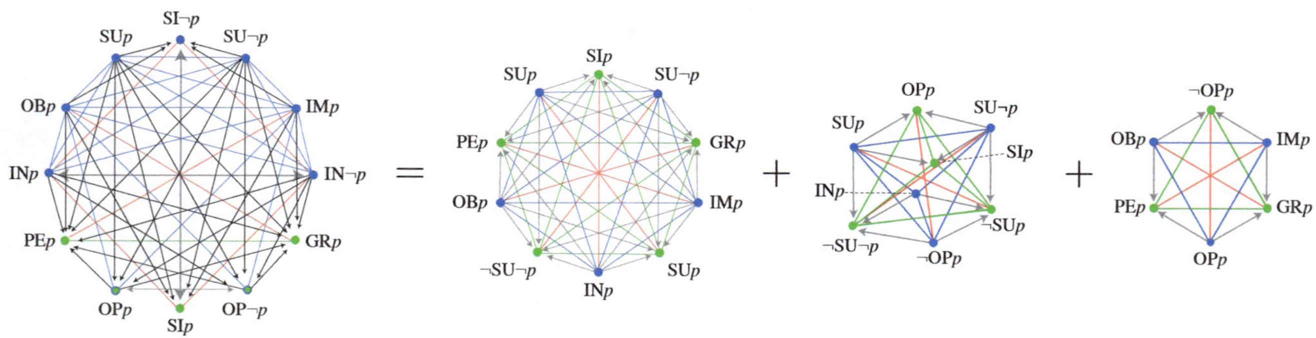

Figure 26: The Conjectural Series of the "Logical Tri-Simplexes" of Dimension m

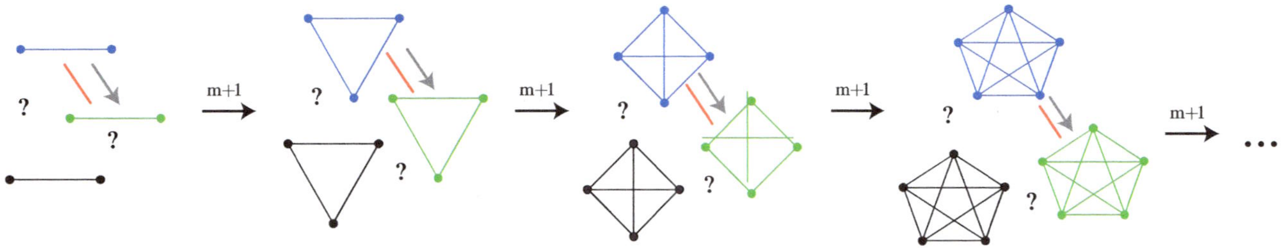

Figure 27: The Backbone of the Logical Bi-Simplexes: The Aristotelian 2²-Semantics and 2²-Lattice

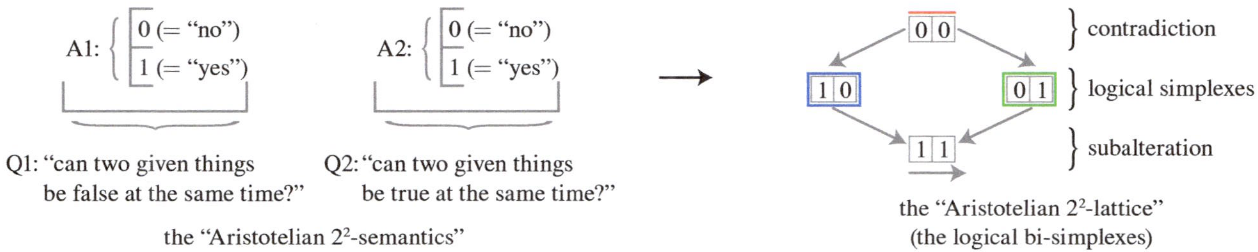

Figure 28: The Backbone of the Logical Tri-Simplexes: The Aristotelian 3²-Semantics and 3²-Lattice

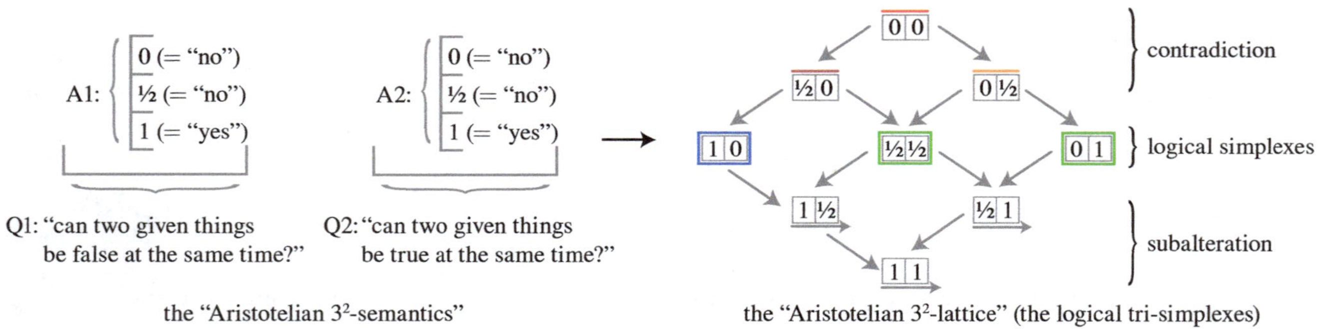

Figure 29: The Logical Tri-Simplexes of Dimension 1 (Logical Tri-Segment) and 2 (Logical Tri-Triangle)

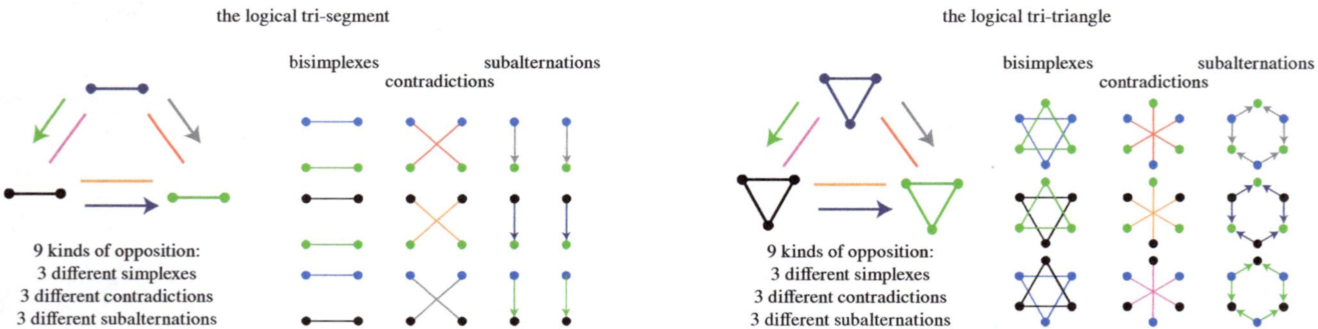

Figure 30: Global View of the Logical Tri-Segment and of the Logical Tri-Triangle

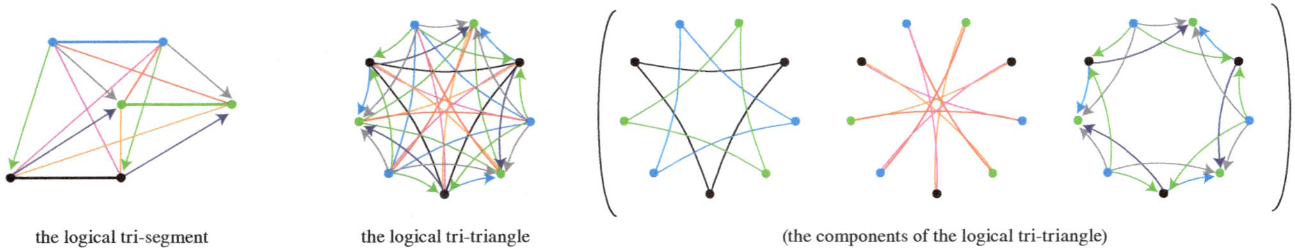

Figure 31: The Logical Tri-Simplexes of Dimension 3 (Logical Tri-Tetrahedron) and 4 (No Name)

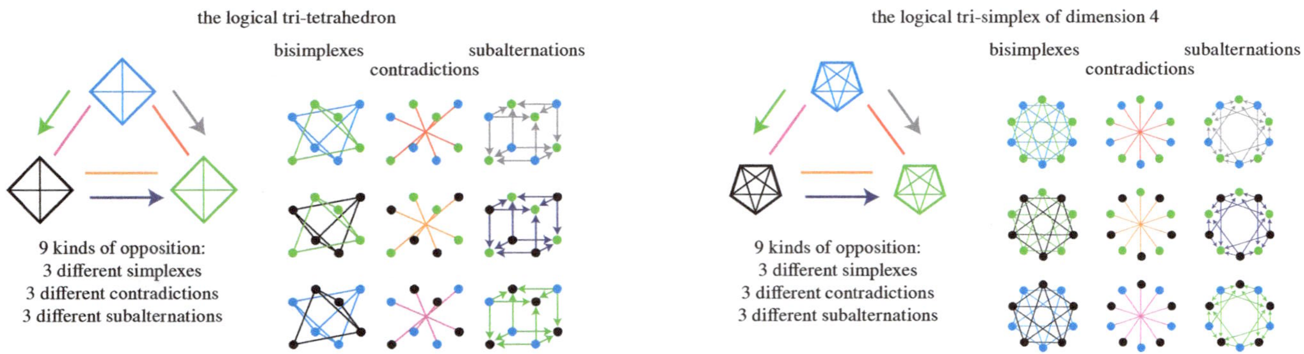

Figure 32: The Series of the Logical p-Simplexes of Dimension 3 (the Logical p-Tetrahedra)

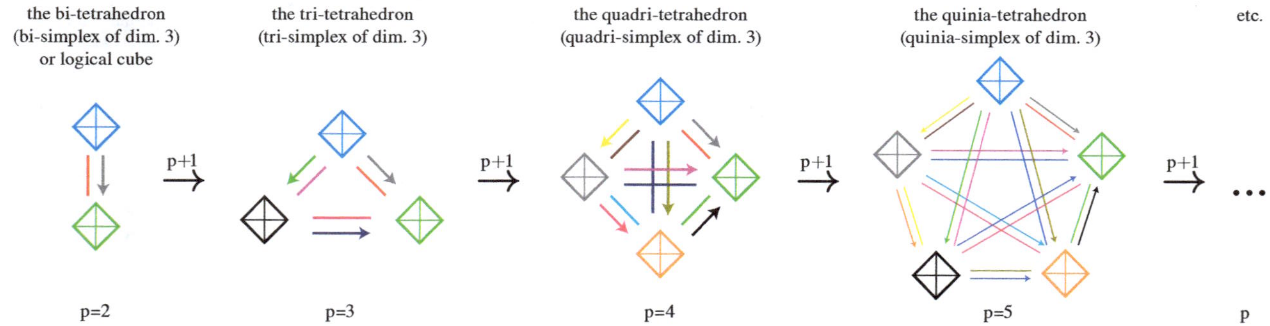

Figure 33: The (Still) Mysterious Aristottelian 2^3-Semantics and 2^3-Lattice (the Backbone of What?)

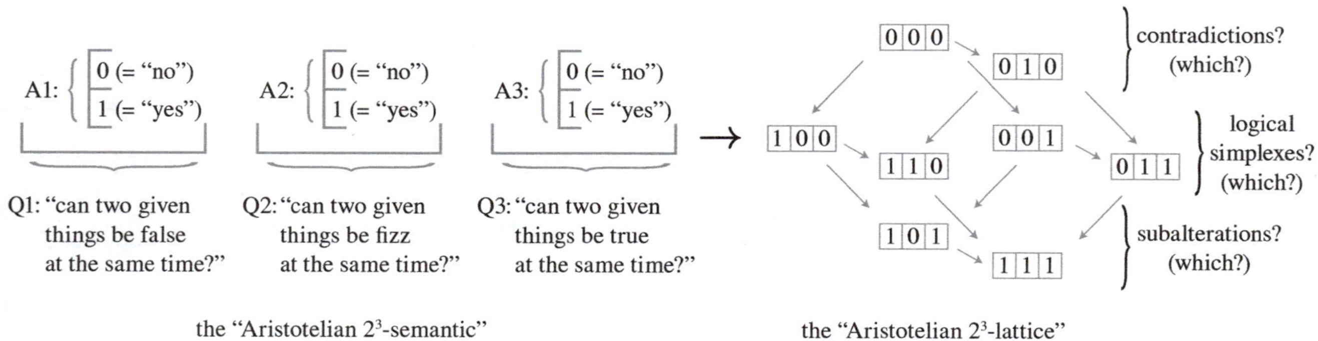

Figure 34: The Series of the Geometrical q-Dimensional Hyper-Cubes Corresponding to the Aristotelian 2^q-lattices

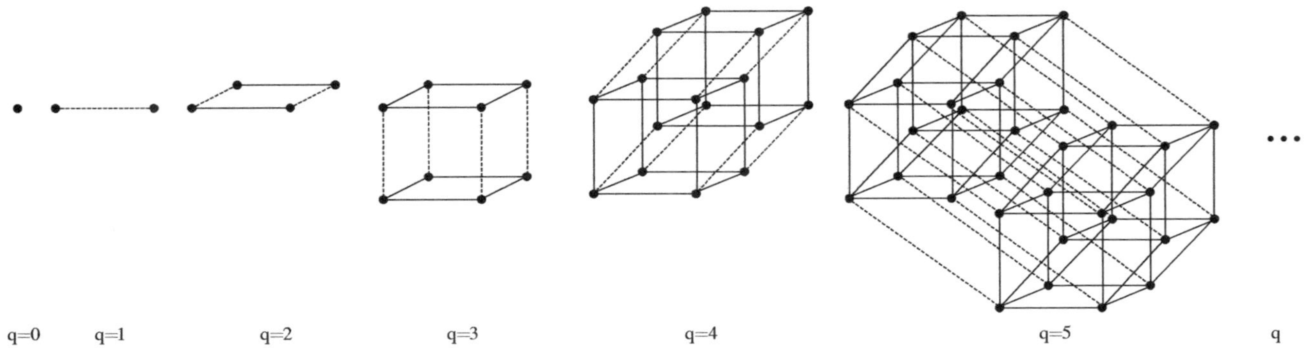

Figure 35: The Infinite *npq*-Parallelepiped of Opposition (Giving All the Known Kinds of Opposition)

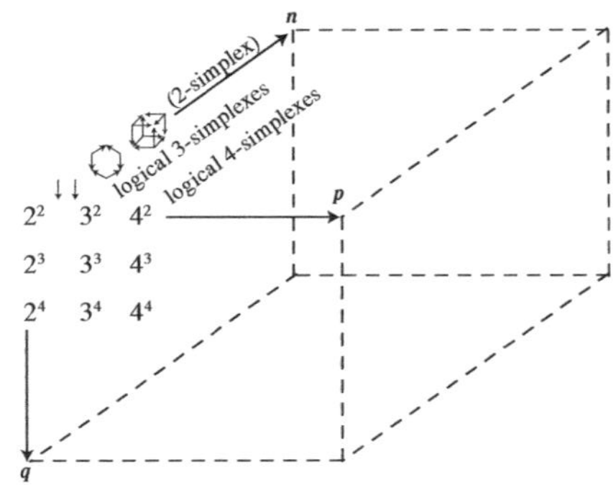

Figure 36: There is also a PG-Geometry, Grounded on Suited Piagetian-Gottschalkian p^q-Semantics and Lattices

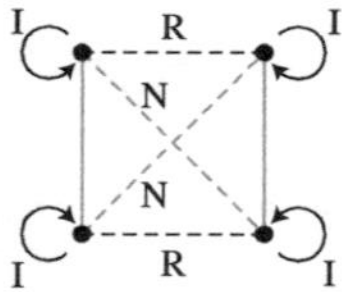

	I	N	R	C
I	I	N	R	C
N	N	I	C	R
R	R	C	I	N
C	C	R	N	I

Piaget's INRC group (its operations)

the logical PG-square, generated
by the Piagetian-Gottschalkian 2^2-semantics

→

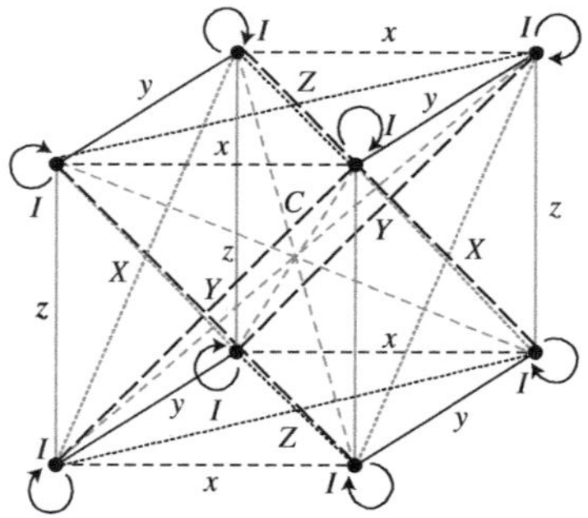

	I	*C*	*X*	*Y*	*Z*	*x*	*y*	*z*
I	*I*	*C*	*X*	*Y*	*Z*	*x*	*y*	*z*
C	*C*	*I*	*x*	*y*	*z*	*X*	*Y*	*Z*
X	*X*	*x*	*I*	*Z*	*Y*	*C*	*z*	*y*
Y	*Y*	*y*	*Z*	*I*	*X*	*z*	*C*	*x*
Z	*Z*	*z*	*Y*	*X*	*I*	*y*	*x*	*C*
x	*x*	*X*	*C*	*z*	*y*	*I*	*Z*	*Y*
y	*y*	*Y*	*z*	*C*	*x*	*Z*	*I*	*X*
z	*z*	*Z*	*y*	*x*	*C*	*Y*	*X*	*I*

the logical PG-cube, generated by the Piagetian-Gottschalkian 2^3-semantics

The Application of Vector Theory to Syllogistic Logic

JONATHAN WESTPHAL[1]

Abstract

Syllogistic arguments can be represented with vectors. If individual premises of a syllogism are diagramed as vectors, then the conclusion of a (valid) syllogism will be the vector sum of the two premises. The result is Leibniz's Scheme VIII Square of Opposition. The application of the vector representation to Rescher and Gallagher's extension of Venn Diagram techniques to plurative propositions is shown.

Keywords: Syllogisms, Vectors, Vector Sums, Valid Argument, Diagrammatic Logic, Leibniz, Plurative Propositions

1. Introduction

This paper shows how syllogistic arguments can be represented with vectors, extending the diagrammatic method of Westphal and Hardy (2005). If individual premises of a syllogism are diagramed as vectors, then the conclusion of a (valid) syllogism will be the vector sum of the two premises. Its aim is to introduce a new diagrammatic expression of syllogistic logic, rather than to suggest a new logic of any sort, such as three-valued logic, or to extend an existing diagrammatic logic. Nevertheless, the suggested diagrams do imply a Boolean view of existential import. The relationship of

1 School of Cognitive Science, Hampshire College, Amherst, MA 01002, USA, email: jontywestphal@gmail.com. This research was carried out as part of a subcontract to US DOD Grant HQ006-05-C-0010. I wish to thank Bill McCurdy for his invaluable help with this line of research.

the suggested representation to Rescher and Gallagher's extension of Venn Diagram techniques to plurative propositions is shown.

I will begin by representing the subject term as a point a unit distance along the horizontal axis from the origin O, the predicate term as a point a unit distance up the vertical axis, and the two term complements as points unit distances in the opposite directions. This gives us four quadrants, which are, traveling anticlockwise from S at 3 o'clock, the SP, –SP, –S –P and S –P quadrants. Propositions can now be represented as vectors to and from the points within these quadrants, including O.

Diagram I: The Four Quadrants

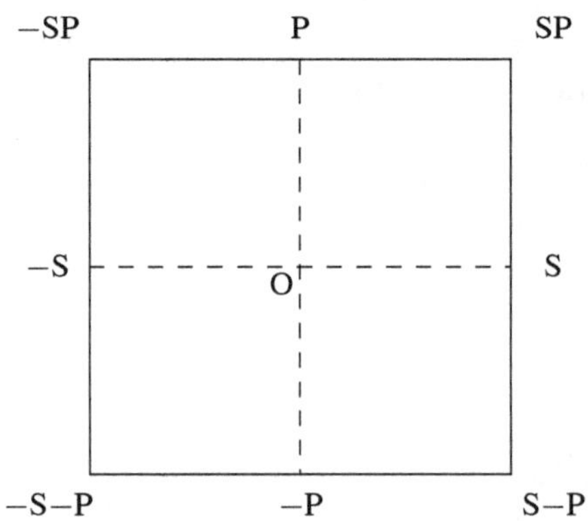

In "Some Logical Difficulties" and elsewhere in other texts Leibniz describes several alternative schedules of standard-form categorical propositions. Consider the one which G.H.R. Parkinson calls "Scheme VIII" (Leibniz, 1966; Kneale and Kneale, 1962).

A: A non-B non-Ens
O: A non-B Ens
E: AB non-Ens
I: AB Ens

We can use this scheme to represent the four standard-form categorical propositions with vectors in the following way.

2. The A-Proposition

The A-proposition 'All S is P' is most naturally represented by the free vector from S to P. (Diagram II). It indicates that whatever is S is translated to P. The translation of this whole vector into the S – P quadrant gives the position vector **S – P → O**. This is to be read as 'No S is – P', taking the origin O as a Leibniz's non-Ens, in some sense a center of *denial*. The E-proposition resulting from the translation of 'All S is P' to the S – P quadrant is the *obverse* of that A-proposition. And when the given A-proposition is translated into the –S – P quadrant, we get the contrapositive, which is 'All –P is –S'. (The vector **O → –SP** is a particular proposition that I will discuss below, under *(4) Particular Propositions.*)

Diagram II: The A-Proposition with Translations

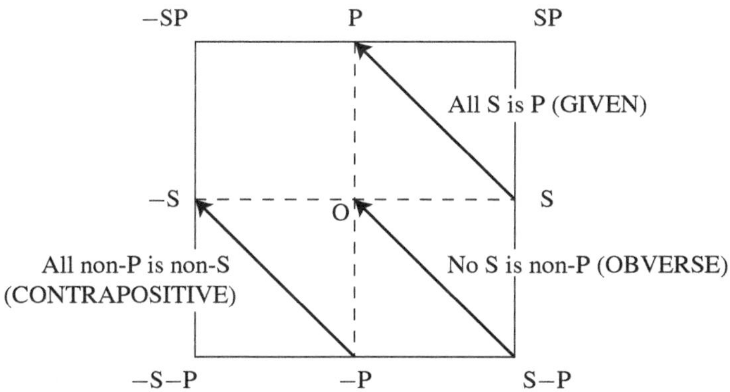

Apart from the "position vectors", which are fixed to their point of origin in the sense that they cannot be translated at all, *translations in the vector logic space turn out to be equivalences.*

3. The E-Proposition

We now know that the vector from S – P to O is an E-proposition, and the same is true of the vector from SP to O. **SP → O** is 'No S is P', or in the Scheme VIII terminology 'SP non-Ens'. Translated into the S – P quadrant it becomes **S → –P**, which is the *obverse* of **SP → O**. In the –SP quadrant this vector becomes a second obverse, the obverse of the converse of the original E-proposition, or **P → –S**. And in the –S – P quadrant, where one might perhaps expect to find the (invalid) contrapositive 'No non-P is non-S', one finds instead the interesting **O → –S – P** (Diagram III), a position vector to be discussed below.

Diagram III: The E-Proposition with Translations

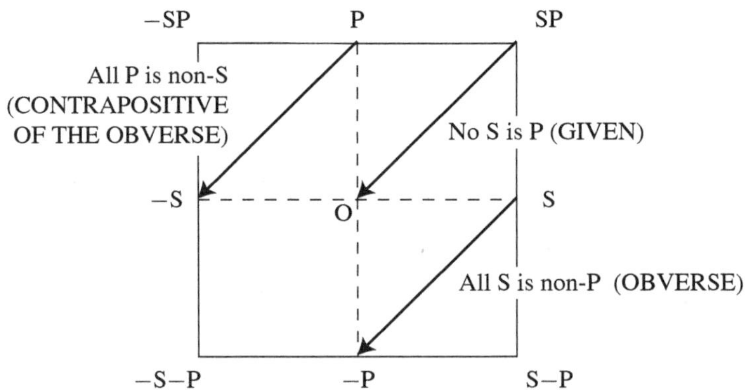

It is interesting to see how the E-form vector **SP → O** travels the diagonal of a parallelogram between the two disjoint terms and separates them, as it were driving between them and carrying them together into the non-Ens, whereas the A-proposition maps the S-term onto the P-term.

It is also significant that the two translations of the given free vector representing the universal proposition have exactly two equivalents, produced by the "immediate inferences" that are valid apart from those shown in the Square of Opposition.

4. Particular Propositions

So far the examples given of vectors in the space have been the universal A- and E- form propositions. How do matters stand with the particular I- and O- forms? In Scheme VIII the I-proposition is 'SP Ens', and this contradicts the universal 'No S is P'. We can take **O → SP** to be 'Some S is P' (Diagram IV). With the standard forms movement *away* from the origin is towards "Ens", just as movement towards it is the tendency to "non-Ens".

The quantity of a standard-form categorical proposition is expressed in the Scheme VIII vector diagram by the fact that the proposition points *away* from the origin. Position vectors which start at the origin therefore cannot be translated, as they would then no longer be bound to take us away from the origin. Position vectors, for example wind velocity, are bound to a point, though it need not be the origin (Borisenko and Trapov 1968). All the other universal propositional vectors are free vectors. In its Scheme VIII representation, therefore, **S → P** is a universal proposition, and **O → SP** is a particular proposition. **S → P** is contradicted by the reverse *position* vector **O → S – P**, but not by **P → S**, which represents 'All P is S' and is the (invalid) converse but not the contradictory of 'All S is P'.

Diagram IV: The I-Proposition

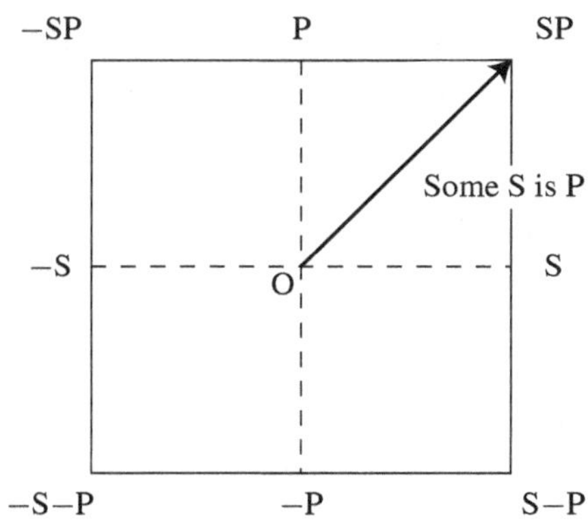

In the same way that the I-proposition is diagramed as the vector from O to SP, the O-form is diagramed as the vector from O to S – P. Its negation is its reverse free vector, 'All S is P.'

Diagram V: The O-Proposition

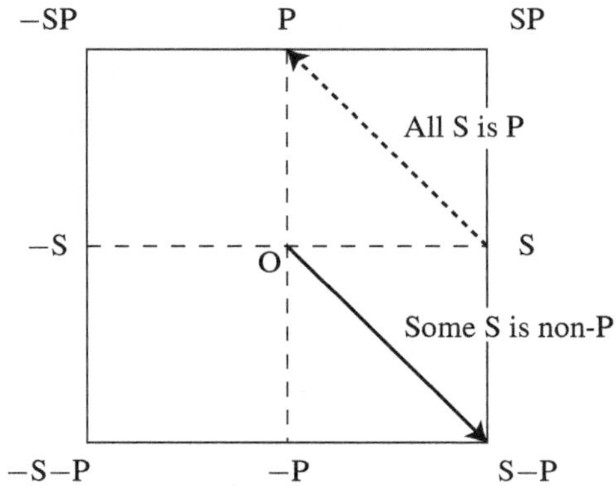

5. The Vector Square of Opposition

We are now able to diagram the Square of Opposition with vectors, using Leibniz's canonical Scheme VIII formulations. The two particular propositions contradict the two universal propositions. The particular propositions are (a) position vectors, and they are (b) the reverse vectors of the vectors representing the two universal propositions. Leibniz's choice of 'No S is nonP' as the canonical form of the A-proposition is significant. His Scheme VIII "Square" gives us two universal propositions, but they share the exclusionary quantifier "No". In addition, both universal propositions are contradicted in explicit form by their negations, the I and O propositions. The result is the symmetries of Diagram VIa, which would be spoiled by an A-proposition from represented by the vector from S to P (shown dashed).

Diagram VIa: Leibniz's Scheme VIII; Square of Opposition as Vectors

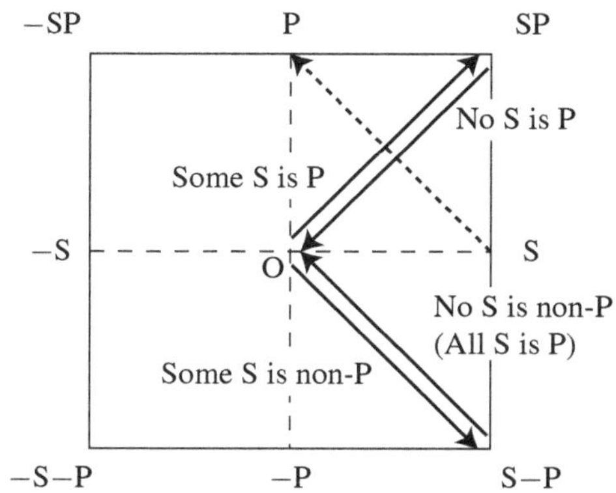

Diagram VIb shows how the Scheme VIII Square of Opposition is related to the canonical Square of Opposition with A, E, I, and O propositions at the four corners. The canonical Square is the Scheme VIII Square without the origin, so that it contains no position vectors. The familiar A, E, I, O scheme in the Square of Opposition is given by a reflection in the A-O axis, and taking the arrowheads or directions as indicating the quantity and quality of the proposition. Universal propositions point North, and particular propositions point South. Affirmative propositions point West, and negative propositions point East.

Diagram VIb: Rotated Scheme VIII Square without Origin

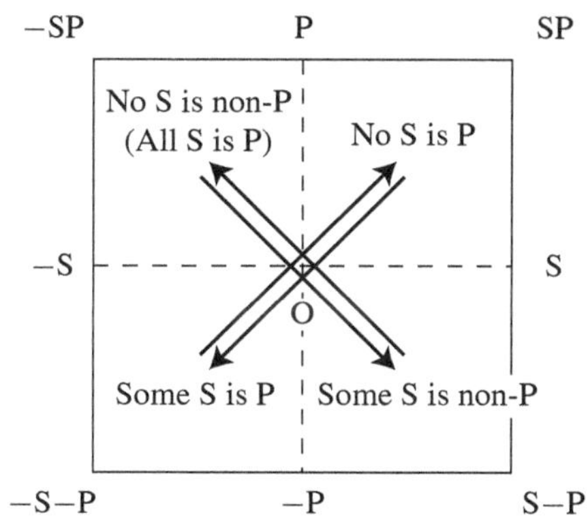

6. Existential Import

It is obvious on inspection that without the extra existential premise the vector representation **S → P** or 'All S is P' does not by any translation yield the vector **O → SP** or 'Some S is P'. The vector representing the conclusion of an argument by limitation from universal premises is only validly constructed with an extra premise. In this way the vector system is Boolean and confirms the non-existential interpretation of the universal proposition. The A-proposition 'All S is P' or **S → P**, together with **O → S**, taken to mean 'Something is S' or 'There is something which is S' ('S is not non-Ens'), does give the resultant **O → P**: 'Something is P.' But it does not do so without the particular premise **O → S**.

Diagram VII: Existential Import

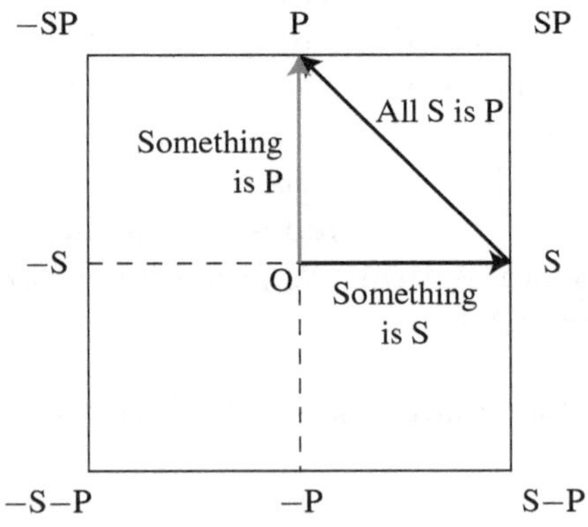

7. Valid Syllogisms

The system described above allows syllogistic arguments to be expressed and tested for validity. The general rule is that *the conclusion of a standard-form categorical syllogism is the resultant or sum of two component vectors which represent the premises.* For the purposes of "vector logic" the right method is to sum the two given vectors by drawing the third side of a triangle which travels from the initial point or tail of the first vectors to the final point or head of the second. This third vector, the directed third side of the triangle, is the sum and the conclusion.

The middle term is represented by "M", and an "M dimension" is added to give the vector diagrams a third dimension. This third dimension can be conceived as orthogonal to the SP plane, and the unit vector **O** → **M** represented as a vector from O to the point M, half way between S and P at the unit distance from O.

Let us begin with AAA-1, 'All M is P; all S is M; therefore all S is P'. Vectors are drawn from S to M and from M to P. The resultant is 'All S is P'.

Diagram VIII: AAA-1 Barbara

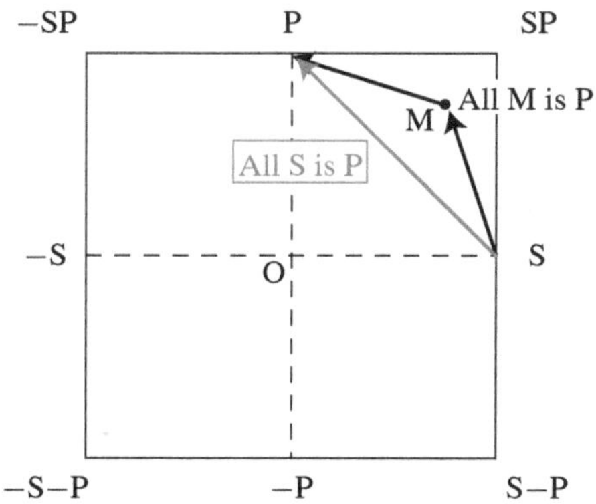

	All M is P
	All S is M
Therefore	All S is P

The same principles apply to EAE-1, for which the vector sum of the premises gives the conclusion. 'No M is P; all S is M; therefore no S is P'. The two premises are $\mathbf{MP} \rightarrow \mathbf{O}$ and $\mathbf{S} \rightarrow \mathbf{M}$, and their sum is $\mathbf{SP} \rightarrow \mathbf{O}$. It is best to translate the $\mathbf{S} \rightarrow \mathbf{M}$ vector to meet the negative premise at the point SP, by translating it to $\mathbf{SP} \rightarrow \mathbf{MP}$. Like all translations, the translation $(\mathbf{S} \rightarrow \mathbf{M}) \rightarrow (\mathbf{SP} \rightarrow \mathbf{MP})$ is an implication, in this case the immediate inference "by added determinants". An example is 'A democracy is a government by the people, therefore a successful democracy is a successful government by the people' (Eatin, 1931). (MP is the point at the head of the vector obtained by summing the unit vectors **M** and **P**.)

Diagram IX: EAE-1 Celarent

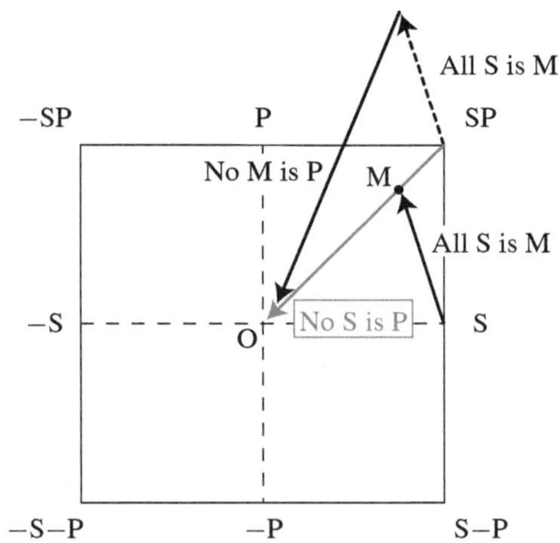

	No M is P
	All S is M
Therefore	No S is P

Consider next an argument with a particular premise, for example AII-1 (Diagram X). We have $\mathbf{M} \rightarrow \mathbf{P}$ and $\mathbf{O} \rightarrow \mathbf{SM}$. The resultant is $\mathbf{O} \rightarrow \mathbf{SP}$. The particular premise is a position vector, and so it cannot be translated from the origin. If a vector sum is to be achieved, then it must be the universal premise which is translated, to **SM → SP**, or 'Everything which is S and M is S and P.'

Furthermore, if one premise is a position vector, then the conclusion is bound to be a position vector as well. Hence there is a vector form of the rule that a valid syllogism with a particular premise must have a particular conclusion.

Diagram X: AII-1 DARII

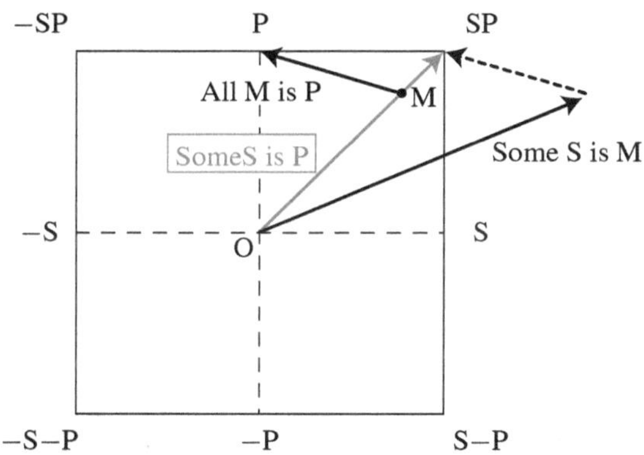

	All M is P
	Some S is M
Therefore	Some S is P

The same principles apply to all the valid syllogisms. Consider EIO-3, which has a particular negative proposition as a conclusion. In this example, as in the previous one, the particular proposition or position vector must not be moved, and the vector sum triangle must be constructed from this vector in its original position. This results in **O → S – P.**

Diagram XI: EIO-3 Ferison

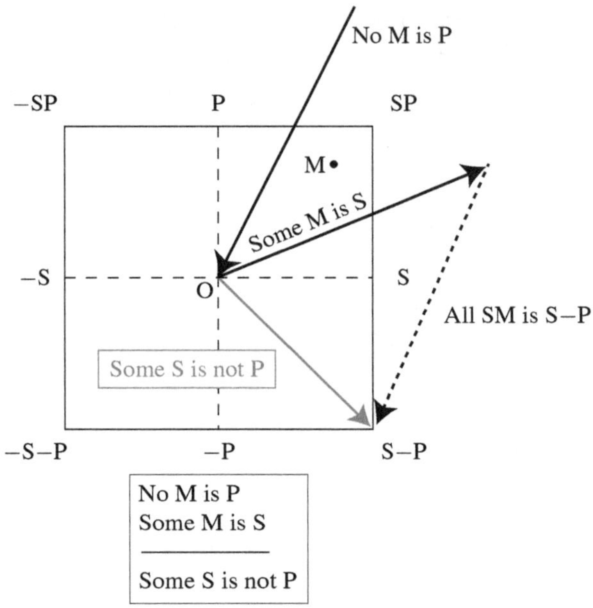

The vector diagrams have something of an advantage in economy over the corresponding Venn diagrams, as the vector diagrams do not use two different symbols, cross and hatching. There is just one symbol, the arrow of the vector or directed line segment, traveling to the origin or away from it, to the anti-origin at infinity. Furthermore, the vector diagrams can be constructed with all the terms and their complements in an array which goes from the square (two terms) to a hexagon (three terms, with "S", "M", "P", "−S", "−M", and "−P") to an octagon (four terms, adding "Q" and "−Q"), and so on without limit, to a circle for an infinite number of terms in spaces with higher numbers of dimensions.

8. Invalid Syllogisms

Just as a valid syllogism consists of premises which, represented as component vectors, yield a resultant which is the conclusion of the syllogism, so the invalid syllogisms are those which offer as a conclusion a vector which is not in fact the resultant or diagonal. Take, for example, the Fallacy of the Undistributed Middle: 'All P is M; all S is M; therefore all S is P' (Diagram XII). The vector diagram also shows, however, that with the appropriate translations one could draw the conclusion 'All SP is M' from the premises of the Undistributed Middle, dropping the redundant "M" in 'All SP is M and M'.

Diagram XII: Undistributed Middle (Invalid)

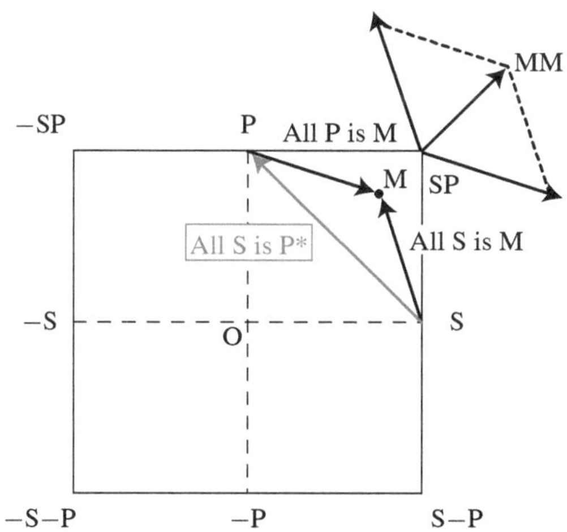

All M is P
All S is M

Therefore All S is P*

Take finally an argument with two particular premises, for example OOO-1. This is a difficult case because the two premises do seem to yield a vector sum that does not correspond to a valid conclusion.

Yet they do so only by the algebraic vector sum method of Galileo, which constructs the diagonal of a parallelogram two of whose sides are the premises. The Galileo construction can be applied to an example like OOO-1, *whereas the triangle or geometrical method cannot!* With the two position vectors, the instruction to join the arrowhead of or endpoint of the first vector to the starting point of the second vector cannot be obeyed, as both vectors are position vectors. Neither can move. This would count as an *ad hoc* restriction only if we did not already have in hand the doctrine of the non-translatability of particular propositions (Diagrams II–V). And the reason is very clear. The two vectors **O → S** and **O → P** in Diagram XIII have the sense, respectively, of (∃x)Sx and (∃x)Px. Clearly these two propositions do not imply (∃x)(Sx & Px). (If something is a cat and something is red, it doesn't have to be the cat that is red, just as if there is something which is red all over and there is something which is green all over, it most certainly does not follow that there is something which is red all over and green all over. Proof: It is impossible for something to be red all over and green all over at the same time.)

Diagram XIII: OOO-1 (Invalid)

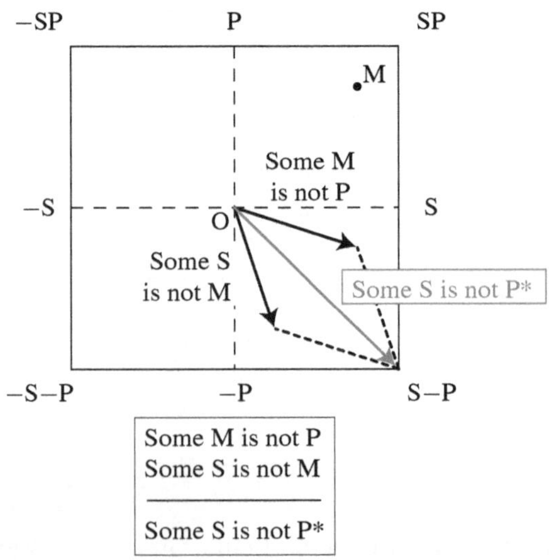

So it is the existence of geometrical vector sum triangles which is the mark of the valid syllogism, rather than the existence of a resultant in an algebraic Galileo construction. (Geometry is not algebra!) It seems to me, therefore, to be just a mistake to suppose that the two methods are equivalent. *The example of a pair of position vectors shows otherwise.* This modest conceptual point, typically overlooked by mathematicians, is of some logical importance, as no valid standard-form categorical proposition has two particular premises.

Diagram XIV: (Invalid)

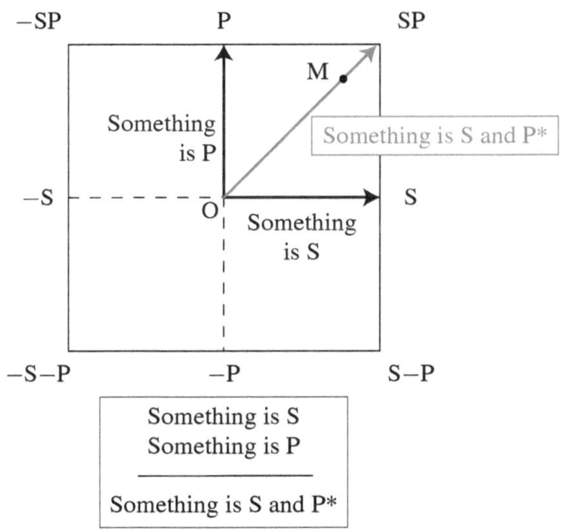

9. Plurative Propositions

In 1965 Rescher and Gallagher (Rescher and Gallagher 1965) showed how to extend the Venn Diagram techniques for validity testing to plurative propositions, as Geach called them, i.e. those employing the non-standard quantifier "most". Rescher and Gallager proposed enriching Venn Diagrams with a vector-like arrow showing the cardinality relationships (of "some" to "most" and to "all") of one area of the diagram to another. Rescher and Gallagher's arrow can be given a similar and related rôle in the present diagrammatic vector logic.

Consider the AUU-1 syllogism, which I have dubbed "AURUM". ("U" is the non-standard-form categorical proposition 'Most S is P'.) As before, All M is P is represented by the vector from M to P, **M → P**. The U proposition 'Most S is M' is represented by a vector starting at a point three-quarters of the way along the line OS, and ending at M. This is shown as the short vector **0.75S → M**. The resultant of the two vectors, **0.75S → P**, is obtained by the usual vector sum method, and automatically computes Rescher and Gallagher's (R2), which states that 'the head of an arrow may always be contracted', where needed. Rescher and Gallagher may be credited with the discovery of vectorial methods for plurative propositions. From this point of view the present paper can be regarded as extending these methods to the standard-form categorical propositions, making the abstraction away from the Venn Diagram circles for classes complete.

Diagram XV: AUU-1, Aurum

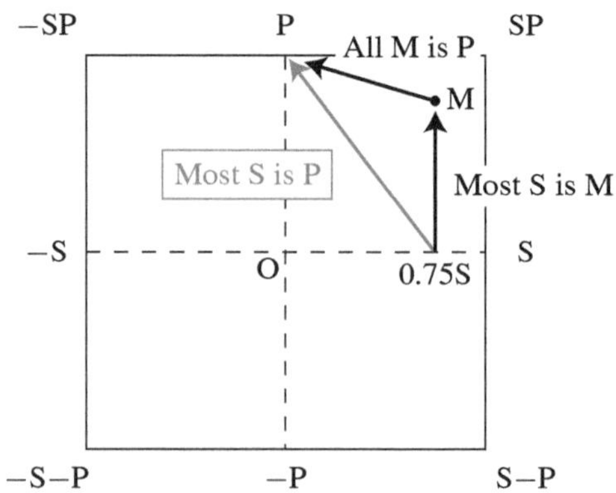

All M is P
Most S is M

Most S is P

10. Further Work

In addition to further work on non-standard quantifiers, including those in plurative propositions, the first steps to an extension of the methods described in this paper to full predicate logic will clearly be the representation in the predicate vector logic of (i) multiple variables; (ii) two- and many-place predicates; (iii) multiple quantifiers with overlapping scope.

Works Cited

Borisenko, A. I., & Trapov, I. E. (1968). *Vector and Tensor Analysis with Applications.* (R. A. Silverman, Trans.). New York: Dover.

Eaton, Ralph M. (1931). *Logic: an Introductory Survey.* New York: Scribner's.

Kneale, W., & Kneale, M. (1962). *The Development of Logic.* Oxford: Oxford University Press.

Leibniz, G. W. F. (1966). Some Logical Difficulties. In G. H. R. Parkinson (Ed. and Trans.), *Leibniz, Logical Papers* (pp. 117 ff.) Oxford: Oxford University Press.

Rescher, N., & Gallagher, N. A. (1965). Venn Diagrams for Plurative Syllogisms. *Philosophical Studies, 16 (4),* 49–55.

Westphal, J., Hardy, J. (2005). Logic as a Vector System, *Journal of Logic and Computation, 15 (5),* 751–765.

Interpreting Squares of Oppositions with the Help of Diagrams

SUMANTA SARATHI SHARMA[1]

Abstract

There are certain logical relations between the four types of categorical propositions, when they have the same subject and predicate terms. The diagrammatic representation of these relations among these propositions is called the Traditional Square of Opposition. Formal rules were developed to test the validity of syllogisms comprising categorical propositions. Apart from these rules, Euler Circles were used to test the validity of syllogisms, diagrammatically. Later, with the advent of modern logic, it was found that certain relations between propositions as expressed in the Traditional Square do not hold true. This paved way for a revised Square. At this juncture, Euler circles became ineffective, to validate the modern interpretation. It was then supplanted by Venn-Peirce diagrams. The impasse is attributed to the problem of existential import. The present paper explores the possibility of developing an alternative diagrammatic technique, which would satisfy both traditional as well as the modern viewpoints.

Keywords: Categorical Syllogisms, Euler Circles, Method of Least Representation, Square of Opposition, Venn-Peirce Diagrams

1 Department of Humanities and Social Sciences, Indian Institute of Technology Kanpur, U.P., India, email: sumantas@iitk.ac.in; School of Philosophy and Culture, Shri Mata Vaishno Devi University Katra, J&K, India, email: ss.sharma@smvdu.ac.in, sumantas@gmail.com

1. Introduction

It is said that Aristotle was concerned with pairs of statements (propositions) where one is the denial of another[2]. Later, Boethius in the middle ages represented the logical relations among the propositions, as found in the *Organon* and its subsequent commentaries, with the help of a diagram. The diagrammatic representation of the logical relations between the four types of categorical propositions (**A**, **E**, **I**, **O**) is known as the Square of Opposition. This paper is divided into five sections. The first section introduces the traditional square of opposition. With the advent of the notion of empty terms, certain difficulties were raised against this square. Thus, in the second section, we discuss empty terms and problems related to empty terms. This problem revised the traditional square and the paved way for the modern square. In the third section, we discuss the modern square of opposition briefly. These changes have made several changes in the rules which were employed to test the validity of syllogisms. In this paper, we concentrate chiefly on one of those syllogisms namely, *Darapti*. In the next section, we show the validity/invalidity of the aforesaid syllogism with the help of Euler circles and Venn-Peirce diagrams. In the fifth section, we attempt to develop an alternative diagrammatic technique to test the validity of syllogism which can test the validity of syllogism in both the traditional and modern formats.

2. The Traditional Square of Opposition

Apuleius in his commentary on *De Interpretione* mentions in detail the logical relations that hold between the four types of categorical propositions. However, he never drew any diagrams to express the relationship between them. Later, Boethius expressed the relations among these propositions based on the formulations of Apuleius with the help of a diagram known as the 'Square of Opposition'. Boethian representation of the relations between

2 Here, denial may be understood as the relations of contradiction, contrary, sub-contrary and sub-altern. See, Kneale and Kneale (1962, p. 52).

propositions which he had used extensively in his writings and teachings is the oldest found in logic texts.

Figure 1

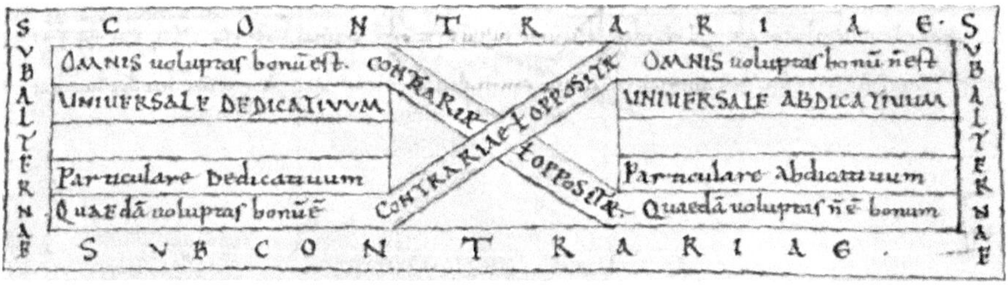

This representation as we can see is in Latin[3]. Later, several logicians followed this work and rendered the same diagram in English. Parsons' representation of the Traditional Square of Opposition is most widely used and referred in logical texts now. In his representation, the four propositions namely, the universal affirmative, the universal negative, the particular affirmative and the particular negative are placed at the four corners of a square, and the relations represented as lines drawn between them. There are exactly four types of relations between these propositions. They are contradictory, contrary, sub-contrary and subaltern. Figure 2 is Parsons' diagrammatic representation.

Two propositions are contradictory if and only if they cannot both be true and they cannot both be false together. There are two pairs of propositions which are contradictories. 'Every *S* is *P*' and 'Some *S* is not *P*' are contradictories. Similarly, 'No *S* is *P*' and 'Some *S* is *P*' are also contradictories. Two propositions are said to be contraries if and only if they cannot both be true but can both be false. 'Every *S* is *P*' and 'No *S* is *P*' are contraries. The contrary relationship holds between the universal affirmative and its corresponding universal negative only. Two propositions are

3 Boethian Square is the oldest of all the representations known as the Traditional Square. However, the later representations of this Square in English arguably are different with respect to its formulation of 'O' type propositions. For my own convenience, I have discussed the Parsons Square as the Traditional Square, since I am unable to follow Latin.

Figure 2

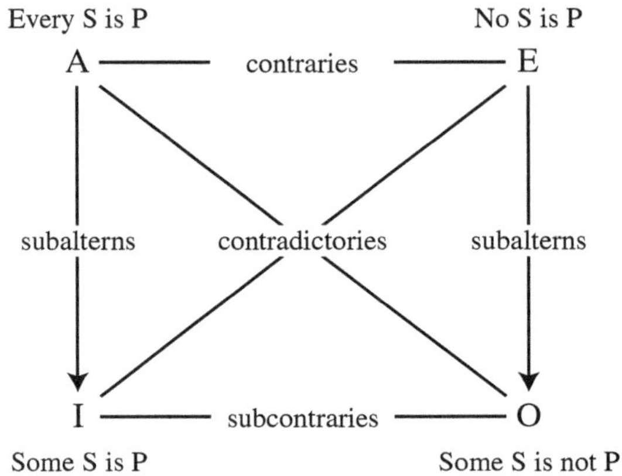

subcontraries if and only if they cannot both be false but can both be true. 'Some *S* is *P*' and 'Some *S* is not *P*' are subcontraries. The sub-contrary relationship holds between the particular affirmative and its corresponding particular negative only. A proposition is a subaltern of another if and only if it must be true if its superaltern is true, and the superaltern must be false if the subaltern is false. There are two pairs of propositions which are subalterns. 'Some *S* is *P*' is a subaltern of 'Every *S* is *P*' and 'Some *S* is not *P*' is a subaltern of 'No *S* is *P*'[4].

The relations expressed in the Traditional Square seem to be intuitively correct. However, there were certain apprehensions, which were raised against this representation of relations between propositions with the passage of time. These concerns were regarding the notion of 'empty terms'. In the next section, I shall briefly introduce the problem of empty terms and its implications for the Traditional Square of Opposition.

4 See, Parsons' entry on Square of Opposition in the Stanford Encyclopedia of Philosophy available online at http://plato.stanford.edu/entries/square/

3. The Problem of Empty Terms

An 'empty term' is a term for which there is nothing existent as its referent. For example, the term 'chimera' is an empty term as 'chimeras' are considered imaginary creatures that have no real existence. Thus, the term 'chimeras' denotes nothing real. The determination of the truth value of propositions that contain empty terms – either as subject or predicate – gives rises to a problem we call the 'problem of empty terms'. Consider the universal affirmative proposition 'all chimeras are animals'. Since 'chimera' refers to nothing that exists, can we say whether the proposition is true or false? There are opposing intuitions with regard to this issue. One is to say that when an empty term occurs, either as the subject or as the predicate of a proposition; the proposition is false since the term does not refer to anything at all. The other is to say the truth or falsity of a proposition does not depend on the existence of the referents of the terms in the proposition.

The first intuition namely, is that propositions with empty terms are false, has certain implications for the Traditional Square of Opposition. Let us take the particular affirmative proposition 'some chimera is an animal'. Since the subject term refers to nothing, the proposition comes out to be false. According to the Square, particular propositions (**I** & **O**) are subcontraries. This means both **I** and **O** types of propositions cannot be false together. Since the **I** proposition 'some chimera is an animal' is false, its sub-contrary, namely the particular negative 'some chimera is not an animal' must be true. However, the question that arises here is: which chimera is not an animal? If the particular negative is true, its subject term must have a reference. But the subject term of the proposition 'some chimeras are not animals' is an empty term, which by definition does not refer to anything at all. Thus, we can say the Square along with the doctrine of empty terms leads us into trouble. The doctrine of empty terms states that when they occur in a proposition, it is false since they don't have a reference. However, the Square of Opposition tells us that the sub-contrary of a false proposition is true. And truth of a proposition demands that the referents of its terms must exist. Thus, it seems that either the entire doctrine embodied in the Traditional Square of Opposition is erroneous, or we are stuck with the problem of empty terms, which in conjunction with the standard doctrines embodied in the *Square*, lead to apparently incorrect conclusions.

Let us first understand what made the modern logicians believe that the universal propositions lack existential import. Thus, here I discuss how modern logicians understand the **A**, **E**, **I** and **O** propositions. Propositions of type **A** (all *S* is *P*) mean, given any individual thing whatever, if it is *S* then it is *P*. Similarly, **E** (no *S* is *P*) type propositions may be paraphrased as, given any individual thing whatever, if it is *S* then it is not P. **I** type propositions (some *S* is *P*) are understood as; there is at least one thing which is both *S* and *P*. Last, the **O** type proposition (some *S* is not *P*) is paraphrased as, there is at least one thing which is *S* but not P. We must note here, that in **A** and **E** types of propositions there is no reference of any object or entity whereas in **I** and **O** types of propositions it is said that there exists at least one individual of the cited type. This paraphrasing was based on the symbolic notation for propositions.

The modern symbolization of the four types of categorical propositions is given as follows:

All S is P (**A**) is symbolized as (**x**) [Sx ⊃ Px]
No S is P (**E**) is symbolized as (**x**) [Sx ⊃ ∼ Px]
Some S is P (**I**) is symbolized as (∃**x**) [Sx ∧ Px]
Some S is not P (**O**) is symbolized as (∃**x**) [Sx ∧ ∼ Px]

Here, (**x**) stands for, all **x**, (∃**x**) stands for, some **x**, **Sx** stands for, **x** belongs to S, **Px** stands for, **x** belongs to P, '⊃' stands for implication, '∧' stands for conjunction and '∼' stands for negation. The above is the symbolic manifestation of the paraphrasing of propositions. (**x**) and (∃**x**) are called the logical quantifiers and '⊃', '∧' and '∼' are called the logical connectives. **Sx** and **Px** are called the antecedent and consequent of the universal propositions respectively whereas they are called the conjuncts of the particular propositions. It is because of the fact that in a conditional proposition (universal propositions); the terms of the proposition are called antecedent and consequent. On the other hand, in a conjunctional proposition (particular propositions) the terms are called conjuncts.

In symbolic logic, an implication (conditional proposition) is true, if its consequent is true or its antecedent is false. An implication is false if an only if its antecedent is true whereas its consequent is false. In case of conjunctional propositions, a proposition is true if and only if both the conjuncts are true and false otherwise.

Let us now suppose if the subject term is empty i.e. **Sx** is empty. According to the analysis of modern logicians, if **Sx** is empty, then the truth-value of **Sx** is false. If **Sx** is false, then $\mathbf{Sx} \supset \mathbf{Px}$ must be true, since the antecedent of the conditional proposition is false. Whereas, $\mathbf{Sx} \wedge \mathbf{Px}$ is false since one of the conjuncts is false. According to the Traditional Square of Opposition, **A** and **I** propositions are subalterns, i.e. if **A** is true, then **I** must be true and if **I** is false, then **A** must be false. The above analysis of modern logicians falsifies the traditional claim of **A** and **I** being subaltern. Similarly, $\mathbf{Sx} \supset \mathbf{Px}$ and $\mathbf{Sx} \supset \sim \mathbf{Px}$ are both true if **Sx** is false. This defies the traditional claim that **A** and **E** cannot be true together. Similarly, $\mathbf{Sx} \wedge \mathbf{Px}$ and $\mathbf{Sx} \wedge \sim \mathbf{Px}$ are both false, in the given condition. This also is against the traditional claim that **I** and **O** cannot be false together. Thus, we find that the relations of subaltern, contrary and subcontraries are all falsified. The only relation in the Traditional Square of Opposition, which holds in accordance with the modern point of view, is that of contradictories between universal affirmative and particular negative as well as universal negative and particular affirmative.

Kneale and Kneale[5] opines, that when the difficulties of Aristotle's theory of opposition were realized, attention was concentrated chiefly on the question whether universal statements could properly be said to entail their subalterns, and the notion of *existential import* was introduced as something required especially for the understanding of universal statements within the Aristotelian scheme. Boger[6] argues that modern logicians are more puzzled by considerations of existence. They fear that Aristotle's logic leads to peculiar violation of the Traditional Square of opposition whereas these matters were not troublesome to Aristotle. Aristotle himself affirms that truth or falsity of a given proposition depends on whether propositions are expressions of state of affairs. The cause or ground of the truth of a sentence is the state of affairs it denotes. The Kneales[7] opine that in order to justify Aristotle's doctrine as a whole it is necessary, then, to suppose that he assumed application for all the general terms with which he dealt. However, for a modern day logician this seems a restriction to the scope of logic, since it is not difficult to construct a general statement in which at least one term lacks application.

5 Quoted from, Kneale and Kneale (1962, p. 59)
6 See, Boger, (2004, pp. 157–160)
7 See, Kneale and Kneale (1962, p. 25)

4. The Modern Square of Opposition

Since modern logicians accept only *contradiction* as a valid form of relation among the universal and particular propositions, the Square of Opposition as revised by modern logicians is as follows:

Figure 3

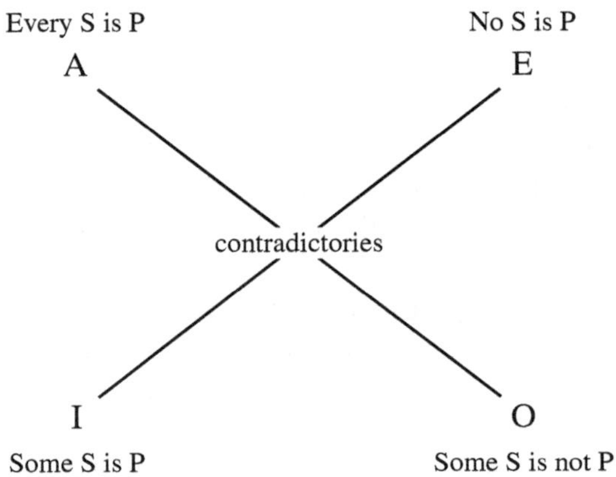

Thus, the Traditional Square of Opposition is revised to what one may call a 'Cross' of Opposition. We have already seen that relations represented by the Traditional Square of Opposition hold only on the assumptions that the terms of the various categorical propositions are non-empty. Modern logicians wanted to develop a system of logic which is not threatened by the problem of empty terms. Their strategy was to free universal propositions of their existential implications. This led them to revise the Square of Opposition. However, this revision had certain implications for the Aristotelian system of deductive inference. Once it is granted that universal propositions do not have existential commitments, many of the immediate inferences do not hold. For example, consider the immediate inference of conversion by limitation. From the proposition 'all unicorns are horses' we cannot draw the conclusion 'some horses are unicorns'. The proposition, 'all unicorns are horses' being a universal proposition is devoid of existential implications is understood as a conditional statement 'if anything is a unicorn it will be a horse'. Since 'unicorns' is an empty term it does not refer to anything. Consequently, the antecedent of the conditional is false making the whole conditional vacuously true. However, the conclusion 'some horses are unicorns' is false, as there are nothing which is both a horses and a unicorn. So this is an invalid argument since from a true premise we draw a false conclusion.

Similarly, if the universal propositions are purged of their existential implications, some of the syllogisms considered as valid, in the Aristotelian logic cease to be valid. Consider for example, an instance of categorical syllogism of the form *Darapti* (AAI-III):

All Cows are Mammals.
All Cows are Herbivorous.
Therefore, some Mammals are Herbivorous.

In the syllogism, the premises and the conclusion are true and it does not violate any of the Aristotelian rules of inference. However, that this argument is not valid can be shown by considering another argument in Darapti itself with true premises and a false conclusion. Consider, for example:

All trespassers are prosecuted
All trespassers are people.
Therefore, some people are prosecuted

In a situation where there are no trespassers both the premises of this argument turns out to be true vacuously. However, the conclusion that 'some people will be prosecuted' will be false in a world, *ceteris paribus*, where nobody trespasses. Hence, it is an invalid argument.

That some of the arguments which are valid in the Aristotelian system is no more valid from the modern point of view, suggested that the Aristotelian rules of syllogistic inference are not sound as they allow inferences that are not really valid. To remedy this, modern logicians suggested that an additional rule regarding existential import must be added to the system of rules of syllogistic inference. This rule which pertain to the quantity of the premises and the conclusion may be stated as follows:

No valid syllogism with particular conclusion has two universal premises[8].

With the addition of this rule, *Darapti* ceases to be a valid form of an argument. Since universal propositions do not have existential import whereas particular propositions do, to infer a particular proposition from two universal premises would be to go beyond what is warranted by the premises[9]. Any syllogism which violates the above rule is said to commit the *existential fallacy*. In the next section, we will consider the same syllogism (*Darapti*) and test its validity with the help of Euler Circles and Venn-Peirce diagrams.

5. Euler Circles and Venn-Peirce Diagrams

As we know, *Datisi* i.e. AII-3 is written as:

Figure 4

All M is P
Some M is S

∴ Some S is P

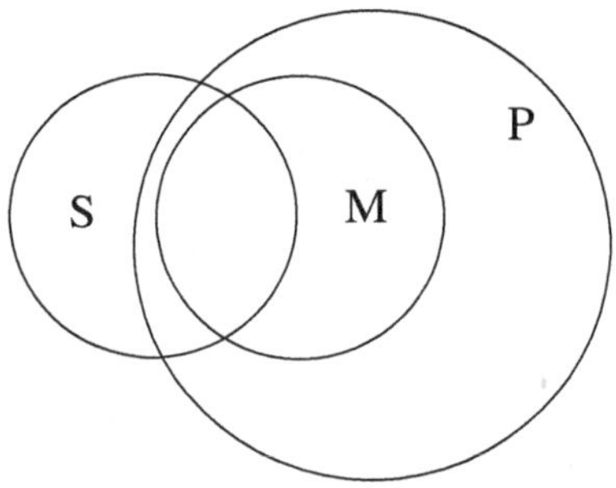

8 The first rule states, that a valid standard form categorical syllogism must contain exactly three terms. The middle term must be distributed at least once. Any term which is distributed in the conclusion must also be distributed in the premise. The syllogism cannot have two negative premises. If one of the premises is negative then the conclusion must be negative and vice-versa.

9 Quoted from Copi (1972 p. 203).

In this diagram, we draw a circle *M* inside *P* and then we draw a circle *S* as intersecting with *M*. It is not possible to draw *S* intersecting *M* without *S* intersecting *P*. As we can see [in Fig.(4)] some part of circle *S*, is inside circle *P*. This is exactly what is claimed by the conclusion. Thus, this syllogism is shown to be valid with the help of Euler Circles. However, according to Venn-Peirce diagrams it drawn as:

Figure 5

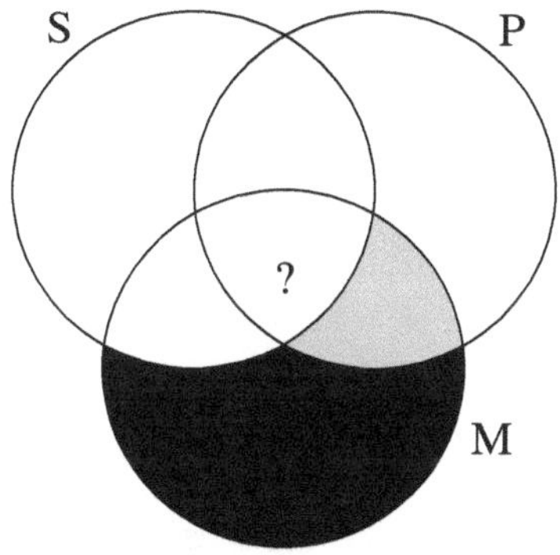

The above diagram [Figure (5)] first draws all M is P by shading off the part of M which is not P. Similarly, the part of M, which is not S, is also shaded off, as given in the second premise. The conclusion of the syllogism states, 'Some S is P', which means that we must get an '**X**' mark in the common part of S and P. This does not happen. Hence, we draw a question mark (?) to show the inability of the premises to imply the conclusion. Therefore, according to Venn-Peirce Diagrams, AAI-3 is an *invalid* syllogism.

We must note that *Darapti* was rendered valid when we check its validity as per the Euler Circles whereas it is invalid here. Euler Circles do not consider the rule of existential fallacy whereas Venn-Peirce Diagrams does consider the above rule and thus *Darapti* is invalid here. In Euler Circles, we use closed curves to show the relationship between the class of objects denoted by the subject and predicate terms of a proposition. Here, we find no principle which distinguishes between the representation of universal and particular propositions. On the other hand, a basic Venn diagram consists of three overlapping circles to show three terms (*S*, *P* and *M*) present in a syllogism. However, in these diagrams, we use 'shading' for universal propositions and an '**X**' to mark particular propositions. The basic Venn Diagrams shows how the terms of the syllogism shall be placed, whereas 'shading' and '**X**' marks expresses the relationship between terms. It is evident that if both the premises are universal propositions then they will be expressed with the help of shading, but the conclusion is a particular proposition, which requires an '**X**' for showing validity. Thus, a particular conclusion from two universal premises cannot follow according to this system of diagrams. This diagrammatic method follows the rules of the Modern Square; thus, all the syllogisms which violate the rule of existential fallacy are invalid here. In the next section, we will attempt to develop an alternative diagrammatic technique to test the validity of syllogisms in both the traditional and modern viewpoints.

6. Method of Least Representation[10]

The proposed representation method is based on the following findings. Traditional understanding (developed by Aristotle and diagrammatically expressed by Euler) seeks to find a representation of circles in the concerned area, i.e. whether the diagram draws the area claimed by the conclusion. However, in the case of the modern viewpoint we look for specific demonstration. In the proposed diagrammatic technique, we use rectangles for universal propositions and right-angled triangles for particular propositions. When we are corroborating the traditional perspective, we seek the representation of the conclusion in the diagrammed figure. However, when we deal with the modern interpretation, the shape of the geometric figure

10 The Method of Least Representation was originally construed when the author has presented his work in First World Congress on the Square of Opposition, Montreux, Switzerland June 2007. Later, this was revised and represented as the Method of Minimal Representation: An alternative diagrammatic technique to test the validity of Syllogisms. See Sharma, (2008, pp. 412–414)). In this paper, the version of diagrammatic technique has been significantly elaborated, expanded and strengthened aiming clarity.

becomes important and necessary differences (if any) are made. The diagrams are explained below:

Universal Affirmative Proposition (**A**): All *S* is *P*. Here, a rectangle is drawn from the right bottom edge containing it and it is shaded. The arrow shows that the possibility of finding *S* is inside *P* only.

Figure 6

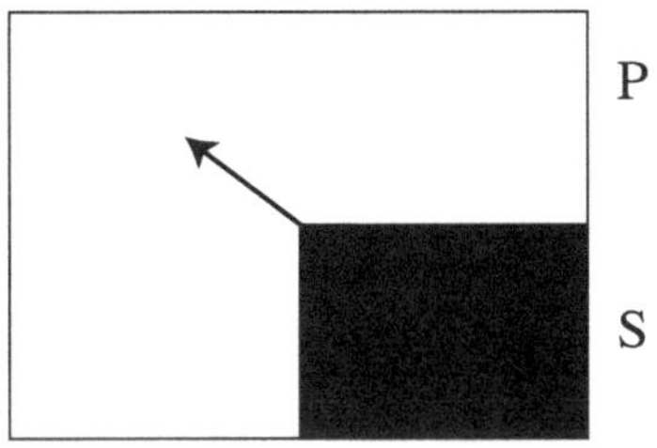

Universal Negative Proposition (**E**): No *S* is *P*. Here, two disjoint rectangles are drawn and the arrow shows that the possibility of finding *S* is outside *P* only.

Figure 7

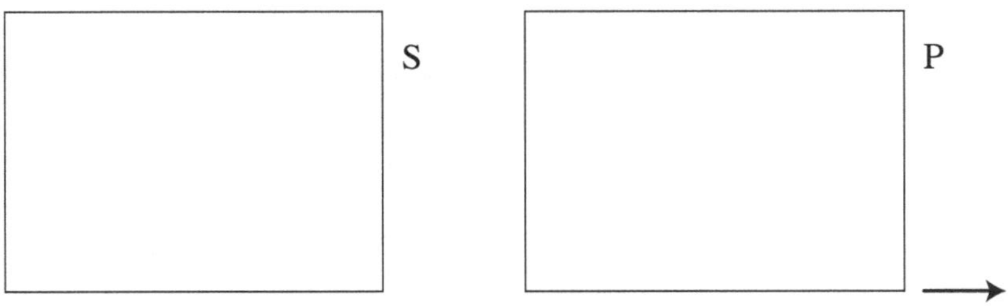

Particular Affirmative Proposition (**I**): Some *S* is *P*. Here, a right-angled triangle is drawn from the right bottom edge containing it and is shaded. The arrows suggest that the possibility of finding *S* is both inside as well as outside *P*.

Figure 8

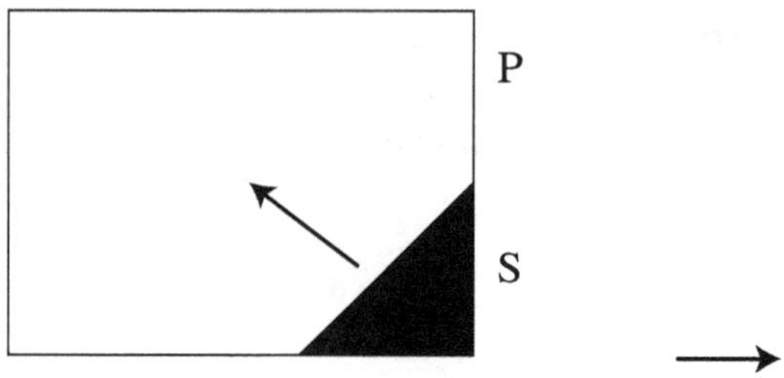

Particular Negative Proposition (**O**): Some *S* is not *P*. Here, a right-angled triangle is drawn from the right bottom edge outside it and is shaded. The arrows suggest that the possibility of finding *S* is both inside as well as outside *P*.

Figure 9

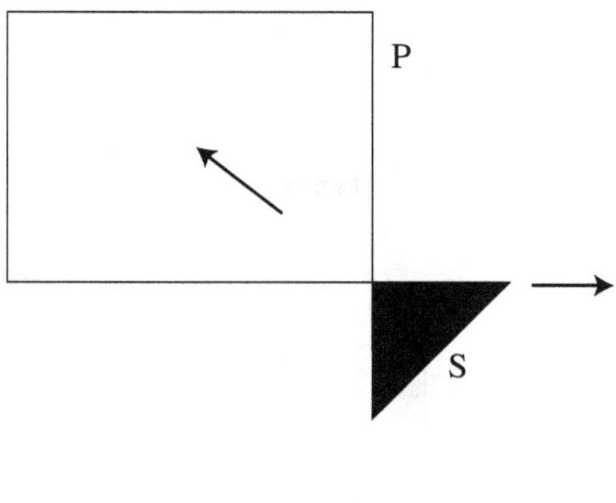

To sum up:

a) The Universal Propositions (**A** & **E**) are represented by rectangles.
b) The Particular Propositions (**I** & **O**) are represented by right-angled triangles.

c) In Universal Affirmative Proposition (A), a rectangle is drawn from the right bottom edge of the greater rectangle containing it and it is shaded.
d) In Universal Negative Proposition (E), two disjoint rectangles are drawn of equal area.
e) In Particular Affirmative Proposition (I), a right-angled triangle is drawn from the right bottom edge the greater rectangle containing it and is shaded.
f) In Particular Negative Proposition (O), a right-angled triangle is drawn from the right bottom edge of the greater rectangle outside it and is shaded.

Now, let us take the syllogism *Darapti* (AAI-3) and test its validity with the help of this method. The diagram is given below:

Figure 10

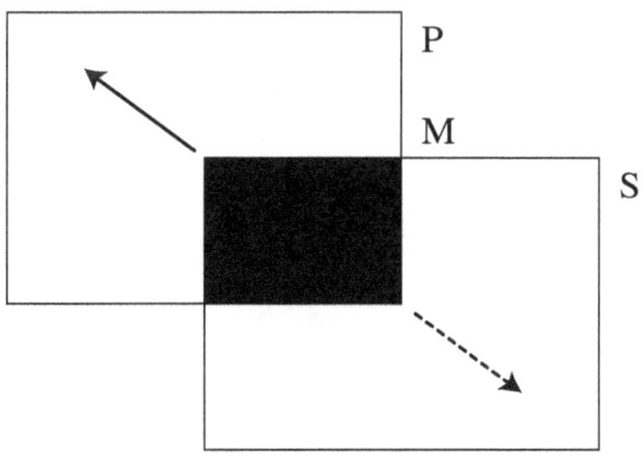

In the above diagram, we first draw small rectangle M inside the larger rectangle P. Similarly, we draw the same small rectangle M in the larger rectangle S as shown above. We shade M and put in the arrows. The solid arrow above is from the first premise, whereas the dotted arrow below is from the second premise, but both depict the possibility of finding M. Now, we will test the validity of the above mentioned categorical syllogism.

In the traditional interpretation, we need to find that there shall be a part common between S and P. In the diagram, it is portrayed by M. Therefore,

AAI-3 is a valid syllogism in the traditional interpretation. Now let us verify the same in the modern interpretation.

In the modern viewpoint, we need to have only a right-angled triangle S in the rectangle P. If we examine the diagram above, we find a rectangle in place of a right-angled triangle. This suffices, to establish that **AAI-3** is an invalid syllogism in the modern viewpoint.

The above example shows us that we can have different set of inference rules for the same diagram. Moreover, these different set of inference rules can test the validity in two different interpretations, and can have dissimilar validity judgments. In the case mentioned above, we found that the same diagram is valid in traditional interpretation, whereas it is invalid from the modern viewpoint. We must note that it is impossible to draw a right-angled triangle from two Universal premises, which are drawn as rectangles. This means, that whenever, we will draw Universal premises, we need to have a universal conclusion or else it will be invalid in modern viewpoint. The *Method of Least Representation* looks for diagramming the conclusion in traditional interpretation whereas a perfect geometrical shape is required for validation in the modern viewpoint. Finding a representation in the mentioned area suffices for the traditional format whereas inability to find the suggested geometric figure indicates that the syllogism is invalid in the modern system. The rationale in using two dissimilar geometric shapes for two different types of propositions namely universal and particular lies here.

7. Conclusion

Valid moods (in both the interpretations) have been tested with the proposed diagrammatic technique and results were found confirming both the points of view. The new technique seeks to demonstrate that we do not have to have two separate systems of diagrams, one representing the traditional point of view and another championing the modern conception regarding the problem of existential import, if we were to compare and contrast the two logical systems. The scope of the paper is limited to checking the efficacy of the illustrated diagramming techniques to categorical syllogisms only. The modern revised square opened one of the perennial debates in validating syllogisms, i.e., the problem of existential import. Venn-Peirce diagrams broadened the issue further with its apt and consistent diagrammatic representations. This paper attempted to develop a unified diagrammatic technique to address the existing stalemate representing syllogistic reasoning. An attempt has been made to present a single diagrammatic technique to test the validity of syllogisms in both the elucidations. If a single depiction is possible to validate syllogisms in both the formats, we may interpolate a single *Square,* which is left as a challenge for future work.

Acknowledgements

I am grateful to Dr. C.A Tomy for correcting the first draft. I thank Prof. R.S Misra, Prof. B.H Boruah, Dr. V. K Tripathi and Dr. Hari Narayanan V. for encouraging me to carry out this work. I am indebted to the anonymous referees for their constructive feedback on an earlier draft. My sincere gratitude to Dr. Gillman Payette for his painstaking final corrections and Dr. Anil Kumar Tewari for the software support, without whom, this paper would not have taken the present form. This work is supported by Junior Research Fellowship Scheme of University Grants Commission, New Delhi, India.

References

Boger, G. (2004). Aristotle's Underlying Logic. In D.M. Gabbay & J. Woods (Eds.), *Handbook of the History of Logic* (pp. 157–160). Amsterdam: Elsevier B.V.

Copi, I. M. (1972). *Introduction to logic.* New York: Macmillan., Edn. IV p. 203.

Kneale, W., & Kneale, M. (1962). *The Development of Logic*. Oxford: Clarendon Press.

Sharma, S. S. (2008). Method of Minimal Representation: An Alternative Diagrammatic Technique to Test the Validity of Categorical Syllogisms. In *Lecture Notes in Artificial Intelligence, 5223,* 412–414. Springer: Heidelberg.

Part 3: Philosophy of Language

Classical vs. Modern Squares of Opposition, and Beyond[1]

DAG WESTERSTÅHL

Abstract

The main difference between the classical Aristotelian square of opposition and the modern one is not, as many seem to think, that the classical square has or presupposes existential import. The difference lies in the relations holding along the sides of the square: (sub)contrariety and subalternation in the classical case, inner negation and dual in the modern case. This is why the modern square, but not the classical one, applies to any (generalized) quantifier of the right type: *all, no, more than three, all but five, most, at least two-thirds of the*, … After stating these and other logical facts about quantified squares of opposition, we present a number of examples of such squares spanned by familiar quantifiers. Special attention is paid to possessive quantifiers, such *Mary's, at least two students'*, etc., whose behavior under negation is more complex and in fact can be captured in a *cube* of opposition.

Keywords: generalized quantifiers, monotonicity, possessive quantifiers, square of opposition

1 I thank the audiences at the 1st Square Conference in Montreux, June 1–3, 2007, and the 10th Mathematics of Language Workshop at UCLA, July 28–30, where earlier versions of this paper were presented, for useful comments. In particular, Larry Horn's remarks were valuable to me. Section 4.4 of the paper builds on joint work with Stanley Peters on the semantics of possessives. The version presented at the MoL Workshop was delivered in honor of Ed Keenan on the occasion of his 70th birthday. In several of his publications over the years, Ed has explored to the general notion of inner negation, which he calls post-complement, and which is a crucial ingredient in the modern square of opposition. Work on this paper was supported by a grant from the Swedish Research Council. A much abbreviated version of it appeared in Westerståhl (2008).

1. Background

1.1 Squares

A square of opposition illustrates geometrically a particular way that negation interacts with certain operators. In the *quantified square* (Figure 1), the operators are the Aristotelian quantifiers, but several other operators follow a similar pattern (Figure 2). In these squares – here in addition a *modal* square, a *temporal* square, and a *propositional* square – you have ordinary, *contradictory* or *outer* negation along the diagonals, and various weaker forms of negation as well as other logical relations along the edges of the square. Precisely what these other relations are is a matter of debate, as we will see presently. But an easily identifiable rough pattern is common to all these squares. For example, the *deontic* square replaces *necessary* by *obligatory*, *impossible* by *prohibited*, and *possible* by *permitted*.

In a clear sense, the quantified square is the fundamental one: it recurs when you spell out the meaning of the operators in the other squares. For example, a conjunction is true iff *all* its conjuncts are true, and a disjunction is true iff *some* of the disjuncts are. Likewise, a proposition is necessary iff it is true under *all* circumstances (in *all* possible worlds), it is possible if true under *some* circumstance, and impossible if true under *none*.[2]

2 One square whose contents were discussed by Aristotle falls partly outside this pattern. This is the *singular* square (Horn, 1989, §§1.1, 7.2–3):

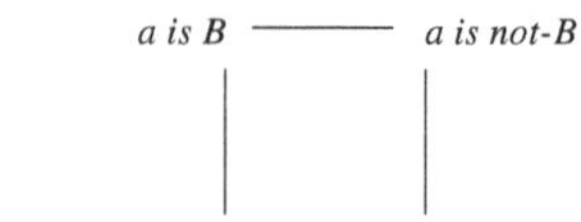

The crucial notion here is that of *predicate negation*. E.g. one may distinguish *not-white* (or *not white*) as a 'logical' contrary of *white*, whereas *black* is a 'polar' contrary (along a color scale), and *brown* yet another kind of contrary (it is a contrary since nothing can be both white and brown). There are intricate linguistic (and possibly logical) issues involved, but they don't have much to do with quantification, and I shall disregard them here.

Figure 1: The Quantified Square

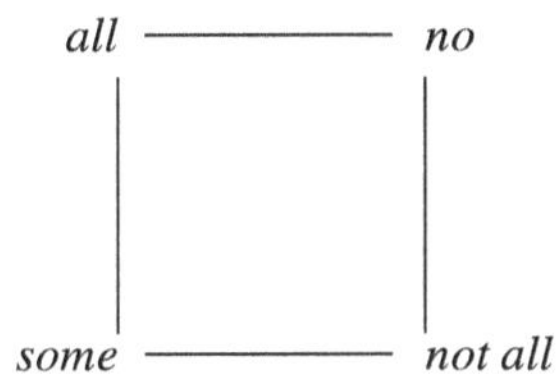

Figure 2: Various Squares

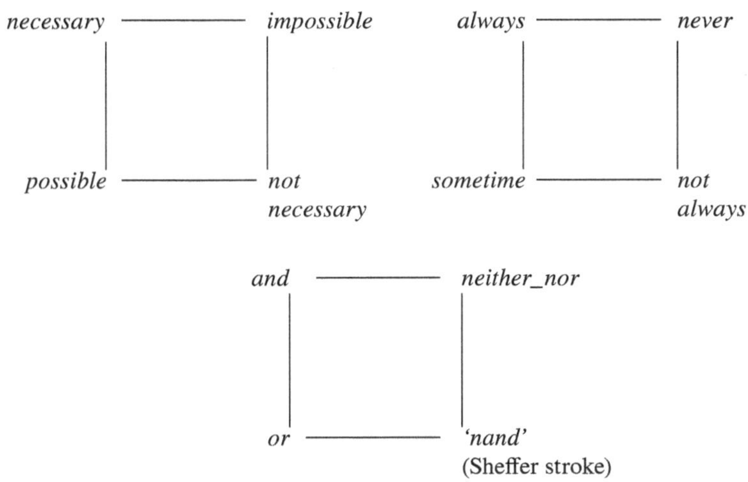

1.2 The Aristotelian Square

In this paper we focus on the quantified square, and more generally on the interaction between negation and quantification. Let us draw the classical quantified square, that I will simply call the *Aristotelian square* (its content was described in words by Aristotle, and it was drawn as a square by Apuleios and Boethius some 800 years later), in somewhat more detail. In Figure 3, the only non-standard notation is all_{ei}, which is the universal quantifer with *existential import*, i.e. such that $all_{ei}(A,B)$ is true iff A is non-empty and included in *B*. This is what Aristotle and most of his medieval followers, in agreement with many philosophers and most linguists today, took words like "all" and "every" to mean. In contrast, Frege and most logicians after him take them to stand for the quantifier *all* without existential import: *all* $(A,B) \Leftrightarrow A \subseteq B$. I come back to this issue presently, but let me for the record first describe the content of the Aristotelian square.

Figure 3: The Aristotelian Square

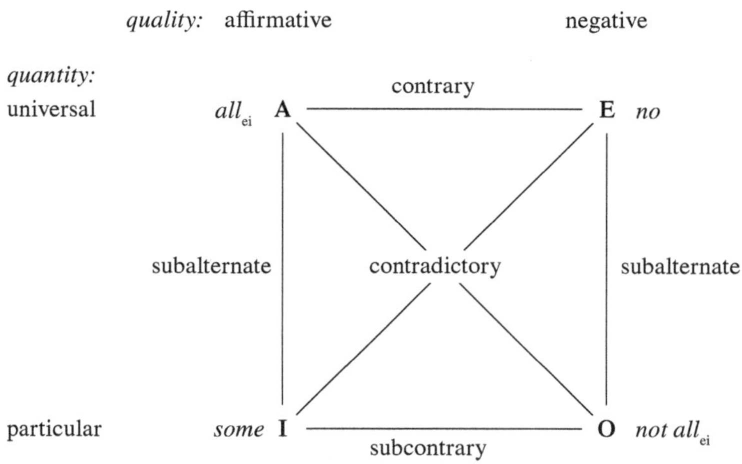

The (medieval) naming **A**, **E**, **O**, **I** of the four corners is handy and will be used consistently in what follows. As to the logical relations depicted in the square, we have, aside from contradictory negation along the two diagonals, the relations of contrariety, subcontrariety, and subalternation along the sides. These are seen as relations not between the quantifiers themselves but the corresponding statements made with them. Thus, φ and ψ are *contrary* iff they cannot both be true, *subcontrary* iff they cannot both be false, and ψ is *subalternate* to φ iff whenever φ is true, so is ψ (i.e. iff φ implies ψ).[3]

3 One often adds the requirement that φ and ψ *can* both be false in the definition of contrariety, and likewise that φ and ψ *can* both be true in the definition of subcontrariety. Everything I say below about these notions holds for the revised definitions as well.

That the **A** and **E** corners are contrary clearly shows that the universal quantifier must be taken to have existential import in this square. The same conclusion follows from the fact that **A** is held to imply **I**. But whereas few linguists or philosophers have had doubts about this, there has been much uneasiness about the consequence that the quantifier at the **O** corner, the contradictory negation of all_{ei}, then must have the truth conditions: *not* $all_{ei}(A,B)$ iff *either A* is empty *or* something in *A* is not in *B*. It doesn't seem very plausible that a sentence of the form "Not all As are B" would mean that. Various remedies have been suggested. One, that may initially appear to have some plausibility, is to say that *all* corners have existential import. Thus, for example, "No As are B" would amount to "There are As, but none of them are B". Alternatively, one might even stipulate that 'empty terms' are simply prohibited.

I will not follow any of these routes. My main reason is a simple (Gricean) distinction between the *truth-conditions* of a sentence and when it is *appropriate to utter* it. If I know that there are no girls in your class, it is inappropriate for me to say

(1) No girls in your class came to the party.

But it surely isn't *false*. Having been at the party, I may utter (1) appropriately, believing there are girls in your class; your informing me that there aren't any doesn't compel me to *retract* (1). Retracting it would amount to saying

(2) Some girls in your class came to the party.

but I know this is false. The distinction is nowadays accepted by most theorists, I think. For another example, if I know that Mary is at home, it is inappropriate for me to say

(3) Mary is either at home or at the movies.

It is even misleading to say so, but it clearly isn't false. No one concludes from the inappropriateness of (3) in that situation that *A* is sometimes true even though $A \vee B$ is false!

So we shouldn't take *no* to have existential import. (The case of *all* is slightly more complex and will be returned to below.[4]) As indicated in the diagram above, statements can be classified according to *quality*: they are either *affirmative* or *negative*; negative statements don't make existential claims in the Aristotelian square, but affirmative ones do. Another classification concerns *quantity*: in an obvious sense, a statement (of the relevant form) is either *universal* or *particular.*

1.3 Plan

A number of square-related issues have been discussed in the literature, some of them since the days of Aristotle. For example,

(a) What (if anything) is less valuable/real/natural/informative about negative statements as compared to positive ones?
(b) What (if anything) makes a statement negative rather than positive?
(c) Why is the **O** corner never lexically realized (in any language)?
(d) Can the corners be distinguished/ordered in terms of the simplicity (conceptual, computational, or other) of the corresponding quantifiers?
(e) Does the **A** corner have existential import?

These issues will not be discussed, other than indirectly, here (except that I make a brief remark on (d) in section 5).[5] Instead, I will focus on the modern

4 Aristotle's own position on empty terms may be unclear, but many medieval philosophers had no inhibitions whatever against them. A nice example is Paul of Venice (c. 1400, quoted from Parsons, 2004), who claimed that the sentence

(i) Some man who is a donkey is not a donkey.

is *true*, since $A =$ the set of men who are donkeys is empty. This in turn is an argument that the truth conditions at the **O** corner are as we described them above, even if one uses the form "some_not" rather than "not all".

5 Horn (1989) is a classic history of negation, to which I refer for discussion of the above and many other issues related to negation. For recent contributions to the discussion, in terms of the idea of a natural logic for language, see Jaspers (2005) and Pieter Seuren's Natural Logic Project (a description is a available at his home page). While Horn (and many others) considers (a) and (b) to be non-issues, he offers an interesting explanation of (c). Löbner (1990) and Jaspers (2005) lay great emphasis on (d) (though their respective answers differ). A pragmatic view on (e) is defended in Horn (1997), and, in a different way, in Peters and Westerståhl (2006, ch.4.2.1). But, as we will see,

variant of the Aristotelian square. But my main aim is not to argue in favor of the Fregean version of those four quantifiers. Rather, my point is that the modern square, in contrast with the Aristotelian one, represents forms of negation that are much more ubiquitous in language than the usual discussion reveals, and that this particular square is just one example among indefinitely many others with the same pattern of negation but with other quantifiers at the corners. Indeed, any (generalized) quantifier (of type ⟨1,1⟩), and thus any interpretation of any simple or complex determiner, spans a modern square (but not a classical one).

So my first task is to show that the 'oppositions' introduced by Aristotle, i.e. (sub)contrariety and subalternation, though not uninteresting, are not quite the ones to focus on in the study of natural language negation. The distinction may seem subtle for Aristotle's square, but it becomes obvious as soon as you consider squares with other quantifiers. The second part of the paper, then, examplifies how these new squares are manifested in language, and revisits some of the old issues in this more general context. I end with a particularly interesting example, that of possessive determiners, whose interpretation involves two quantifiers. This allows a new form of negation, and is seen to be representatble in a *cube* of opposition rather than a square.

2. Classical vs. Modern Forms of Opposition

The modern or Fregean version of the Aristotelian square is drawn in Figure 4. At first sight, one may get the impression that the only difference is that *all* has replaced all_{ei} at the **A** and **O** corners (I continue to use these names of the corners, but without writing them in the diagrams). Likewise, if one disregards empty terms, the two squares seem to coincide. So the difference might appear to boil down simply to different views about existental import. But this impression is *completely misleading*, and it stems from considering only the four Aristotelian quantifiers. Instead, the main

there are more interesting issues of existential import for quantifiers other than those in Aristotle's square.

difference consists in the different choice of relations along the sides (not the diagonal) of the square. Existential import is a side issue in this context.

Figure 4: The Modern Version of Aristotle's Square

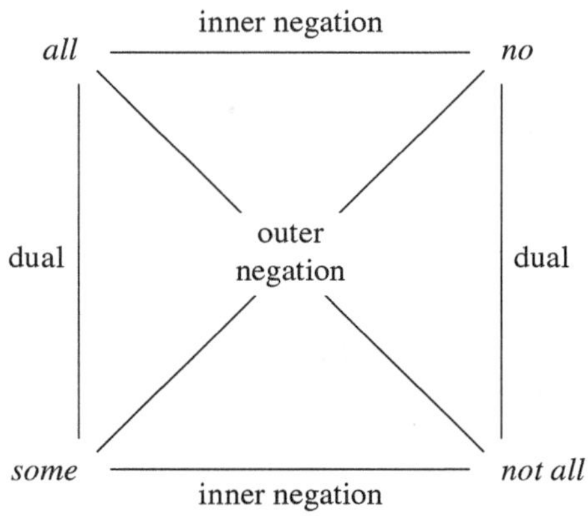

In the modern square there are only two relations along the sides: inner negation and dual, and like outer negation they can be seen as operations on the quantifiers themselves:

(4) a. The *outer negation* of Q, $\neg Q$, is defined by: $\neg \mathrm{Q}(A,B) \Leftrightarrow$ not $\mathrm{Q}(A,B)$
   b. The *inner negation* of Q, $Q\neg$, is defined by: $\mathrm{Q}\neg(A,B) \Leftrightarrow \mathrm{Q}(A,M-B)$ (where M is the universe)
   c. The *dual* of Q, Q^{d}, is defined by: $Q = \neg(Q\neg) = (\neg Q)\neg$

One sees that these relations do indeed hold as indicated in Figure 4. So it is the same relation along both horizontal sides, in contrast with the Aristotelian square, but more importantly, this relation has little to do with (sub) contrariety. In general, as we will see, nothing prevents Q and $Q\neg$ from being both true, or both false, of the same arguments. Likewise, the relation along the vertical sides has little or nothing to do with subalternation, since we may easily have Q true and Q^{d} false of the same arguments. The fact that $all_{\mathrm{ei}}(A,B)$ implies *some*(A,B), and hence $\neg$*no*(A,B), is just incidental to that particular square.

To appreciate these points, we need a technical concept.

2.1 Type ⟨1,1⟩ Quantifiers

I only give the bare definitions necessary for what follows; for (much) more about quantifiers, see Peters and Westerståhl (2006).

(5) **Definition.** A (generalized) *quantifier* Q *of type* ⟨1,1⟩ associates with each universe M a binary relation Q_M between subsets of M.

Many such quantifiers interpret simple or complex determiners, and we can name them by the corresponding determiner phrase in italics (in, say, English). Here are a few examples, starting with some already mentioned ($|X|$ is the cardinality of the set X). For all M and all $A,B \subseteq M$,,

$all_M(A,B) \Leftrightarrow A \subseteq B$
$(all_{\text{ei}})_M(A,B) \Leftrightarrow \varnothing \neq A \subseteq B$
$no_M(A,B) \Leftrightarrow A \cap B = \varnothing$
$\textit{at least two}_M(A,B) \Leftrightarrow |A \cap B| \geq 2$
$\textit{exactly five}_M(A,B) \Leftrightarrow |A \cap B| = 5$
$\textit{all but three}_M(A,B) \Leftrightarrow |A - B| = 3$
$\textit{more than two-thirds of the}_M(A,B) \Leftrightarrow |A \cap B| > 2/3 \cdot |A|$
$most_M(A,B) \Leftrightarrow |A \cap B| > |A - B|$
$\textit{the ten}_M(A,B) \Leftrightarrow |A| = 10 \ \& \ A \subseteq B$
$\textit{John's}_M(A,B) \Leftrightarrow \varnothing \neq A \cap \{a : \text{John `possesses' } a\} \subseteq B$
$\textit{no\_except John}_M(A,B) \Leftrightarrow A \cap \mathrm{B} = \{\text{John}\}$
$\textit{some but not all}_M(A,B) \Leftrightarrow A \cap B \neq \varnothing \neq A - B$
$\textit{infinitely many}_M(A,B) \Leftrightarrow A \cap B$ is infinite
$\textit{an even number of}_M(A,B) \Leftrightarrow |A \cap B|$ is even

Also,

(6) let **1** (**0**) be the trivially true (false) quantifier (*at least zero*, *fewer than zero*).

The following properties of quantifiers will be relevant:

(7) a. Q is *conservative* (CONSERV) iff $Q_M(A,B) \Leftrightarrow Q_M(A, A \cap B)$
b. Q satisfies *extension* (EXT) iff for $A,B \subseteq M \subseteq M'$, $Q_M(A,B) \Leftrightarrow Q_{M'}(A,B)$

c. Q is *closed under isomorphism* (ISOM) iff for $A,B \subseteq M$ and $A',B' \subseteq M'$, $|A-B| = |A'-B'|$, $|A \cap B| = |A' \cap B'|$, $|B-A| = |B'-A'|$, and $|M-(A \cup B)| = |M'-(A' \cup B')|$, entails $Q_M(A,B) \Leftrightarrow Q_{M'}(A',B')$

Call the first argument of a type $\langle 1,1 \rangle$ quantifier its *restriction* and the second its *scope*. CONSERV and EXT together mean that quantification is in effect restricted to the restriction argument (Peters and Westerståhl, 2006, ch. 4.5). All quantifiers interpreting natural language determiners, in particular all those listed above, satisfy these two properties. Many satisfy ISOM as well; in the list above, all do except those mentioning particular individuals, i.e. all except *John's* and *no_except John*.

EXT entails that the universe is irrelevant, so we may drop the subscript $_M$ for such quantifiers, as we in effect did in Definition 4 of the various negations. One easily sees that the combination CONSERV + EXT is *preserved* under inner and outer negation (and hence under duals),[6] and so is ISOM. Also, note that CONSERV + EXT entails that the definition of inner negation may be expressed as follows:

(8) $Q\neg(A,B) \Leftrightarrow Q(A,A-B)$

In what follows, we assume that these two properties hold of the quantifiers mentioned.

2.2 *Modern vs. Classical Squares*

Every type $\langle 1,1 \rangle$ quantifier *spans* a (modern) square. Define:

(9) $square(Q) = \{Q, \neg Q, Q\neg, Q^{\mathrm{d}}\}$

The following facts are easily verified.

Fact 1
(a) $square(\mathbf{0}) = square(\mathbf{1}) = \{\mathbf{0},\mathbf{1}\}$.
(b) *If Q is non-trivial, so are the other quantifiers in its square.*

6 But this fails for e.g. contrariety: it is easy find quantifiers Q and Q' which are contraries and such that Q satisfies CONSERV + EXT but Q' doesn't.

(c) *If* $Q' \in$ *square*(Q), *then square*(Q) = *square*(Q').
(d) *square*(Q) *has either two or four members.*

By (c), any two squares are either identical or disjoint. As to (d), a square normally has four members, but it can happen that $Q = Q\neg$ (and thus $Q^d = \neg Q$).

Example 2
This holds when Q expresses identical conditions on $A \cap B$ and $A - B$ (cf. (8) above), such as the following quantifier in the case when $k = n$,

$$Q_{(k,n)}(A,B) \Leftrightarrow |A - B| = n \;\;\&\;\; |A \cap B| = k$$

or the quantifier

$$\textit{exactly half}(A,B) \Leftrightarrow |A \cap B| = |A - B|$$

A more spectacular example is due to Ed Keenan, who noted that the sentences

(10) a. Between one-third and two-thirds of the students passed.
b. Between one-third and two-thirds of the students didn't pass.

are logically equivalent, i.e. that if Q = *between one-third and two-thirds of the*, then $Q = Q\neg$.[7]

Thus, applying these kinds of negation to type $\langle 1,1 \rangle$ quantifier results in a rather robust 'square behavior'. And the main point here is that nothing similar holds for classical squares of opposition. To state this, we need to say what a classical square is. The following notion seems natural.

7 See, for example, Keenan (2005). This observation is no mere curiosity, but a consequence of the following two general facts:

(i) a. (*at most p/q of the*)$\neg$ = *at least* $(q-p)/q$ *of the* $(0 < p < q)$
b. $(Q_1 \wedge Q_2)\neg = Q_1\neg \wedge Q_2\neg$

Using this and the fact that $Q\neg\neg = Q$ one can verify that *between* $(q-p)/q$ *and* p/q *of the* (for $q - p \leq p$) is identical to its inner negation. Note that the observation above about the quantifier *exactly half* is an instance of this.

(11) **Definition.** A *classical square* is an arrangement of four quantifiers as traditionally ordered and with the same logical relations – contradictories, contraries, subcontraries, and subalternates – between the respective positions.

Now each position determines the diagonally opposed quantifier, i.e. its outer negation, but *not* the quantifiers at the other two positions. For example, one may check the following:

Fact 3
The square

[**A**: *at least five*; **E**: *no*; **I**: *some*; **O**: *at most four*]

is classical. More generally, for $n \geq k$,

[**A**: *at least n*; **E**: *fewer than k*; **I**: *at least k*; **O**: *fewer than n*]

is classical.

The classical squares in Fact 3 look completely unnatural. There is no interesting sense, it seems, in which *no* is a negation of *at least five* or *at most four.* I conclude that the debate over issues pertaining to the Aristotelian square, such as those listed at the end of section 1.2, is to a large extent restricted to that particular configuration, and does not generalize to other quantifiers. In contrast, the modern square exhibits a general pattern of negation for type ⟨1,1⟩ quantifiers, and thereby for the interpretation of simple and complex natural language determiners.[8]

8 I am of course not claiming that these observations about negation and the square of opposition are new. The identification of the three forms of negation is made, for example, in Barwise and Cooper (1981) and Keenan and Stavi (1986) (as noted, Keenan calls inner negation *post-complement*), and it is fairly obvious that they generate squares of arbitrary quantifiers. Brown (1984) compares these to Aristotle's square, and Löbner (1990) also studies modern squares for other quantifiers (he calls them *duality squares*), although he disagrees with much of the generalized quantifier analysis of noun phrases. But it does seem to me that the behavior of modern squares of opposition in the context of natural language semantics has not been sufficiently explored in the literature. Exactly what I mean by this will be apparent in the following sections.

3. Identifying the Corners

In view of the preceding remarks, it looks like a mildly interesting task to investigate how various (modern) squares of opposition are 'manifested' in a language like English. In the next section I look at a few examples. But first we should deal with another matter.

While *square*(Q) uniquely specifies the *quantifiers* involved, it says nothing about how to distinguish the *corners*. Can we also find quantitative and qualitative aspects in the squares? Is it possible to identify **A**, **E**, **O**, and **I** corners in an arbitrary quantified square?[9]

In general, the answer is No. But in many cases we can obtain an identification, or at least a partial one, by suitably generalizing features of those corners in the Aristotelian square. The features I have in mind are purely semantic ones. Thus, I am not here attempting to say anything about which quantifiers in a square are more basic in terms of, say, conceptual or computational simplicity.

3.1 Monotonicity

A striking feature of the quantifiers in the Aristotelian square is their monotonicity behavior. Indeed, these quantifiers are *doubly monotone* (with a small *caveat* for the **A** and **O** corners).

(12) a. Q is *right monotone increasing* (MON↑) iff
$Q(A,B)$ & $B \subseteq B' \Rightarrow Q(A,B')$
b. Q is *right monotone decreasing* (MON↓) iff
$Q(A,B)$ & $B' \subseteq B \Rightarrow Q(A,B')$
c. *Q is left monotone increasing* (↑MON) iff
$Q(A,B)$ & $A \subseteq A' \Rightarrow Q(A',B)$
d. *Q is left monotone decreasing* (↓MON) iff
$Q(A,B)$ & $A' \subseteq A \Rightarrow Q(A',B)$

9 Thanks to Larry Horn for directing my attention to this question.

Q is doubly monotone if it has both a left and a right monotonicity property. For example, *all* is ↓MON↑.[10]

Fact 4

(Peters and Westerståhl, 2006, ch. 5) *The monotonicity behavior of Q determines the monotonicity behavior of all elements of square(Q):*

1. Q is MON↑ iff $Q\neg$ is MON↓ iff $\neg Q$ is MON↓ iff Q^d is MON↑
2. Q is ↑MON iff $Q\neg$ is ↑MON iff $\neg Q$ is ↓MON iff Q^d is ↓MON
3. *So if Q is doubly monotone, all four combinations are exemplified in its square.*

This means that right monotonicity could be seen as *quality*, with MON↑ as *affirmative* and MON↓ as *negative*. Also (but less naturally), left monotonicity could be seen as *quantity*, with ↑MON as *particular* and ↓MON as *universal*. Thus, we can identify the exact position in the square of any doubly monotone quantifier.

However, many quantifiers are only *right monotone*, like the proportional quantifiers. So we can say, for example, that *at least two-thirds of the* is affirmative: it is either **A** or **I**, but we cannot say which. And this is not unreasonable: the dual of *at least two-thirds of the* is *more than one-third of the*, and it seems arbitrary which of these two should go into the **A** position. In any case, monotonicity cannot decide it (but see section 3.3 below). We cannot *require* a certain monotonicity behavior of the quantifiers in *square*(Q), since many quantifiers are neither left nor right monotone, such as *an even number of* or *exactly ten* (though the latter is a conjunction of a ↑MON↑ and a ↓MON↓ quantifier). Rather, the rule should be that *when Q* exhibits monotonicity, the corners of *square*(Q) can be (partially) identified as described.

10 all_{ei} is MON↑ but only *weakly* ↓MON, in the sense that $Q(A,B)\ \&\ \varnothing \neq A' \subseteq A \Rightarrow Q(A',B)$. See section 4.4 for other weak monotonicity properties. As shown in Peters and Westerståhl, (2006, ch.5), the four properties listed above do not exhaust the monotonicity behavior of the quantifiers in Aristotle's square, but they are the only ones that will be used in this paper.

3.2 Symmetry

We can get some more help from a property that was identified and discussed already by Aristotle, who noted that the *order* between the two arguments of the quantifier is irrelevant at the **I** and **E** corners.

(13) a. Q is *symmetric* iff $Q(A,B) \Rightarrow Q(B,A)$.
     b. Q is *co-symmetric* iff $Q\neg$ is symmetric.

Fact 5
(Peters and Westerståhl, 2006, ch. 6.1) *The symmetry behavior of* Q *determines the symmetry behavior of all elements of square(Q):*

1. Q is symmetric iff $Q\neg$ is co-symmetric iff $\neg Q$ symmetric iff Q^{d} is co-symmetric
2. Also, under CONSERV and EXT, symmetry is the same as *intersectivity*: if $A \cap B = A' \cap B'$ then $Q(A,B) \Leftrightarrow Q(A',B')$. So co-symmetry = co-intersectivity: if $A - B = A' - B'$ then $Q(A,B) \Leftrightarrow Q(A',B')$.

Again, we cannot require that two quantifiers in *square*(*Q*) be (co-)symmetric, since e.g. squares of proportional quantifiers contain no symmetric quantifiers. But *when* there is symmetry behavior, Fact 5 says that we can use it to distinguish the **I** and **E** from the **A** and **O** corners.

Thus, if Q is *right monotone* and *either symmetric or co-symmetric*, then we can again pinpoint its exact position in the square, given that the **I** and **E** positions are symmetric, and the **A** and **O** positions co-symmetric. For example, *at most ten* is at the **E** position. But we already knew that, since *at most ten* is ↓MON↓. Indeed, if Q is right monotone and symmetric, it is clearly also left monotone. In fact, the cases where symmetry would give extra information are somewhat limited. This is also illustrated by the next result. Let FIN mean that attention is restricted to finite universes.[11]

11 Results like the next fact which rely on CONSERV, EXT, ISOM, and FIN are often easily proved with a 'number triangle argument'; see Peters and Westerståhl (2006) for explanations and several examples. I will not give these proofs here.

Fact 6
(CONSERV, EXT, ISOM, FIN) *If Q is* MON↑ *and symmetric, then Q = at least n, for some $n \geq 0$.*

Thus, we only get extra information for cases like *an even number of*, which satisfies all assumptions of Fact 6 except monotonicity, and *no _ except John*, which is CONSERV, EXT, and symmetric but not ISOM or right monotone. This gives us two possible configurations of *square*(*an even number of*) and *square*(*no_except John*).

We should also ask, however, if the two criteria for positioning quantifiers in squares, monotonicity and symmetry, can ever conflict with each other. After all, the intuitions behind them are rather different. A conflict would occur if we found a symmetric quantifier that was also either ↓MON↑ or ↑MON↓ (and correspondingly for co-symmetry). Fortunately, this cannot happen:

Fact 7
(CONSERV, EXT) *If Q is symmetric and either* ↓MON↑ *or* ↑MON↓, *then Q is trivial* (*either* **0** *or* **1**).

Proof. Suppose Q is symmetric and ↓MON↑. If $Q \neq \mathbf{0}$, there are A, B (by EXT, we don't have to worry about the universe) such that $Q(A,B)$ holds. We show:

(14) For all sets C, $Q(C,C)$

This is enough, since then, for any sets D,E, we have $Q(D \cap E, D \cap E)$, hence $Q(D \cap E, D)$ by CONSERV, so $Q(D, D \cap E)$ by symmetry, and so again by CONSERV, $Q(D,E)$. This means that $Q = \mathbf{1}$.

To prove (14), take any set C. We have the following chain of implications: $Q(A,B) \Rightarrow Q(A \cap C, B)$ (by ↓MON) $\Rightarrow Q(A \cap C, A \cap B \cap C)$ (by CONSERV) $\Rightarrow Q(A \cap C, C)$ (by MON↑) $\Rightarrow Q(C, A \cap C)$ (by symmetry) $\Rightarrow Q(C,C)$ (by MON↑).

If Q is instead ↑MON↓, then $\neg Q$ is ↓MON↑ and symmetric (Facts 4 and 5), hence trivial by the above, and therefore so is Q. □

This is reassuring, and indicates that the intuitions behind identifying the corners by means of monotonicity and symmetry are quite robust.

Just as the selection of monotone and symmetric quantifiers is rather restricted (Fact 6), so is the choice of doubly monotone quantifiers, at least when ISOM is presupposed. One can show the following (Westerståhl, 1989, sect. 4.3):

Theorem 8
(CONSERV, EXT, ISOM, FIN) ↓MON↑ *quantifiers are finite disjunctions of quantifiers of the form at least n of the k or more, i.e.* $|A| \geq k$ & $|A \cap B| \geq n$ $(0 \leq n \leq k)$. *More generally,* ↑MON *quantifiers are finite disjunctions of quantifiers of the form* $|A \cap B| \geq n$ & $|A - B| \geq k$.

However, there is an interesting class of non-ISOM quantifiers with significant monotonicity properties: the possessives. For example, *at least five of John's* is ↑MON↑, hence belongs to the **I** corner. And *(all of) John's* is MON↑ and *weakly* ↓MON (you can decrease *A* as long as something belonging to John remains; cf. note 10), so it goes in the **A** corner. On the other hand, *most of John's* is MON↑ but not left monotone, so it is affirmative, but there seems to be no logical indication of whether it is **A** or **I**. We come back to this in section 4.4.

3.3 Summing Up

In addition to monotonicity and symmetry, one can sometimes use a more pragmatic and loose criterion, at least as a rule of thumb for drawing squares: *square(Q)* should reduce to the (modern) Aristotelian square in *limiting cases*. What a limiting case is varies with the type of quantifier considered, but we will see several examples below where this notion is quite natural.

In sum, two semantic features of the Aristotelian square generalize to arbitrary quantifiers, and they can often be used to identify, at least partly, **A**, **E**, **O**, and **I** corners in quantified squares. Surprisingly many quantifiers in natural languages are monotone and/or symmetric.[12] Sometimes other criteria seem natural as well. On the other hand, the endeavor of finding

12 See also Peters and Westerståhl (2006, ch.5.5), where it is shown that symmetry can in fact be seen as a monotonicity property too. So basically, it is in terms of monotonicity that I am trying to cash in the Aristotelian and medieval ideas (with the somewhat misleading labels) of quality and quantity.

unique properties of each corner in squares of opposition should not be taken too far, I think. There is no obvious semantic reason why each such corner should always be identifiable in this way. After all, *square*(*all*) is just one of infinitely many squares.

It is now high time to have a look at some examples of squares.

4. Examples

4.1 Numerical Quantifiers

By a *numerical* quantifier I intend one of the form *at least n* ($n \geq 0$) or Boolean combinations (including inner negation) thereof, i.e. also, for example, *at most n*, *more than n*, *all but n*, *exactly n*, *between k and n*. Here is a typical square in this class, *square*(*at least six*) (Figure 5). In this diagram and the following ones, I use the convention that italics mark a quantifier interpreting a corresponding English determiner, so that one can see directly which corners of the square are 'realizable' as simple or complex determiners.

The quantifiers in Figure 5 are doubly monotone, so there is no question about the identification of the appropriate corners. In the limiting case, when *six* is replaced by *one* (or *five* by *zero*) we get *square*(*all*).

Before making some comments, let us look at another case (Figure 6). This time there is no monotonicity, but we have symmetry and co-symmetry, so *exactly five* should be either at the **E** or the **I** corner. The choice made in Figure 5 to place in at the **E** corner is dictated by the fact that with *five* replaced by *zero*, we again obtain *square*(*all*).

These squares are perhaps not very exciting, but there is nothing wrong with them. The truth conditions at each corner are clear, and one sees how English 'realizes' at least five of the eight corners by means of determiners. It is somewhat doubtful whether "all but at least six" is a well-formed English determiner; see Peters and Westerståhl (2006, p. 132) for some discussion. No determiners seem to correspond to the **I** and **O** corners of *square*(*exactly five*).

Figure 5: ***square(at least six)***

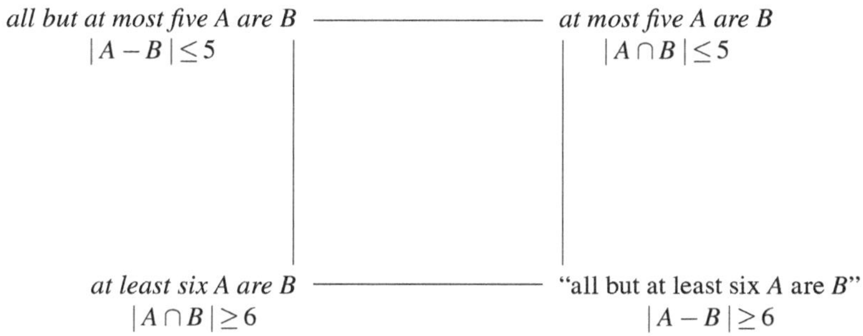

Figure 6: ***square(exactly five)***

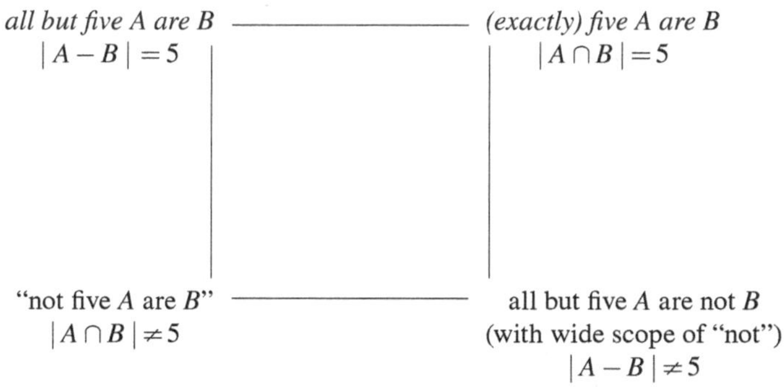

But the main point here is that these squares are far from classical. For example, $|A \cap B| \leq 5$ and $|A - B| \leq 5$ are compatible (provided $|A| \leq 10$), so they are not contraries. And $|A \cap B| = 5$ does not imply $|A - B| \neq 5$ (unless $|A| \neq 10$), etc. Will the squares become classical under suitable presuppositions, just as *square(all)* becomes classical if one presupposes that the restriction argument is non-empty? They will, but the presuppositions are not ones that anyone would deem reasonable in contexts where these expressions are used. This is seen from the next (easily verified) fact.

Fact 9
(a) *square*(at least $n+1$) *is classical iff* $|A|>2n$ *is presupposed.*
(b) *square*(exactly *n*) *is classical iff* $|A|\neq 2n$ *is presupposed.*

Obviously, it makes no sense at all to have

(15) Five students passed the exam.

presuppose that the number of (salient) students was distinct from ten. *Exactly five* simply doesn't fit in a classical square of opposition.

4.2 Proportional Quantifiers

Figure 7: *square(at least 2/3 of the*)

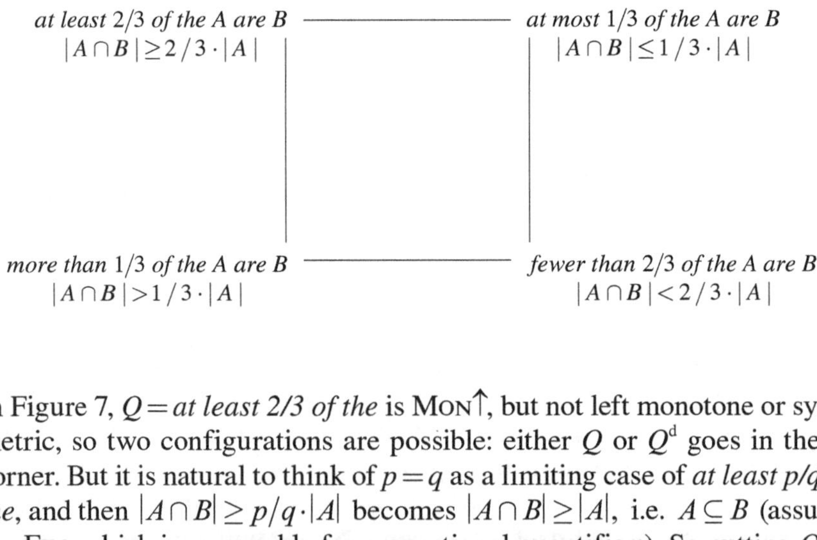

In Figure 7, $Q=$ *at least 2/3 of the* is MON↑, but not left monotone or symmetric, so two configurations are possible: either Q or Q^{d} goes in the **A** corner. But it is natural to think of $p=q$ as a limiting case of *at least p/q of the*, and then $|A\cap B|\geq p/q\cdot|A|$ becomes $|A\cap B|\geq|A|$, i.e. $A\subseteq B$ (assuming FIN, which is reasonable for proportional quantifiers). So putting Q in the **A** corner, we again get *square*(*all*) in the limiting case.

We note that all four corners are 'realized' as English determiners, and that the square is not classical. However, in this case it is at least possible to argue that there is existential import at each corner, namely, if one gives a compositional analysis of determiners of the form [Det of the], and notes that *the* does have existential import.[13] But if $A \neq \varnothing$ is added to the truth conditions, we no longer have a (modern) square. (But we do get a classical one.) This could be taken to favor a presuppositional analysis of proportional quantifiers as regards existential import.

Alternatively, we can of course consider *square(at least 2/3 of the*$^+$), where *at least 2/3 of the*$^+(A,B) \Leftrightarrow |A \cap B| \geq 2/3 \cdot |A| > 0$. In this square (which incidentally is both classical and modern), we get the same problem at the **O** and **I** corners that the Aristotelian square had at the **O** corner, i.e. that the truth condition is a disjunction, one of whose disjuncts is $A = \varnothing$. The whole matter boils down to understanding exactly what happens when proportional determiners occur in negation contexts. This might be interesting to investigate, but I will not pursue it here.

4.3 Exceptive Quantifiers

Exceptive quantifiers, i.e. quantifiers involved in the interpretation of exceptive noun phrases, like "every professor except Mary" or "No students except foreign exchange students", have been studied extensively in the literature, (see Peters and Westerståhl 2006, ch.8, for an overview of the issues involved and a proposed general analysis). Their interaction with negation is not without interest. Here I just exemplify with the very simplest case (Figure 8).

Figure 8: ***square(every_except Mary)***

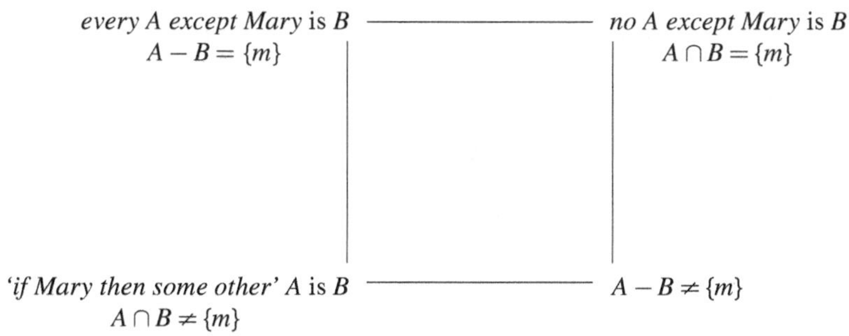

13 Such an analysis is given in Peters and Westerståhl (2006, ch.7.11).

In this square there is (co-)symmetry and no monotonicity. But in the limiting case when the exception set is *empty* we obtain *square(all)*. *square(every_except John)* is both modern and classical. The **O** corner appears to be unrealized. A suggestion (from Peters and Westerståhl, 2006, ch.4.3) has been made in Figure 8 for the **I** corner; it should be taken as *possible* English determiner with the desired interpretation.

4.4 Possessive Quantifiers

The final example comes from possessive constructions. It relies on the account of possessives given in chapter 7 of Peters and Westerståhl (2006), and on further development of that work in Peters and Westerståhl (2010).[14] But my remarks here, which are meant only to illustrate how these quantifiers interact with negation, are mostly independent of the fine details of the semantics of possessives. I will just state the bare essentials needed to present the examples.

The main observation is that possessives, in prenominal as well as postnominal form, involve *two quantifiers*, one (Q_1) over the 'possessors' and the other (Q_2) over the 'possessions'. This holds for all possessives,[15] even though Q_2 is sometimes implicit. The following examples illustrate this.

14 There is a vast literature on possessives, and it may seem strange that I have to refer to a relatively new account and not to some standard theory. But the fact is that this literature has been mostly concerned with the simplest cases of possessive constructions, and in particular not paid attention to their doubly quantificational nature, which is crucial here, but easily missed in the simplest cases. An exception, however, is Barker (1995), who notes that possessives involve quantification both over the possessors and the possessions. He attempts to handle both at the same time with a form of unselective binding à la Lewis (1975). Using two monadic quantifiers instead of one polyadic quantifier simplifies matters; the relation between Barker's account and ours is discussed in Peters and Westerståhl (2010).

15 That is, if one disregards *modifying* or *descriptive* uses of the genitive. This use is most clear in compound noun-like constructions like "child's toys" or "men's room", but it can also occur with quantifiers, and there can be an ambiguity between the modifying use and what we call the *quantifying* use, as in

(i) Five students' tennis rackets were left in the locker room.

In the modifying use, *five* quantifies over tennis rackets: 'five tennis rackets of the kind used by students were left'. In the quantifying use, it quantifies over students:

(16) a. Mary's pupils are bright.
b. No car's tires were slashed.
c. Most of John's friends came to the party.
d. Some pupils of most teachers failed the exam.

In (16d) both quantifiers are clearly visible: $Q_1 = most$ quantifies over teachers (the 'possessors') and $Q_2 = some$ quantifies over pupils (the 'possessions'). In (16c), Q_2 (*most*) is again explicit; the 'possessor' is John, and we can think of *John* as a Montagovian style type $\langle 1 \rangle$ quantifier. But in (16a) and (16b), Q_2 is implicit. In (16a) it is presumably *all*. But in (16b) it is unlikely to be *all*; the sentence doesn't normally express that no cars had all of their tires slashed (which is consistent with lots of cars having a few tires slashed); rather, it says that no car had *any* tire slashed, i.e. $Q_2 = some$.

This leads us to take the genitive suffix "'s" as well as the genitive preposition "of" to denote a higher-order operator *Poss*, which takes Q_1 and Q_2, as well as a set *C* and a binary relation *R* as arguments, and yields a type $\langle 1,1 \rangle$ quantifier as output:

(17) Three	athletes	of	each	country	paraded.
Q_2	A	$Poss$	Q_1	C	B

Note that "Three of each country's athletes paraded" means exactly the same. The only part of the semantics which is not explicit in these sentences is the possessor relation *R*. That is precisely correct (we claim), since that relation is 'free' in the sense that even though a default relation exists in many cases, one can always think of contexts where a completely different relation would be intended. Even "John's brothers" can in a suitable context refer to some brothers that John is, say, responsible for, but who are not brothers of John. We thus leave *R* as a parameter in the semantics.

The definition of *Poss* is as follows:[16]

(18) $Poss(Q_1,C,Q_2,R)(A,B) \Leftrightarrow Q_1(C \cap dom_A(R),\ \{a : Q_2(A \cap R_a, B)\})$

'five students were such that their tennis rackets (which together may be more than five) were left'. I consider only quantifying uses here.

16 It can be shown that when Q_1 and Q_2 are CONSERV and EXT, so is $Poss(Q_1, C, Q_2, R)$, which is why the universe *M* has been left out in (18).

I will not be concerned here with further motivation of this definition, just explain that R_a is the set of things 'possessed' by a, $\{b : R(a,b)\}$, and $dom_A(R)$ is the set of things 'possessing' something in A, $\{a : A \cap R_a \neq \varnothing\}$.

Now let us see how *Poss* behaves under negation. The following fact is easily verified.

Fact 10

(a) $\neg Poss(Q_1,C,Q_2,R) = Poss(\neg Q_1,C,Q_2,R)$
(b) $Poss(Q_1,C,Q_2,R)\neg = Poss(Q_1,C,Q_2\neg,R)$
(c) $Poss(Q_1,C,Q_2,R) = Poss(Q_1\neg,C,\neg Q_2,R)$

So we know what the inner and outer negation of $Poss(Q_1, C, Q_2, R)$ is, and thus its square of opposition. For example, for the universal reading of *Mary's*, i.e. the one with $Q_2 = all$, one verifies that according to (18),[17]

(19) $Mary's(A,B) \Leftrightarrow \varnothing \neq A \cap R_m \subseteq B$

where $m =$ Mary. In other words, "Mary's As are B" says that Mary 'possesses' at least one A (a kind of possessive existential import), and all of the As she 'possesses' are B. We get the square of Figure 9.

We see from (19) that *Mary's* is MON↑ and *weakly* ↓MON (you can decrease A as long as $A \cap R_m \neq \varnothing$), so it should be in the **A** corner. Likewise, at the **E** corner we have MON↓ and weak ↓MON. But because these left downward monotonicities are weak, we do not have double monotonicity at the remaining two corners. More exactly, the **I** corner is MON↑ but not ↑MON (since we can have $A \cap R_m = \varnothing$, $A \subseteq A'$, and $\varnothing \neq A' \cap R_m \subseteq \overline{B}$), and the **O** corner is MON↓ but not ↑MON. But the existing monotonicities suffice to uniquely name the corners of *square*(*Mary's*).

17 Actually, (18) is formulated for a type $\langle 1,1 \rangle$ Q_1 with a 'frozen' noun argument C, as in *each country*, but *Mary* is a type $\langle 1 \rangle$ quantifier I_m (defined by $I_m(B) \Leftrightarrow m \in B$). So to apply 4.4 we need to rewrite I_m as a 'frozen' type $\langle 1,1 \rangle$ quantifier; this can be done, for example, with $Q_1 = all_{ei}$ and $C = \{m\}$. In Peters and Westerståhl (2006, ch.7) we argue that this apparently *ad hoc* move (a) gives correct truth conditions in all cases where it is needed, and (b) is necessary if one wants a unified account of possessives.

Figure 9: ***square(Mary's)***

Mary's A are *B* $\varnothing \neq A \cap R_m \subseteq B$	*none of Mary's A* are *B* $\varnothing \neq A \cap R_m \subseteq \overline{B}$
Mary has no *A* or some of her *A* are *B* $A \cap R_m = \varnothing \vee A \cap R_m \not\subseteq \overline{B}$	Mary has no *A* or not all of her *A* are *B* $A \cap R_m = \varnothing \vee A \cap R_m \not\subseteq B$

Now, while in this case the **E** corner provides a natural kind of negation of "Mary's As are B", the **O** corner does not, and the **I** corner too seems odd.[18] That is, it is unlikely that

(20) Mary's pupils aren't bright.

could mean that *either* Mary has no pupils *or* some of her pupils are not bright. Even for the logician's

(21) It is not the case that Mary's pupils are bright.

one would be hard pressed to get that reading. Still, (20) is ambiguous. It can mean what the inner negation yields, that she has pupils but none of them are bright. But it also seems possible to use (20) to express that she has pupils and some of them aren't bright. This becomes clearer if we start instead from a version of the positive statement, i.e. the universal reading of (16a), where Q_2 is explicit:

(22) All of Mary's pupils are bright.

18 Note that the **I** corner of a square, i.e. the dual of the **A** corner, should not be seen as a form of negation on a par with the **E** and **O** corners. It is rather a kind of *double* negation. Still, it is often 'realized' in English, but the truth condition at that corner in Figure 9 seems distinctly odd.

Now it is rather clear that

(23) All of Mary's pupils aren't bright.

has both readings (though this time with a preference for the second one), whereas

(24) Not all of Mary's pupils are bright.

seems to have only the second reading. But none of these negative statements expresses the truth conditions of the **O** corner.

It might seem natural to deal with this situation similarly to a strategy hinted at for proportional quantifiers earlier, i.e. to always presuppose that at all corners in Figure 8, $A \cap R_m \neq \varnothing$ is assumed. But this overlooks two important facts. First, although contradictory negation doesn't seem to occur for (16a), it occurs for other possessives. For example, consider

(25) Not everyone's needs can be satisfied with standard products.

This seems simply to deny that everyone is such that (all) his/her needs can be satisfied with standard products, i.e. it says that someone has at least one need that cannot be so satisfied. Note that when *everyone's* is analyzed with *Poss* according to (18), there is no 'possessive' existential import, because $Q_1 = every$ (on our account) doesn't have existential import. And indeed it seems that (25) can be truly uttered in the case when no one in fact has any needs (even if it would be odd to for someone to utter it knowing that).

The second, perhaps even more compelling, fact is that our semantics using *Poss* provides a straightforward way of expressing the desired truth conditions. It is just a new form of negation, applicable in cases when a quantifier is composed out of two other quantifiers as in *Poss*. Call this *middle negation*:

(26) $\neg^m Poss(Q_1,C,Q_2,R) =_{\text{def}} Poss(Q_1,C,\neg Q_2,R)$

(By Fact 10 (c) and obvious laws of negation we also have $\neg^m Poss(Q_1,C,Q_2,R) = Poss(Q_1\neg,C,Q_2,R)$.) There is a corresponding 'middle dual':

(27) $Poss(Q_1,C,Q_2,R)^{d^m} =_{\text{def}} Poss(Q_1,C,(Q_2)^{d},R)$

We now have a 'square of middle opposition':

(28) $square^m(Q) =_{\text{def}} \{Q, Q\neg, \neg^m Q, Q^{d^m}\}$

This square is easily seen to have the same crucial square property as the standard one:

Fact 11
If Q' belongs to $square^m(Q)$, then $square^m(Q') = square^m(Q)$.

For *Mary's*, we can now check that the relations in the diagram of Figure 10 obtain. Here all four corners are realized as possessive determiners, and we have in one diagram the two natural ways to negate (16a). Further, the semantics itself gives all corners 'possessive' existential import, and we have double monotonicity of the expected kind at each corner.

Figure 10: ***square^m*****(*****Mary's*****)**

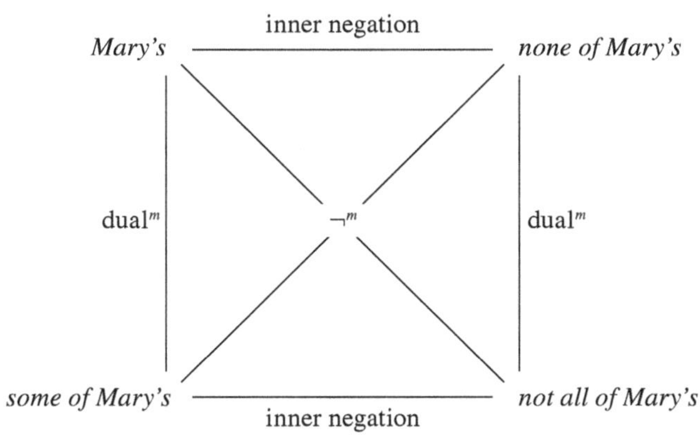

But we shouldn't stop there. Ordinary contradictory negation and ordinary dual also exist for possessive quantifiers. Moreover, *square*(*Q*) and *square*^{m}(*Q*) are closely related. In fact, they represent all the possible ways to negate quantifiers of this form. In principle, considering the four standard possibilites for Q_1, i.e. the four members of *square*(Q_1), and similarly for Q_2, there would be 16 corresponding cases for *Poss*(Q_1, C, Q_2, R). But, using Fact 10 (c) and standard laws of negation, one sees that

Fact 12
Of the 16 cases for putting, or not, inner and outer negations and duals in Poss(Q_1, C, Q_2, R), *only 8 yield distinct quantifiers.*

Furthermore, these 8 quantifiers are naturally represented in a *cube of opposition*, that we can define by means of the diagram in Figure 11, for *Poss*(Q_1, C, Q_2, R). Here, the front side of the cube is *square*(Q), and the top side is *square*^{m}(Q). Also, the back side is *square*($\neg^m Q$), and the bottom side is *square*^{m}($\neg Q$). This means that the four horizontal edges of the cube (oriented as in the diagram) stand for inner negation, the four vertical edges stand for dual, and the four remaining, slanted, edges for middle dual. Likewise, the diagonals of the front and back faces represent outer negation, and those on the top and bottom faces middle negation. The remaining diagonals, those on the remaining sides and the four diagonals inside the cube, also stand for easily identifiable operations on quantifiers of the form *Poss*(Q_1, C, Q_2, R).

Figure 11: *cube*(Q)

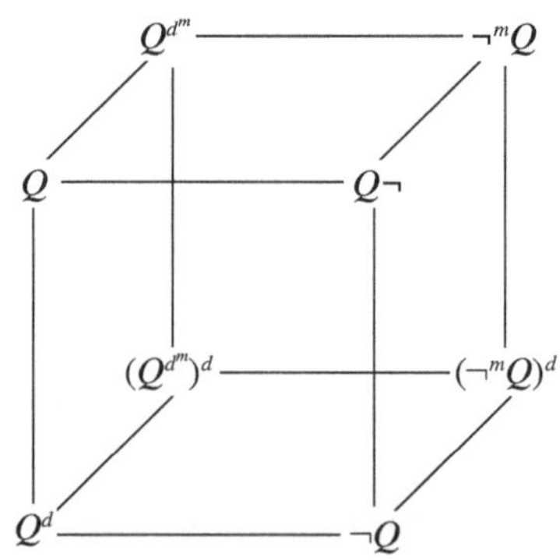

The robustness of this representation is indicated by the following fact.

Fact 13
(a) *If $Q' \in cube(Q)$, then $cube(Q') = cube(Q)$.*
(b) *The same 'cubic behavior' is exhibited by any binary operation F from quantifiers to quantifiers satisfying the conditions in Fact 10:*
- $\neg F(Q_1,Q_2) = F(\neg Q_1,Q_2)$
- $F(Q_1,Q_2)\neg = F(Q_1,Q_2\neg)$
- $F(Q_1,Q_2) = F(Q_1\neg,\neg Q_2)$ [19]

Can we distinguish corners also in the cube? Continuing to use monotonicity as a guide, and following the example *Mary's*, we calculate:

Fact 14
The monotonicity behavior of cube (Mary's), oriented as in Figure 10 with Q=Mary's, is as depicted in Figure 12.

Interestingly, the monotonicity facts for the two ordinary squares, i.e. the front and back sides, of this cube are similar but not quite the same. At the front (*square(Mary's)*), we only have weak left monotonicity at the **A** and **E** corners, and consequently only right monotonicity at the **O** and **I** corners. At the back (*square($\neg^m$Mary's)*), on the other hand, we have full double monotonicity at each corner. Still, this is enough to uniquely name each corner as we have done for ordinary squares. In the cube, then, we will have *two* **A** corners, *two* **E** corners, etc. Specifically, at the back side we have, going clockwise and starting from $\neg^m$*Mary's*: **O**, **E**, **A**, **I**. Thus, the square in the

19 Another example of an operation of the kind mentioned in (b), very different from *Poss*, is *iteration*, i.e. the natural way to combine the two quantifiers occurring in simple sentences with a quantified subject and object (see, for example, Peters and Westerståhl, 2006, ch.10.1). The operation *It* takes two type $\langle 1 \rangle$ quantifiers and yields a type $\langle 2 \rangle$ quantifier (or two type $\langle 1,1 \rangle$ quantifiers yielding a type $\langle 1,1,2 \rangle$ quantifier), so that, for example, the interpretation of

(i) At least two critics reviewed four films.

is obtained by iterating the type $\langle 1 \rangle$ quantifiers Q_1=*at least two critics* and Q_2=*four films*, and applying $It(Q_1,Q_2)$ to the relation *reviewed*. I am not saying that the corresponding cube of opposition has the same interest as in the possessive case, but the formal similarity between the ways these operators interact with negation is at least striking.

Figure 12: Monotonicity in *cube(Mary's)* ($_w$ means 'weak')

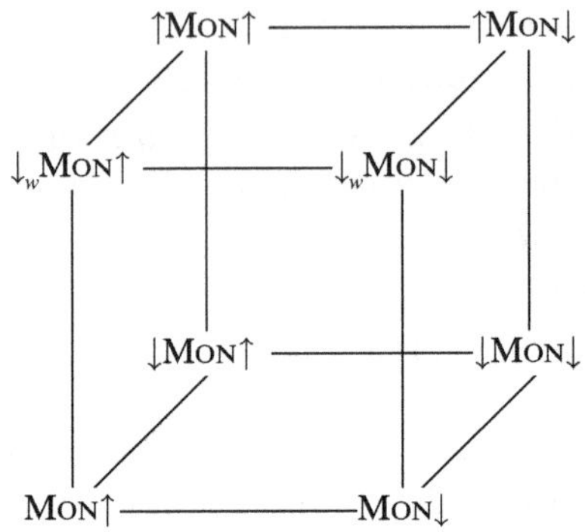

Figure 13: Corners in *cube(Mary's)*

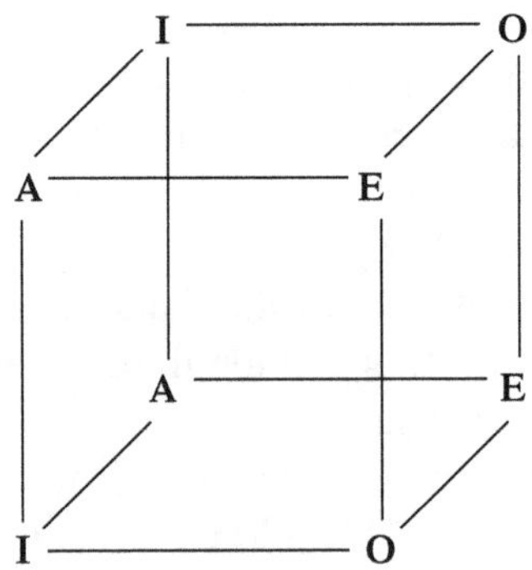

back is 'upside down' compared to the one at the front; see Figure 13. Note that this is also consistent with the naming suggested earlier for the corners of *square*^{m} (*Mary's*) (and *square*^{m}($\neg$*Mary's*)).

Another thing emerging from the examples discussed so far is that the left monotonicity behavior of a possessive quantifier Q does *not* determine that of the other quantifiers in *square*^{m}(Q), in contrast with the situation for ordinary squares. More precisely, we have the following:

Fact 15 *If Q is a possessive quantifier, then Q is* Mon↑ *iff $Q\neg$ is* Mon↓ *iff $\neg^m Q$ is* Mon↓ *iff* Q^{d^m} *is* Mon↑. *But one cannot, without extra information, draw any conclusions from the left monotonicity of Q, except for $Q\neg$ as described in Fact 4.*

Proof. The positive claims are easily verified directly from definition (18). The negative claims are illustrated by the top and bottom sides in Figure 12. □

In conclusion, let us look at one more possessive example. We already noted that in sentence (16-b), *no car's* normally has an existential reading ($Q_2 = some$). Other sentences with *no C's* can be ambiguous:

(29) No students' library books were returned in time.

This can mean that no student returned *any* of his/her books in time, for example, as uttered by someone who then concludes that the school needs to find stronger inducements for students to return at least some library books on time. But it could also have the universal reading that no student returned *all* of his/her books in time, e.g. as said by a librarian noting that every student who checked out library books had to pay at least one late fee.[20] Not surprisingly, both these readings occur in the same cube. Instead of drawing the whole cube with all the information, I content myself with the ordinary and the middle squares of the universal reading of *each car's* (equivalently, the existential reading of *no car's*): Figures 14 and 15. The truth conditions are all calculated from the definition of *Poss* in (18). One sees that *each C's* is ↓Mon↑, and so we have double monotonicity at each corner (Fact 4) in Figure 14. But, as indicated in Fact 15, this does not hold in Figure 15. The **I** and **O** corners can still be uniquely identified in this diagram, since we have Mon↑ and Mon↓ there, respectively, *and* since the **A** and **E** corners were already identified, but one can verify that left monotonicity fails at these corners. The universal reading of *no C's*, as in one interpretation of (29), is found at the **O** corner of $square^m$(*each car's*).

20 The example is from Barbara Partee (p.c.).

Figure 14: *square(each car's)* (universal reading)

each car's A were (all) *B* $C \subseteq \{a : A \cap R_a \subseteq B\}$	*no car's A* were *B* (exist.) $C \subseteq \{a : A \cap R_a \subseteq \overline{B}\}$
at least one of some car's A was *B* $C \not\subseteq \{a : A \cap R_a \subseteq \overline{B}\}$	*not all of some car's A* were *B* $C \not\subseteq \{a : A \cap R_a \subseteq B\}$

Figure 15: *square^m(each car's)* (universal reading)

each car's A were (all) *B* $C \subseteq \{a : A \cap R_a \subseteq B\}$	*no car's A* were *B* (exist.) $C \subseteq \{a : A \cap R_a \subseteq \overline{B}\}$
each car's A were *B* (exist.) $C \cap dom_A(R) \subseteq \{a : A \cap R_a \cap B \neq \varnothing\}$	*no car's A* were *B* (univ.) $C \cap dom_A(R) \subseteq \{a : A \cap R_a \cap \overline{B} \neq \varnothing\}$

Thus, we have identified all eight corners of *cube(each car's)* (universal reading). To find the truth conditions at the two corners not present in Figure 14 or 15, recall that the back side of this cube is an ordinary square with outer negation along its diagonals. Using this, Figure 15, and Fact 10(c), one sees that the remaining **A** corner expresses the universal reading of *some car's A* were *B*, and the remaining **E** corner expresses that *none of some car's A* were *B*. So we also see that, due to the flexibility of the formation rules for English possessive determiners, each corner of the cube is in this case 'realized' by such a determiner.

However, the monotonicity behavior of possessive quantifiers is somewhat complex (Peters and Westerståhl 2006, ch. 7.13) and we cannot in general expect all corners of cubes from possessive quantifiers to be uniquely identifiable in terms of monotonicity, even when both component quantifiers Q_1 and Q_2 are monotone.

5. Concluding Remarks

The square of opposition is a useful conceptual tool for understanding how negation interacts with quantification (and thereby several other operators). It doesn't represent a particular quadruple of quantifiers, such as the four Aristotelian ones, but rather a pattern that recurs for *all* quantifiers. This pattern, I have argued, uses the modern square, with the two basic forms of negation, and their combination with each other, the dual. Characteristically, further combination of these three operators yields nothing new; it stays within the square.

I have not discussed which of the three operators is most basic. But such a discussion, if one wants to embark on it, should start from the general square pattern, not from any particular square of opposition. We defined dual in terms of inner and outer negation, but one can take any two of the three operators and define the third in terms of them: $Q\neg = \neg(Q^{d})$ and $\neg Q = (Q^{d})\neg$. One cannot define any of the operators, seen as operators on quantifiers, in terms of just one of the others (they are all idempotent). In this sense, they are all on a par. But other notions of definability can perhaps be invoked; I leave to others to ponder whether this is a worthwhile idea.[21] In this paper I have, aside from a few general observations, used examples to illustrate the variety of negation-quantification interaction in natural language. In particular, complex determiners built from simpler ones, such as the possessive determiners, constitute a rich source of examples, a source it may be worth exploring further.

21 If pressed, I would argue that *outer* negation is most basic, along the following (loose) lines: Outer negation is sentence negation. Inner negation is also sentence negation, though applied to an 'interior' sentence. Dual combines the two.

References

Barker, C. (1995). *Possessive Descriptions*. Stanford: CSLI Publications.

Barwise, J., & Cooper, R. (1981). Generalized quantifiers and natural language. *Linguistics and Philosophy, 4,* 159–219.

Brown, M. (1984). Generalized quantifiers and the square of opposition. *Notre Dame Journal of Formal Logic, 25:4,* 303–22.

Horn, L. R. (1989). *A Natural History of Negation*. Chicago: Chicago University Press. (Republished by CSLI Publications, Stanford, 2001.)

Horn, L. R. (1997). All john's children are as bald as the king of france: existential import and the geometry of opposition. In K. Singer, R. Eggert, and G. Anderson (Eds.), *CLS 33, Papers from the Main Session, 155/79.* Chicago: The University of Chicago.

Jaspers, D. (2005). *Operators in the Lexicon. On the Negative Logic of Natural Language*. Utrecht: LOT.

Keenan, E. (2005). How much logic is built into natural language? In P. Dekker et al. (Eds.), *Proceedings of 15th Amsterdam Colloquium* (pp. 39–45). Amsterdam: ILLC.

Keenan, E., & Stavi, J. (1986). A semantic characterization of natural language determiners. *Linguistics and Philosophy, 9,* 253–326.

Lewis, D. (1975). Adverbs of quantification. In E. Keenan (Ed.), *Formal Semantic of Natural Language* (pp. 3–15). Cambridge: Cambridge University Press.

Löbner, S. (1990). *Wahr neben Falsch*. Tübingen: Niemayer.

Parsons, T. (2004). The traditional square of opposition. In E. N. Zalta (Ed.), *The Stanford Encyclopedia of Philosophy (Summer 2004 Edition)*. URL = http://plato.stanford.edu/archives/sum2004/entries/square/.

Peters, S., & Westerståhl, D. (2006). *Quantifiers in Language and Logic*. Oxford: Oxford University Press.

Peters, S., & Westerståhl, D. (2010). *The semantics of possessives*. [In preparation.]

Westerståhl, D. (1989). Quantifiers in formal and natural languages. In F. Guenthner and D. Gabbay (Eds.), *Handbook of Philosophical Logic*, Vol. IV (pp. 1–131). Dordrecht: Reidel. 2nd ed. (2007): Vol. 14 (pp. 223–338). Dordrecht.

Westerståhl, D. (2008). The traditional square of opposition and generalized quantifiers. *Studies in Logic 1:3,* 1–18. Beijing.

From Logical Intuitions to Natural Logic

PIETER A. M. SEUREN

Abstract

Standard modern predicate calculus (SMPC) does not fit natural logical intuitions at all well. Traditional predicate logic, known as the Square of Opposition, does a much better job in this respect, but, as is well-known, it is logically unsound because it suffers from undue existential import (UEI). Yet even the Square is somehow 'unnatural', in that it has an entailment from all to some, whereas basic-natural intuition says that some should imply not-all. Considering that every logic is defined by the meanings of its operators or constants (in this case all, some, not) and by the range of its lexical variables, it makes sense to investigate the possibility of reconstructing (a) the most basic-natural predicate logic that human beings are, apparently, equipped with at birth, and (b) the predicate logic that emerges when subjects reflect more closely on the meanings of the operators concerned, trying to achieve consistency. Two levels of natural set theory and natural logic are thus distinguished, a BASIC-NATURAL and a STRICT-NATURAL level. The strict-natural level of predicate logic is taken to be identical with the traditional Square of Opposition. SMPC, created mainly by Peano, Frege and Russell and here called CONSTRUCTED, is based not on natural intuitions but on the notions and theorems of standard Boolean set theory. The enterprise thus competes with the Gricean 'maxims' aiming to replace them with more formal principles rooted in human cognition.

Basic-natural predicate logic (BNPC) is closely related to but differs in a single detail from the predicate logic developed by Hamilton (1860) and to some extent also by Jespersen (1917). All three have the three quantifiers (operators) all, some and no, which are mutually exclusive, so that all entails both not-some and not-no, some entails both not-all and not-no, and no entails both not-all and not-some. A newly developed technique for X-raying logical systems with the help of VALUATION-SPACE (VS) MODELLING and POLYGONAL REPRESENTATION shows that BNPC, with its three quantifiers, is the richest and most powerful system of predicate logic unearthed so far, and is also logically sound. Its main defect is that it requires complete knowledge of the verification domain before existentially quantified expressions can be said to be true or false. The Square and SMPC allow for existentially quantified expressions to be assigned a definitive truth value before complete knowledge of the verification domain is achieved.

The Square does not represent Aristotle's original predicate logic but rather the way Aristotle's logic was moulded by his later commentators. Peter Abelard (1079–1142) restored the original Aristotelian logic This Aristotelian-Abelardian predicate logic (AAPC) solves UEI by weakening the Conversions to one-way entailments. This logic is sound but weaker than the Square. SMPC, though logically sound, turns out to be extremely impoverished as regards logical power.

In order to get at the natural logic of mankind as expressed in natural language, it is necessary to develop a presuppositional logic, which probably functions best in connection with the Square. This aspect is discussed in the final section.

Keywords: Abelard, Aristotle, Boole, Hamilton, intensionality, logical intuitions, predicate logic, natural logic, natural set theory, presupposition, presuppositional logic, Russell, square of opposition, undue existential import, valuation space modeling

1. The Problem Stated

The origins of logic lie with Aristotle (384–322 BCE), who, while a student in Plato's Academy, discovered, first, that often two sentences cannot be simultaneously true (they are CONTRARIES), and, secondly, that this is systematically so with sentences containing the words *all*, *some* and *not* (or their Greek equivalents). Sentences of the type *All F is G*, where F and G stand for predicates, cannot be simultaneously true with sentences of the type *No (=not some) F is G* or *All F is not-G*. It follows that when *All F is G* is true, *No F is G* must be false, so that its nonnegative version *Some F is G* must be true. This led to a logical system of what is now known as QUANTIFICATION THEORY or PREDICATE CALCULUS. (Aristotle also developed a modal logic based on the words corresponding to *can* (*may*) and *must*, but we leave this out of account here.) His still defective original system of predicate logic was streamlined and completed by later commentators, especially the Latin author Apuleius (±125–180 CE), the Alexandrian Greek Ammonius (±440–520 CE) and the late Roman and early Christian aristocrat Boethius (±480–524 CE). The resulting system, Aristotelian-Boethian Predicate Calculus, which we may call ABPC but is commonly known as "the Square of Opposition", was the standard form of predicate calculus from the sixth till the beginning of the twentieth century.

On the basis of his predicate logic, Aristotle also developed his SYLLOGISTIC, a taxonomy of reasoning schemata of the type "All philosophers are humans; all humans are mortal; ergo, all philosophers are mortal" or "Some philosophers are Greeks; all Greeks are humans; ergo, some philosophers are humans". Nowadays, syllogistic is no longer much practised, because it has long been known that a simplified version of set theory (Venn diagrams) will do the job in a much more direct and perspicacious fashion. We will pay no further attention to syllogistic in the present paper.

A further system of logic was developed by philosophers belonging to the Stoa, the greatest philosophical school of Antiquity, which started soon after Aristotle's death and continued to flourish for eight hundred years till the very end of the Roman Empire. This Stoic system dealt not with quantification but with the concatenation of sentences through the words *and*, *or* and *if* in combination with the negation word *not*. The system developed by the Stoics, now known as PROPOSITIONAL CALCULUS, has remained unchanged through the centuries till the present day. Since the 1920s it has mostly been taught with the help of the well-known truth tables.

As regards predicate logic, there was a problem, right from the Aristotelian beginnings, consisting in the fact that on a few points the logical analyses presented did not correspond entirely with natural intuitions of speakers regarding the meanings of the sentence types at issue. It has always been felt, for example, that *Some F is G* somehow implies *Not all F is G*, but in the Square the implication goes the other way: *All F is G* implies (entails) *Some F is G*. The logicians' answer is that one should call upon one's reasoning faculties: *on reflection* one sees that *Some F is G* does not imply *Not all F is G*, because, when checking through a number of Fs and having found that so far all Fs are Gs, one must leave open the possibility that it will result in the end that all Fs are Gs. But no-one seems to have bothered much about the question of *how* people come by the persistent intuition that *Some F is G* implies *Not all F is G.*

This clash with natural intuitions suddenly got much worse with the introduction of standard modern predicate calculus (SMPC) around 1900 by men like Gottlob Frege (1848–1925) and Bertrand Russell (1872–1970). This logic is based not on natural intuitions but on standard set theory developed in the wake of Boolean algebra. What is relevant here is the mere fact that SMPC severely clashes with natural intuitions, much more than the Square. The standard, but far from only, example is the case of universally quantified sentences with a null F-class. For SMPC, a sentence of the type *All F is G* counts as true when there are no Fs. Thus, *All mermaids have a bank account* counts as true in this world, where, perhaps unfortunately, mermaids are no more than figments of the imagination. It will be explained in a moment what brings the logicians to such a conclusion.

Right now we want to know how logicians, philosophers and linguists reacted to this undeniable clash of intuitions. The logicians, led by Richard Montague (1931–1971), put all their money on standard modern logic and developed a model-theoretic formal semantics based on the notion of 'possible worlds', the details of which are not important at this point. What counts here is that the clash with natural intuitions was, and still is, brushed aside with an appeal to mathematical necessity. If speakers don't realize that fact, it is their fault, not that of the logical semanticists. To soften the pain, the logicians are quite prepared to allow for palliative therapy, but they are not really concerned to find a principled answer to the predicament. I consider the logicians' answer unsatisfactory.

But what did the linguists say, or rather the linguistic philosophers? (The linguists, on the whole, didn't say anything.) The main figure here is Herbert Paul Grice (1913–1988), an Oxford philosopher who moved to Berkeley in 1967. He is considered the father of PRAGMATICS. Grice's answer came down to (a) accepting SMPC without critique (though with erratic reversions to the Square) and (b) seeking to fill the gap by claiming that the formal semantic machinery by itself is insufficient to explain all the phenomena observed in the use of language. He introduced his famous MAXIMS, which are in effect principles of good social practice, deemed to be naturally taken into account by humans when they operate with the formal systems at work in language, especially standard logic. These principles are based on, and formulated in terms of, common sense and common decency, honesty, and sincerity. Their application is taken to produce the well-known Gricean

IMPLICATURES –noncogent inferences drawn by the listener on the basis of the implicit assumption that the speaker observes the maxims. Formalization is, on the whole, not striven for (the few attempts in this direction that were made by some of Grice's followers have been signally unsuccessful).

The paradigm of classical pragmatics has proved remarkably resistant. Apart from the fact that, especially in the beginning, it provided a home for the many practitioners who turned away in despair from the technical complexities and the obvious empirical inadequacies of existing semantic formalisms, its main strength seems to lie in the picture it presents of humans handling an abstract and intrinsically rigorous system in the rough reality of forever varying everyday life. The abstract system can perhaps be formalized but "life", in its infinite variety, cannot, or so it is believed.

This answer I also consider unsatisfactory. Apart from the overall lack of precision in Gricean pragmatics, it is my view that (a) the Gricean maxims are too superficial to account for the clash of intuitions and (b) Gricean pragmatics fails to address the issue of the alleged inviolability of established standard logic.

Why are the Gricean maxims too superficial? Surely, common sense, decency, honesty and sincerity are deeply ingrained driving forces of human behaviour. Why can't they be called upon to help explain features of linguistic behaviour? Of course, they can, but the question is: do they do the job they were hired for? And here the answer has to be negative.

Consider, for example, the implicatures "not all" from the use of *some* and "not and" from the use of *or*. If these are to be explained by what Grice (1975: 45) calls the maxim of quantity ("Make your contribution as informative as is required", where an entailing sentence is considered more informative than an entailed sentence), then why is there no implicature 'John has not been killed' from the use of *John is dead*? No-one will deny that *John has been killed* entails, and is hence more informative than, *John is dead*, just as *all* induces an entailment to *some* (in the Square) and *and* to *or*, which makes the former more informative than the latter. The answer normally given by pragmaticists is that the semantics of *kill* and *dead* does not involve any appeal to the notion of quantity. But then, what notion of quantity is involved in *and* and *or*? And why should my "contribution" be required to be maximally informative regarding quantity and not regarding other aspects of the world, such as manner of death? I find it hard to avoid the conclusion that the maxim in question fails to explain the implicatures at hand.

A further clear case showing the explanatory deficiency of the Gricean maxims for the conflict between standard logic and natural intuitions is provided by the sentences in (1) and (2):

(1) a. She doesn't like trains and she doesn't like planes.
b. She doesn't like trains or planes.
(2) a. She doesn't like trains or she doesn't like planes.
b. She doesn't like trains and planes.

In terms of standard logic, (1a) and (1b) are equivalent for the same reason as (2a) and (2b) are: both pairs look like, and perhaps are, instances of De Morgan's laws. Yet the equivalence of (1a) and (1b) is immediately recognized by speakers, while that between (2a) and (2b) requires careful reflection. I don't know of any psychological experiment to test this difference, but I am confident that any such experiment will bear out the intuition. The point is that the Gricean maxims, which are all about social propriety and communicative functionality, have no bearing at all on these examples, but they should have if they are to explain the clashes between standard logic and natural intuitions.

This leads up to our second criticism of Gricean pragmatics: why is one so afraid, when studying the semantics of natural language, to nibble at the status of inviolability that adheres to standard modern logic? Even if one accepts that logic enjoys a metaphysical status of mathematical necessity and thus stands above the contingencies of the actual world, the empirical question of what the natural logic of human cognition and human language amounts to still remains open. No professional logician denies that a logic, any logic, is essentially defined by the semantics of its CONSTANTS or OPERATORS, which are represented in each language as lexical items, and by the ranges of its variables. That being so, the operators may well have different meanings in different formal languages. And their meanings in natural language, which, until proven otherwise, we suppose to be universally fixed, may well differ from those defined in various formal languages.

To see this, we must first get a few basic notions right. To begin with, we say that *a logic is a formal calculus of entailments*. But what is an entailment?

When we say that a sentence (or set of sentences) ENTAILS a sentence B we mean that if A is true, then B is necessarily also true in virtue of the linguistic meaning of A (analytical necessity).

We distinguish *semantic* from *logical* entailment. Semantic entailment is based on linguistic meaning tout court. Logical entailment has the additional property of being derivable through a formal logical system. For example, the sentence *Jack has been murdered* entails the sentence *Jack is dead* in virtue of the meanings of both sentences, regardless of whether Jack has really been murdered or is really dead. What counts is that it is not possible for Jack to have been murdered without him being dead, because the linguistically fixed meaning of the predicate *have been murdered* implies that the victim in question is dead. We say that it is in the SATISFACTION CONDITIONS of the predicate *have been murdered* that its subject-term referent must also be dead.

Entailments are thus part of the *type-level* machinery of language, which makes it impossible to override them. By contrast, implicatures are, in principle, operative at the *token level* of language use –though Grice also has his "conventional" implicatures, which are taken to have become part of the type-level machinery, even if neither Grice nor any other pragmaticist has ever indicated how conventional implicatures are to be integrated into the semantic machinery of language.[1]

The next question is: how do we get a formal calculus of entailments? The answer to this question was discovered by Aristotle. To get a formal calculus of entailments, we determine, in the lexicon of any language that has a logic, which words have meanings that allow for the *formal*, or *rule-driven*, derivation of entailments. These words we call "operators" or "logical constants". The remaining words of the language in question are represented as variables with well-defined ranges. If a language has no logical constants, it has no logic.

Aristotle discovered that in Greek one finds as logical constants the words corresponding to *all*, *some* and *not* (and also *may* and *must*). Soon after Aristotle, the Stoics added the words corresponding to *and*, *or* and *if*, thus creating propositional logic. This inventory of logical constants has not been essentially extended since Antiquity. Nor has it ever been doubted

1 The new school of Radical Pragmatics has raised the same question. Its answer is to deny the validity of the concept "type-level machinery' altogether and to try to reduce language as a whole to token-level phenomena requiring no mental machinery beyond general cognition. I will refrain from comment on this new development, as it appears to transgress the boundaries of rationality, given what is known about language and linguistic phenomena.

that all natural languages contain analogs of these classic logical constants, which means that it is universally agreed that human languages have a logic.

When Aristotle, and all logicians after him till the end of the nineteenth century, developed or investigated logic, they did so with regard to the logic of natural language, there being no other source of logical information available. Predicate logic (the theory of quantification) in particular was based on the intuitions of speakers of natural languages concerning plural objects. This changed during the nineteenth century, when the English mathematician George Boole (1815–1864) presented what is now known as Boolean algebra, the foundation of later set theory. Though it took some time, it was gradually discovered that predicate logic could be seen as an application of set theory and hence of Boolean algebra. This insight became ever clearer during the twentieth century. Nowadays, with the benefit of our knowledge of generalized quantifiers (Mostovski 1957), we simply say that the Aristotelian sentence type *All F is G* can be interpreted as "the set of all *F*s is contained in (is a subset of) the set of all *G*s", or $[[F]] \subseteq [[G]]$. And *Some F is G* can be read as "there is a nonnull intersection of the set of all *F*s and the set of all *G*s", or $[[F]] \cap [[G]] \neq \varnothing$. *All* and *some* are thus seen as binary higher-order predicates in logical structure, taking set-denoting terms as arguments. This gives *standard modern predicate calculus* (SMPC), which now reigns supreme and is considered inviolate.

It thus came about that it was no longer the logical intuitions of natural-language speakers that determined the form of predicate logic but the theorems of standard set theory. In principle, SMPC is insensitive to natural logical intuitions and actually conflicts rather badly with them. For example, in SMPC the sentence type *All F is G* is considered true when there are no *F*s ($[[F]] = \varnothing$), because it is a theorem in standard set theory that the null set $\varnothing$ is a subset of any set: $\varnothing \subseteq A$, for any set A. This makes *All F is G* trivially true in all cases where $[[F]] = \varnothing$. For natural language this means that, for example, a sentence like *All mermaids have a bank account* counts as true in this world precisely because there are no mermaids in it. For the same reason, *All mermaids have no bank account* also counts as true, because $\varnothing$ –the set of all mermaids or [[Mermaid]] –is a subset not only of the set of bank-account holders but also of the set of those condemned to a life without a bank account. In my experience, natural speakers find this hard to accept. Curiously, it is much easier for them to accept that the sentence *No mermaid has a bank account*, which is standardly considered equivalent with *All mermaids have no bank account*, is true in the absence of mermaids. Apparently, natural speakers do not consider the two sentence types to be as strongly equivalent as modern logicians do. (We note in passing that pragmatic theory is no better at explaining this difference than modern logic is.) Consequently, we have no choice but to deem the position of modern logic highly counterintuitive.

The logicians' typical answer is: "too bad for natural language." Many logicians, who pity speakers of natural languages, advise palliative therapy: just define *All F is G* not just as "$[[F]] \subseteq [[G]]$" but as "$[[F]] \subseteq [[G]]$ and $[[F]] \neq \varnothing$", adding the clause "and $[[F]] \neq \varnothing$", and all is well. But all is not well, because if one does that, *Not all F is G* comes out as meaning "either $[[F]] \not\subseteq [[G]]$ or $[[F]] = \varnothing$", which again violates natural intuitions: *Not all F is G* clearly implies intuitively that *Some F is not-G* (in fact, both SMPC and the Square take the two to be equivalent). But *Some F is not-G* stipulates the existence of at least one *F* and thus rules out the possibility that $[[F]] = \varnothing$, which disqualifies the disjunct "or $[[F]] = \varnothing$". The addition of the clause "and $[[F]] \neq \varnothing$" to the definition of the universal quantifier thus merely makes the clash of intuitions rear its head elsewhere. Therefore, whichever way one takes it, SMPC does not sit at all well with the facts of language.

We are thus faced with a dilemma, as neither pragmatics nor modern predicate logic appear to be able to account for the logical intuitions of natural-language speakers. Traditional predicate logic –that is, the Square – scores much better in this regard, but even that is not quite adequate, as we have seen.

This brings us to the point of the present section. Strict mathematical formalism places language on a procrustean bed cutting off limbs wherever that is found necessary –a method that is hardly conducive to a better insight into the inner workings of language and cognition. The pragmatic answer is unsatisfactory, as it fails to explain the discrepancies between the logical and semantic intuitions of natural speakers on the one hand and modern standard logic on the other.

My view has always been that room must be made for a theory that is both formal and ecological in the sense that it studies language in its natural habitat. In such a theory, the discrepancies are seen as following from an innate system of logic which again derives from an innate set theory. It is to be expected that innate, or natural, logic and set theory differ considerably from their mathematically based standard versions, and one hopes that the differences can be shown to arise from the functional needs of cognition and language.

To implement this programme it is necessary, first, to investigate if the words (or their equivalents in other languages) *all*, *some*, *not*, *and* and *or* (*if* is too doubtful to be included right away) behave in such a way in natural language as to allow for a logical calculus and, if so, what calculus. This calculus is then to be tested for empirical adequacy as well as for logical soundness. Then, the underlying natural set theory is to be defined, and again tested for empirical adequacy in that it must be checked if indeed humans deal with plural objects the way the theory predicts they do. The Gricean maxims are thus replaced with general principles of human cognition, which is taken to be subject to restrictions of naturalness regarding the way it deals with objects in general and with plural objects (sets) in particular.

2. The Logical Problem of Undue Existential Import (UEI)

It might be submitted that the Square is a good candidate for the status of natural logic. This proposal is not entirely unreasonable, in so far as it is clear that the Square scores a great deal better than SMPC at accounting for natural intuitions. But, as has been shown, there are still too many clashes for it to be declared the winner straight away. Apart from that, however, there is a very serious logical problem with the Square, known as UNDUE EXISTENTIAL IMPORT (UEI). The problem is that the Square fails to cater for cases where $[[F]] = \varnothing$, as with the sentence *All mermaids have a bank account.* the Square only functions when $[[F]] \neq \varnothing$, that is, when there is at least one *F*. This is very serious, because it means that the Square as a whole needs as a precondition for any *F*-predicate that there is at least one object satisfying the conditions of *F*. But this is a condition that cannot be imposed on any logic because, as we have seen, a logical system is defined as being based solely on meanings that are fixed in the language at hand, so that it is by definition independent of actual states of affairs. Put differently, if the Square were a valid logical system, it would entail the actual existence of objects satisfying whatever predicate serves as a value for the variable term *F*. But no logical system can entail actual existence: actual existence of anything is *per se* contingent, never logically necessary. During most of the history of logic, this problem was unknown, though Aristotle himself was probably aware of it. So was the medieval French philosopher Abelard (1079–1142) and later medieval logicians, until, roughly, the year 1500. But it was not until the middle of the nineteenth century that it began to play a major role.

Let us have a closer look. The Square, like SMPC, is subject to the PRINCIPLE OF THE EXCLUDED THIRD (PET), which implies that (a) all sentences formed according to the formation rule "(NOT) ALL/SOME F IS (NOT-) G" always have a truth value, and (b) there are no truth values other than TRUE and FALSE. In section 8, which deals with presuppositions, we will tinker with this principle. In its traditional form, the Square is represented as the Square of Opposition, shown in fig. 1 and no doubt known to many readers. The symbols **A**, **I**, **E** and **O** were introduced by Boethius. They are the first vowels of the Latin ***A****ff****I****rmo* (I affirm) and *n****E****g****O*** (I deny), respectively –clearly a device invented for classroom use. Each symbol stands for two basic sentence types which are considered logically equivalent: the one is always free to CONVERT into the other, so that one speaks of the CONVERSIONS. Thus, as shown in fig. 1, *All F is G* is considered equivalent with *No F is not-G*, *Some F is G* with *Not all F is not-G*, and so forth.

Two sentences are CONTRARY ("C") just in case they cannot both be true simultaneously. Two sentences are SUBCONTRARY ("SC") just in case they cannot both be false simultaneously. Two sentences are CONTRADICTORY ("CD") just in case they cannot both be either true or false simultaneously: the truth of the one entails the falsity of the other and vice versa. This is how predicate logic was taught throughout the Middle Ages and later till the beginning of the twentieth century. As has been said, it is the combined product of Aristotle and his later commentators.

The operator typically used for the assertion of existence is the existential quantifier *some*. As has been said, *Some F is G* says by definition, in both the Square and SMPC, that there is a *nonnull* intersection between [[F]] and [[G]], and a "nonnull set" is defined in set theory as a set that contains at least one element. Logicians, following the ontological prescriptions issued by Quine (1953), have restricted this by stipulating that nothing is an element unless it has actual existence (leaving the nature of mathematical elements open). Likewise, *Some F is not-G* says that there is a nonnull intersection between [[F]] and $\overline{[[G]]}$ –that is, the set of those objects in the universe **OBJ** of all objects that are not G, or: the complement of [[G]] in

Figure 1: The Traditional Square of Opposition

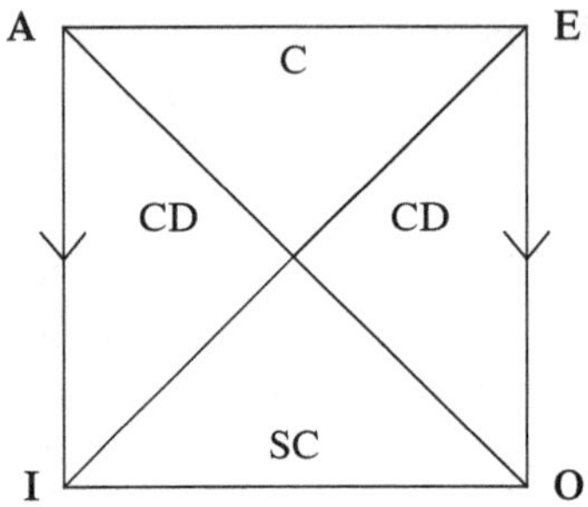

A: All F is G / No F is not-G
I: Some F is G / Not all F is not-G
E: No F is G / All F is not-G
O: Some F is not-G / Not all F is G

CD: contradictories
C: contraries
SC: subcontraries
∨: entailment

OBJ. Since, in set theory, $\varnothing \cap A = \varnothing$ for any set A, both [[F]] and [[G]], must be nonnull for there to be a nonnull intersection between [[F]] and [[G]]. That is why we say that the quantifier *some* induces an entailment of exitence for both [[F]] and [[G]]. Therefore, the **I**-corner and the **O**-corner both entail actual existence for elements in the *F*-class. The question is: does the same hold for **A**- and **E**-sentences? The answer is, obviously, "yes", because there is an entailment from **A** to **I** (in accepted formal notation: $\mathbf{A} \vdash \mathbf{I}$) and from **E** to **O**.

To prove that the entire system of the Square suffers from UEI, it is necessary to show that the negations of the four sentence types likewise entail existence of at least one *F*. This is easily shown. Take *not*-**I** (in formal notation: ¬**I**). ¬**I** is equivalent with **E** and, as has been shown, $\mathbf{E} \vdash \mathbf{O}$ in the Square and thus carries an entailment of actual existence of at least one *F*. Analogously for ¬**O**, which is equivalent with **A**, and $\mathbf{A} \vdash \mathbf{I}$. Now take ¬**A**. Since $\neg\mathbf{A} \equiv \mathbf{O}$, ¬**A** again entails nonnullness for the *F*-class. Analogously for ¬**E**, which is equivalent with **I**. It thus follows that the entire system of the Square suffers from UEI. And, as has been said, this makes the system logically faulty.

It has been found only recently that Aristotle himself was probably aware of the danger of UEI. Careful reading of his texts (mainly his *On Interpretation*) shows that he stopped short of positing the Conversions (*All F is G* ≡ *No F is not-G* and *Some F is G* ≡ *Not all F is not-G*). All he did was allow for an entailment from *All F is G* to *Some F is not-G* but not vice versa, which breaks the chain of existential import.[2] His commentators, however, were unaware of the danger of UEI and "completed" Aristotle's original predicate calculus in such a fashion that it included the Conversions, thus unwittingly introducing the error of UEI.

Abelard was in all likelihood the first to spot the problem after Aristotle. He reconstructed Aristotle's original logic, producing what I have called Aristotelian-Abelardian predicate calculus (AAPC), which is elucidated and commented on below.[3] As is now gradually being discovered by Edward Buckner,[4] medieval philosophers and logicians after Abelard raised a dispute not so much about as related to the UEI question, and not so much from a strictly logical as from a doctrinal-metaphysical point of view. The school of essentialists, headed by Thomas Aquinas, maintained that a sentence like *All humans are animals* must be true even if there are no humans, because animalness is an essential property of humanness. By contrast, the nominalists, taking their clue from Abelard, rejected this argument maintaining that *All humans are animals* is false when there are no humans. The Church took sides with the essentialists, with the result that the nominalists were marginalized and sometimes even excommunicated. We say, of course that it is logically irrelevant whether the *G*-property is in any way 'essential' for the *F*-property.

After the Middle Ages, the debate fizzled out. As far as is known, there are no signs of logicians between, say, 1500 and 1850 being aware of the UEI problem. In any case, if there were, they had no influence on mainstream logical teaching. Also, the fierce opposition between essentialists and nominalists came to an inconclusive end. The first to show awareness of the UEI problem in modern times was the London mathematician-logician Augustus De Morgan (De Morgan 1847). His answer consisted in restricting predicate logic to a 'universe of discourse' with nonnull *F*-classes, but he failed to show what logic would be valid outside such a universe of discourse. Then, of course, came Frege and Russell, who established standard modern predicate calculus (SMPC), which has since been considered to be

2 See Kneale & Kneale (1962: 57), who state correctly that Aristotle did not admit the Conversions. Yet they apparently failed to see that this prevented UEI from creeping into the system. For more comment, see Seuren (2010, Chapter 5).

3 Independently, I hit on the same solution (Seuren 2002). Soon after that publication I became aware of the fact that Abelard (and Aristotle himself) had already presented it.

4 See http://maverickphilosopher.powerblogs.com/posts/1177106402.shtml.

the final and definitive answer. It is this solid bulwark of opinion that the present enterprise aims at undermining or at least putting into perspective.

3. A Further Analysis of the Square

It is my opinion that natural cognition and in particular natural language embody a solution to the UEI problem that differs from the solution proposed in SMPC and is, in fact, a great deal more functional for the purposes of acquiring knowledge and for linguistic interaction. But before we can go into this in greater detail, we must have a closer look at ABPC (the Square).

First, we must improve on the traditional Boethian notation in terms of **A**, **E**, **I** and **O** sentence types, because this notation makes the role of external and internal negation in the system invisible and covers up the implicit assumption of the Conversions. From now on, we will no longer use the symbols **E** and **O** and use only **A** for *All F is G* and **I** for *Some F is G*. The external negation is symbolized by means of the standard bivalent negation sign "¬", while the internal or predicate negation is rendered as *. For both external and internal negation it holds that double negation cancels out. This gives the following eight BASIC SENTENCE TYPES (without unnecessary multiple negations):

A	stands for	*All F is G*	A*	stands for	*All F is not-G*
I	stands for	*Some F is G*	I*	stands for	*Some F is not-G*
¬A	stands for	*Not [all F is G]*	¬A*	stands for	*Not [all F is not-G)*
¬I	stands for	*Not [some F is G] (No F is G)*	¬I*	stands for	*Not [some F is not-G)*

This makes it possible to express the Conversions explicitly in the Square notation. But there is more. The traditional Square fails to express the fact that it is actually defined by (i) the mere relation of contrariety between **A** and **¬I** as shown in fig. 2-a, (ii) the bivalent negation (¬) and (iii) the Conversions. For if **A** and **¬I** are contraries (which was Aristotle's point of departure), it follows that when **A** is true, **I** must be true because **¬I** is by definition the contradictory of **I**. Therefore, **A** ⊢ **I**. This gives rise to the simple logical triangle <**A**, **I**, **¬I**> as part of fig. 2-b and, by contraposition, the corresponding logical triangle <**¬I**, **¬A**, **A**>, together forming a 'natural' logical square <**A**, **I**, **¬I**, **¬A**> in fig. 2b. One sees that, as a result of the linking up of the two triangles in fig. 2b, a new logical relation comes about, the relation of subcontrariety between **I** and **¬A**.

Then, there is the principle that the logical properties of *all* and *some* are insensitive to whether or not the *G*-predicate is positive or negative. That is, given that *All F is G* is defined as saying that $[[F]] \subseteq [[G]]$, it follows that *All F is not-G* is defined as $[[F]] \subseteq \overline{[[G]]}$, as has been said above. Although this principle is hardly ever mentioned explicitly in the literature, it deserves a name. I have named it the MODULO-*-PRINCIPLE. In virtue of this principle, given that **A** ⊢ **I**, it follows that **A*** ⊢ **I***. And since **I*** is the contradictory of **¬I***, which makes **¬I*** and **A*** contraries, we can set up, besides the triangle <**A**, **I**, **¬I**>, the corresponding, logically isomorphic, triangle <**A***, **I***, **¬I***>. Finally, since **¬I** is considered equivalent with **A*** (the Conversions), the two triangles can be linked up so as to form the new improved square notation shown in fig. 2-c, which expresses both the roles of the internal and external negations and the Conversions explicitly. The 'bonus' of the extra relation of subcontrariety in fig. 2b now turns up again in fig. 2c as subcontrariety between **I** and **I***.

Figure 2: The Building Up of the Improved Square Notation

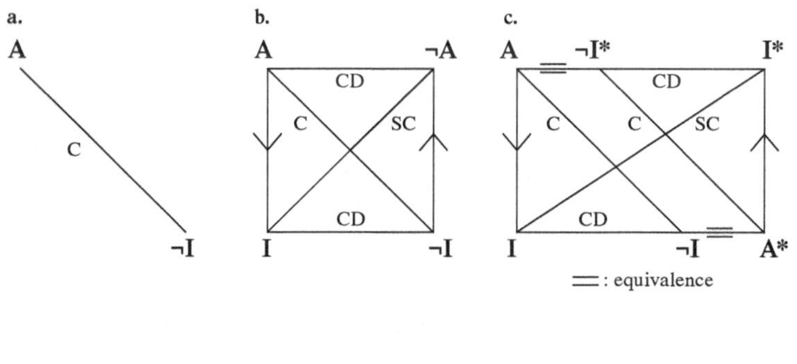

This improved square notation has the further advantage of showing that the Square really consists of two logically isomorphic triangles, whose equivalence is warranted by the Modulo-*-Principle. The *logic* of the classic Square and the improved square is the same, namely that of the Square, but the new *notation* is more transparent and more revealing.

(One can also construct a triangle <A, I, ¬A> and map that onto the isomorphic triangle <A*, I*, ¬A*>, as shown in fig. 3.[5] Here, however, subcontrariety is constitutive of the triangles and contrariety comes in as a bonus when the two are combined. Logically, there is no difference with fig. 2-c. Empirically, however, fig. 2-c is preferable because the relation of contrariety is about the most natural logical relation there is, whereas subcontrariety is notoriously hard to grasp, as teachers of logic know well.)

Figure 3: The Nonnatural Isomorphic Square

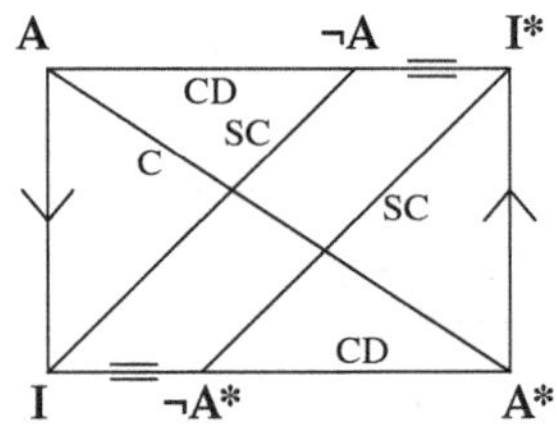

4. Valuation-Space Modelling and Polygonal Representations of Predicate Logics

There is a second way of graphically representing the Square, or indeed any system of predicate logic. Consider a language **L** to be a, possibly infinite, set of sentences. In principle, if a sentence S is to be interpreted and assigned a truth value, S must be referentially linked up with, or, as we say, KEYED TO, one or more given individuals in a given situation at a given time. Situational and objectual keying is, in principle, a necessary condition for interpretation and truth-value assignment. The few so-called "eternal" sentences, such as *All humans are mortal*, which contain no definite referring term and no indexicals, can do without specific keying, are regarded as marginal cases whose key equals the universe of all situations.

5 I am indebted to Dany Jaspers (2005: 34–35) for calling this fact to my attention.

Now consider the set, or universe, **U** of all possible situations *sit* such that each *sit* makes one or more sentences of **L** true. Each sentence S of **L** thus has a space in **U** containing precisely those situations *s* in **U** that make S true, *modulo key*. This space is called the VALUATION SPACE (VS) of S, or /S/.[6]

This allows us to set up a VS-model for the Square in the way shown in fig. 4 (the oblique strokes have been omitted to avoid symbol crowding). Space 1 is reserved for cases where $[[F]] \subseteq [[G]]$; space 2 for cases where either [[F]] ⊘⊘ [[G]] (I use the symbol "⊘⊘" for mutual partial intersection: X ⊘⊘ Y just in case $X \cap Y \neq \varnothing \neq X \neq Y$) or $[[G]] \subset [[F]]$; space 3 for cases where [[F]] ○○ [[G]] ([[F]] and [[G]] are mutually exclusive). One sees that /**A**/ = {1}, /**I**/ = {1,2}, /**A***/ = {3} and /**I***/ = {2,3}. The external negations, of course, select the corresponding complements: /¬**A**/ = {2,3}, /¬**I**/ = {3}, /¬**A***/ = {1,2} and /¬**I***/ = {1}.

Figure 4: The VS-Model for the Square

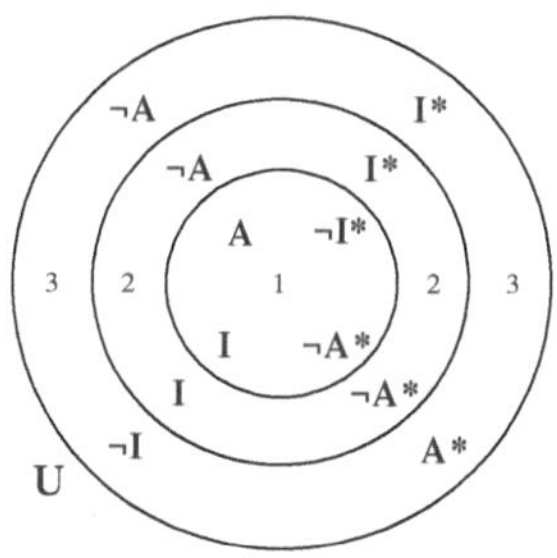

A VS-diagram reflects the logical relations in the following way:
For all sentences P and Q:

Equivalence:	$P \equiv Q$	$/P/ = /Q/$
Entailment:	$P \vdash Q$	$/P/ \subseteq /Q/$
Contrariety:	$P >< Q$	$/P/ \cap /Q/ = \varnothing$
Subcontrariety	$P \geq\leq Q$	$/P/ \cup /Q/ = \mathbf{U}$
Contradictoriness	$P \# Q$	$/P/ \cap /Q/ = \varnothing$ and $/P/ \cup /Q/ = \mathbf{U}$

6 The notion of valuation space was first introduced in Van Fraassen (1971), who, however, did not develop VS-models and polygonal representations.

We can now read off immediately the logical relations that are valid in the Square:

$\mathbf{A} \equiv \neg\mathbf{I}^*$ because $/\mathbf{A}/ = /\neg\mathbf{I}^*/ = \{1\}$.
$\mathbf{A} \vdash \mathbf{I}$ because $/\mathbf{A}/ = \{1\}$ and $/\mathbf{I}/ = \{1,2\}$ so that $/\mathbf{A}/ \subseteq /\mathbf{I}/$.
$\mathbf{A} >< \neg\mathbf{I}$ because $/\mathbf{A}/ = \{1\}$ and $/\neg\mathbf{I}/ = \{3\}$ so that $/\mathbf{A}/ \cap /\neg\mathbf{I}/ = \varnothing$.
$\mathbf{I} \underline{><} \mathbf{I}^*$ because $/\mathbf{I}/ = \{1,2\}$ and $/\mathbf{I}^*/ = \{2,3\}$ so that $/\mathbf{I}/ \cup /\mathbf{I}^*/ = \mathbf{U}$.
$\mathbf{I} \# \neg\mathbf{I}$ because $/\mathbf{I}/ = \{1,2\}$ and $/\neg\mathbf{I}/ = \{3\}$ so that $/\mathbf{I}/ \cap /\neg\mathbf{I}/ = \varnothing$ and $/\mathbf{I}/ \cup /\neg\mathbf{I}/ = \mathbf{U}$.
$\mathbf{A} \# \neg\mathbf{A}$ because $/\mathbf{A}/ = \{1\}$ and $/\neg\mathbf{A}/ = \{2,3\}$ so that $/\mathbf{A}/ \cap /\neg\mathbf{A}/ = \varnothing$ and $/\mathbf{A}/ \cup /\neg\mathbf{A}/ = \mathbf{U}$.

VS-models allow one to set up a third way of representing any system of predicate logic, including the Square. Since the Square has eight basic sentence types, we can set up an octagon, each vertex being reserved for one sentence type and the connecting lines representing the logical relations said to hold between any two vertices in any given logical system. For the Square this is shown in fig. 5.

Figure 5: The Octagonal Representation of the Square

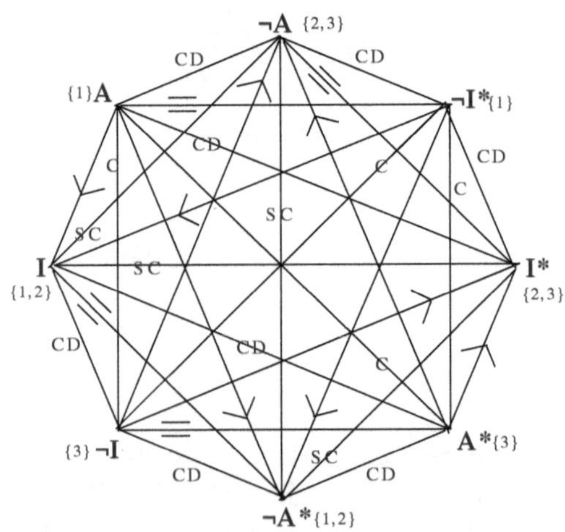

One notes that the octagonal graph of fig. 5 is *complete* in the sense that each vertex stands in some logical relation to all other vertices. Absence of a logical relation means that the two vertices involved are logically independent in that they can be simultaneoulsy both true or both false or the one can be true while the other is false. We say that a logical system is more *powerful* to the extent that it has fewer absentees among its logical relations.

It is now even clearer that the Square suffers from UEI, since the octagonal representation shows that each vertex, and hence each of the eight sentence types, entails (or is equivalent with) an **I**-type or an **I***-type sentence. What the VS-model of the Square needs is a fourth space for situations where $[[F]] = \varnothing$. But this space cannot be added because when **A** and **A*** are taken to be true when $[[F]] = \varnothing$, the entailments $\mathbf{A} \vdash \mathbf{I}$ and $\mathbf{A^*} \vdash \mathbf{I^*}$ will fall, and when **A** and **A*** are deemed false, the Conversions will bite the dust.

For AAPC, the system originally proposed by Aristotle and elaborated by Abelard (and subsequently forgotten in the history of logic), the needed fourth space for situations where $[[F]] = \varnothing$ can be added without a problem. Again, space 1 is reserved for cases where $[[F]] \subseteq [[G]]$; space 2 for cases where either [[F]] ↀ [[G]] or $[[G]] \subset [[F]]$; space 3 for cases where [[F]] ○○ [[G]]. The newly added space 4 is reserved for cases where $[[F]] = \varnothing$. In space 4 of fig. 6, **A** and **A*** count as false, just like **I** and **I***. As a result, the Conversions have been weakened to one-way entailments: $\mathbf{A} \vdash \neg\mathbf{I^*}$ and $\mathbf{I} \vdash \neg\mathbf{A^*}$ but not vice versa.

Figure 6 shows that AAPC is less powerful than the Square, because there are a number of vertex pairs that are unconnected and thus logically independent.

When we set up a VS-model and an octagonal representation for SMPC, the result is fig. 7. One notes that SMPC is identical to AABP but for space 4, where SMPC assigns truth to **A**- and **A***, whereas AAPC considers them false. The consequences of this seemingly small difference, however, are drastic. The twenty-eight logical relations of the Square have dwindled to a paltry twelve contained in two isomorphic quadruples **<A,I*,¬I*,¬A>** and **<A*,I,¬I,¬A*>**, which owe their logical isomorphy to the Modulo-*-Principle. This great loss of logical power of SMPC compared with the Square has never been noted in the literature.[7]

7 A distinct oddity in SMPC lies in the fact that, since **A** counts as true when $⟦F⟧ = \varnothing$, $\neg\mathbf{A}$ ($\equiv \mathbf{I^*}$) entails that $[[F]] \neq \varnothing$. This is odd because normally the negation cancels all entailments except necessary truths, but in this case the negation introduces a highly

Figure 6: The VS-Model of AAPC and its Octagonal Graph

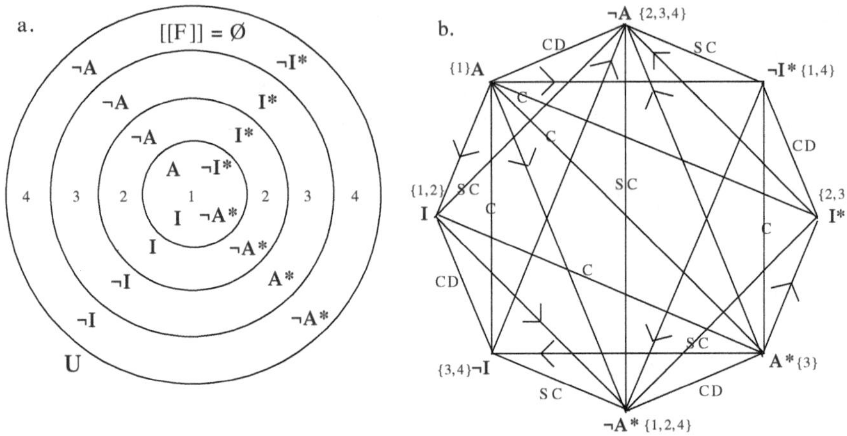

Figure 7: The VS-Model of SMPC and the Poor Remnants of its Octagonal Graph

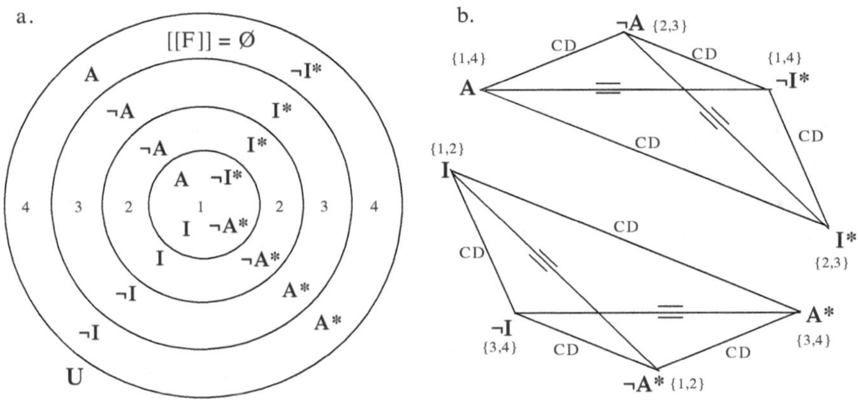

Does this now mean that AAPC is a stronger contender for the status of natural logic than the Square or SMPC? Not quite. True, we are in search of a natural logic that is logically sound rather than one which is not. And of the three systems investigated so far AAPC is sound, unlike the Square, and more powerful than SMPC, which would put AAPC ahead of the others. But there are further contenders, not so far investigated.

5. The Principles of Natural Set Theory: The Bootstrapping Hypothesis

The present study relies on the general hypothesis that there is a gradient in the analytical powers of humans ranging from the 'rawest' or most basic level of unsophistication to the most advanced levels of abstract thinking as found in the modern centres of science. The degree of achievement is taken to depend on the cultural and other conditions that trigger the use and development of available cognitive reserves at the right age interval. Let us call this the BOOTSTRAPPING HYPOTHESIS. Some extremely interesting and challenging work has been done over the past few decades regarding the arithmetical capacities of infants, schoolchildren, and members of illiterate cultures (in particular, Ginsburg et al. 1984, Dehaene 1997, Butterworth 1999, Pica et al 2004). Pica et al. (2004) investigated the arithmetical ability of speakers of Mundurukú, an Amazonian language spoken by some 7000, mostly illiterate, Brazilian Indians. The language has no numerals beyond five (the word for 'five' being the equivalent of 'hand', as also in many South-East Asian languages), reflecting the fact that the speakers were unable to count or do sums beyond five. Yet after some training, they quickly extended their counting and computing abilities both cognitively and lexically, creating expressions like 'two hands', even though their achievements never matched those of humans born into culturally more developed societies (Pierre Pica, p.c.).

contingent entailment of existence. For **I*** this is perfectly acceptable, but for **¬A** it is at least surprising.

Butterworth states (1999: 7):

> What makes human numerical ability unique is the development and transmission of cultural tools for extending the capability of the Number Module. These tools include aids to counting, such as number words, finger-counting, and tallying; and also the accumulated inventions of mathematicians down the centuries –from numerals to calculating procedures, from counting-boards to theorems and their proofs.

Unfortunately, I must disagree with Butterworth. Far from being unique to the human numerical ability, the power to extend a naturally given ability through cultural development appears to be more general. Although the exact boundaries of this human power are unknown, it does seem that, along with numeracy and, apparently, also the reading ability (Dehaene 2005) and geometrical abilities (Dehaene et al. 2006), logic and its underlying set theory are worthy candidates to be included. This is what is investigated in the present paper.

Given this, let us now look briefly at the principles of basic-natural set theory (BNST), viewed as constraints upon standard set theory (for a more detailed discussion see Seuren 2006 and Seuren 2010, Chapter 3). To begin with, let it be assumed that BNST entails that the mind, though naturally capable of processing and operating with sets or 'plural objects', does not naturally represent a set as either the null set (∅) or the universe of all objects (**OBJ**). It also seems unnatural, for unsophisticated humans, to regard singletons, or sets consisting of just one element, as 'sets', as is borne out by the well-known difficulty of explaining to beginning students of set theory the difference between, say, the individual object *a* and the set {a}. For natural cognition, sets are *per se* plural sets, whose cardinality runs from 2 upward.

This means that, in BNST, the null set is not a set at all: the cognitive counterpart of ∅ is the absence of any set, something which is cognitively real and may be called "null" but cannot play the role of a set. The opposite notion of **OBJ**, as known in standard modern set-theory, is too nondescript to be cognitively real to formally untrained minds. What does seem to play a role is the notion of RESTRICTED UNIVERSE OF OBJECTS or $\mathbf{OBJ}^{R}$, involving the totality of all objects within a contextually defined universe of discourse. Therefore, all standard set-theoretic definitions involving **OBJ** should, for natural set theory, be redefined as involving the notion of $\mathbf{OBJ}^{R}$, which does count as a natural set.

Given these assumptions, we posit the first principle of natural set theory, PNST–1. Sets that are neither ∅ nor **OBJ** nor a singleton are called "natural sets". PNST–1 expresses the fact that NST is a theory of *plural objects*:

PNST–1: ∅, OBJ, AND SINGLETONS ARE NOT NATURAL SETS
Sets are never cognitively represented as having an EXTREME VALUE –that is, as the null set (∅) or as the totality of objects (**OBJ**). Nor are they represented as containing just one element. The null set is not a set and **OBJ** is replaced with $\mathbf{OBJ}^{R}$.

In the absence of any experimental data, and hence of any precise scale of naturalness, we posit hypothetically that *PNST–1 is both basic-natural and strict-natural* (see below): it strongly resists intellectual construction. Only at a much more advanced level will human cognitive powers be able to override PNST–1.

A further principle of natural set theory requires the nonidentity of any two or more natural set representations. When two or more sets are cognitively represented, they are naturally taken to be distinct from each other: the identity relation is considered constructed. This gives the second principle of natural set theory, which eliminates the relation of identity:

PNST–2: NATURAL SETS ARE DISTINCT
When sets are distinguished cognitively, they are represented , at the basic but not necessarily at the strict-natural level, as being extensionally DISTINCT, differing as regards their membership. There is no natural cognitive relation of identity between sets.

Among other things, PNST–2 accounts for the fact that natural intuition reads "inclusion" as "proper inclusion" (⊂), and not as "included in or identical with" (⊆). This principle is taken to be valid at the basic-natural but not necessarily at the strict-natural level. It seems to be more easily overridden in the sphere of object-language predicate logic than in the much more abstract metatheory of logic where the logical relation of entailment occupies a central place. Thus, whilst it seems relatively easy to gain the insight that *All F is G* is true when [[F]] = [[G]] (in a logic where *All F is G* is defined as [[F]] ⊆ [[G]]), it seems a great deal harder to convince one that every sentence *P* entails itself, which amounts to saying that /P/ ⊆ /P/. It takes a considerable amount of advanced analytical thinking – more, apparently, than natural cognition and common natural language can bear – for the latter to be accepted.

In actual fact, PNST–2 is more general. It applies to any kind of object representation, not only of the plural objects we call sets but also of individual objects: basic-natural cognition allows no two objects of any kind to stand in a RELATION OF IDENTITY, since that would make them one single object. Yet it does allow for a mental OPERATION OF IDENTIFICATION, given distinct levels of representation. This is what underlies the predicate of identity in natural language: what was thought to be distinct becomes one. Such an operation involves two levels of cognitive representation, one at which the two virtual objects or sets are distinct and one at which they have merged into one. If truth is claimed for the latter, two virtual objects or sets have been identified. If truth is claimed for the former, one virtual object or set has been cognitively split up into two.

In the wake of PNST–1 and PNST–2, we hypothetically posit the following three principles, applying to the set-theoretic functions of *union*, *subtraction* and *intersection*.

PNST–3: BASIC-NATURAL UNION REQUIRES TOTAL DISTINCTNESS OF NATURAL SETS

When two (or more) sets *A* and *B* undergo union, *A* and *B* are natural sets and are, at the level of basic, but not strict, naturalness, totally distinct, with no element in common, so that $|A \cup B| = |A| + |B|$ ("$|X|$": the cardinality of set X).

At the basic-natural level, union is undefined for any two (or more) sets that are not totally distinct. (This accounts for the exclusive nature of natural language *or*. When PNST–3 is overridden, making it possible for *A* and *B* to intersect, exclusive *or* is upgraded to inclusive *or*.)

PNST–4: BASIC-NATURAL SUBTRACTION REQUIRES PROPER INCLUSION OF NATURAL SETS

When a set *A* is subtracted from a set *B*, *A* and *B* are natural sets and, at the level of basic, but not strict, naturalness, *A* is a proper subset of *B* so that $|B - A| = |B| - |A| \neq 0$.

Failing this condition, there is, at the basic level, no subtraction. At the strict level, one may subtract a set of ginger cats from a set of male cats and find that no male cat is left.

PNST–5: BASIC-NATURAL INTERSECTION REQUIRES MUTUAL PARTIAL INTERSECTION OF NATURAL SETS

When two (or more) sets *A* and *B* are intersected, *A* and *B* are natural sets and are, at the level of basic, but not strict, naturalness, in mutual partial intersection, with some, but not all, elements in common.

Basic natural intersection is thus defined only for sets *A* and *B* (and possibly more) such that (a) $A \cap B \neq \varnothing$ and (b) $A \cap B \neq A \neq B \neq \mathbf{OBJ}^R$. For the combination of these two conditions we use the symbol "ↀ" (mutual partial intersection): X ↀ Y just in case $X \cap Y \neq \varnothing \neq X \neq Y \neq \mathbf{OBJ}^R$. Condition (a) cannot be overridden at any level of set-theoretic or logical sophistication, but condition (b) is overridden at all levels above basic. (This overriding allows for the upgrading of *some* excluding *all* to *some* including *all*, and hence for the strict-natural subaltern entailment schema in the Square and in AAPC from *All F is G* to *Some F is G*.)

Thus restricted, the natural intersection function ensures *set-theoretic independence*. That is, if $A \cap B$ has a value under PNST–5, then *A* and *B* are set-theoretically independent and vice versa. It is then ensured that, for any element $o \in \mathbf{OBJ}^R$, it is possible that $o \in A$ and $o \notin B$, or $o \in B$ and $o \notin A$, or $o \in A$ and $o \in B$, or $o \notin A$ and $o \notin B$.

There is a final principle, relating to the recursive application of the set-theoretic functions. Whereas in mathematical set-theory one can apply the functions recursively to one's heart's content, this is not so in basic-natural set theory. This applies especially to the function *complement*, which is hypothetically taken to be nonrecursive in basic-natural, and only once-recursive in strict-natural set theory:[8]

PNST–6: BASIC-NATURAL COMPLEMENT IS NONRECURSIVE AND IS DEFINED AS THE RESULT OF BASIC-NATURAL SUBTRACTION

At the level of basic naturalness, complement is restricted to one application. Further, recursive, applications are either strict-natural or constructed.

8 Or, as a friend of mine once said: "Negatio simplex negat; negatio duplex affirmat; negatio triplex confundit."

Moreover, "$\overline{A}$" is defined as B – A, where *A* and *B* are natural sets and *B* functions as a restricted universe of objects (**OBJ**^{R}).

This hypothetical basic-natural set theory translates into basic-natural predicate logic.

6. Basic Natural Predicate Calculus (BNPC)

The present section builds on the hypothesis that nonsophisticated humans have a primitive, innate "theory" for dealing with plural objects, manifested by bare set-theoretic intuitions. They also have bare, immediate logical intuitions giving rise to a logic that differs from the Square, which, as has been said, is not in total agreement with what we take to be bare logical intuitions. We will now try to bring the putative natural intuitions regarding plural objects in agreement with bare logical intuitions, thereby creating a basic-natural predicate calculus (BNPC). There seems to be an opposition between bare intuitions, whether set-theoretic or logical, and what happens when one reflects upon them. I submit that when the bare logical intuitions are subjected to further reflection, the result is the Square. Without further reflection, we have a logic that differs from, and is more "primitive" than, the Square. But what does BNPC look like?

Before we answer this question, we must have a look at the predicate logic devised by the Edinburgh philosopher Sir William Hamilton (1788–1856) , in particular in his posthumous (1866), which may be seen as an attempt to formulate a "natural" predicate logic, both more natural and more complete than the traditional Square.[9] A similar attempt was made by the Danish linguist Otto Jespersen (1860–1943) in his (1917). Jespersen was a first-rate linguist but lacked any knowledge of or training in logic and he had probably never heard of Hamilton. His predicate logic was merely the result of his finely tuned logical intuitions. Entirely in the spirit of his day

9 On account of his logical proposals a bitter polemic developed between him and De Morgan mentioned earlier. The conflict ran to such heights that it was taken to the courts with charges of plagiarism. (A spicy collection of their exchanges is found in De Morgan 1847.) Since De Morgan came out as the winner, nobody except historians really looked at Hamilton's system with much interest.

(at least among linguists), he rejected "official" logic as linguistically and cognitively irrelevant. His predicate logic, however, was not worked out to a sufficient detail for it to count as a separate logical proposal, so that we only look at Hamilton's logic.

Hamilton was ahead of his time in that he saw that quantification involves two sets, now called the matrix or predicate set and the restrictor set, thereby anticipating the notion of generalized quantifiers. Following an already existing tradition, he maintained that, as regards predicate logic, not only the subject term should be quantified but also the predicate term. He thus adds a quantifier, *all* or *some* (in the sense of 'some but not all'), to both the subject term and the predicate of quantified expressions. Accordingly, he does not use anything like the modern quantifiers, but prefixes each set denotation with the operators **t** (total) or **p** (partial), using = for inclusion and || for exclusion.

The expressions of his formal language are summarily translated as follows:

tF = **tG**: *All F includes all G* (*All F is G* and vice versa); [[F]] = [[G]].
tF = **pG**: *All F includes some but not all G* (*All F is G* but not vice versa), *Some but not all G includes all F* (*Some but not all G is F*); [[F]] ⊂ [[G]].
pF = **tG**: *Some but not all F includes all G* (*Some but not all F is G*), *All G includes some but not all F* (*All G is F* but not vice versa); [[G]] ⊂ [[F]].
pF = **pG**: *Some but not all F includes some but not all G*; [[F]] ⚭ [[G]].
tF || **tG**: *All F excludes all G* (*No F is G*); [[F]] ⚭ [[G]].
tF || **pG**: *All F excludes some but not all G* (*All F is G* or *Some but not all G is F*); [[F]] ⊂ [[G]] or [[F]] ⚭ [[G]].
pF || **tG**: *Some but not all F excludes all G* (*All G is F* or *Some but not all F is G*); [[G]] ⊂ [[F]] or [[F]] ⚭ [[G]].
pF || **pG**: *Some but not all F excludes some but not all G*; [[F]] ⚭ [[G]].

A few points are to be noted. First, there is no negation in this system. Then, **pF** = **pG** is equivalent with **pF** || **pG**, both being true just in case [[F]] ⚭ [[G]]. Third, the relation = (includes) is symmetrical, but || (excludes) is not. Fourth, **tF** = **pG** ⊢ **tF** || **pG**, because both are true when [[F]] ⊂ [[G]] but only the latter is true when [[F]] ⚭ [[G]]. Fifth, *Some (but not all) F is G* is equivalent with *Some (but not all) F is not-G*, both excluding *All F is G*. But *Some (but not all) F is G* is not equivalent with its converse *Some (but not all) G is F*. Sixth, *No F is G* (**t**F || **t**G) is not equivalent with *Not-some F is G*, because *No F is G* is true only when [[F]] ○○ [[G]], while the negation of *Some (but not all) F is G* is true when [[F]] = [[G]] or [[F]] ⊂ [[G]] or [[F]] ⚭ [[G]].

Despite the possible logical advantages this system may have, its notation, largely inspired by what is taken to be the need for quantification of the predicate, sits uneasily with natural language, whether in a basic-natural or a strict-natural interpretation. Besides the absence of negation, what stands in the way of considering the Hamiltonian system as a candidate for natural logic is the fact that its translations into a natural language such as English result either in highly artifical and perhaps even ungrammatical sentences or in sentences that are highly ambiguous in terms of the Hamiltonian system. An ordinary sentence type like *All F is G* becomes multiply ambiguous, as it is true when **t**F = **t**G, **t**F = **p**G or **t**F || **p**G. And *Some F is G* is true when **p**F = **t**G, **p**F = **p**G, **p**F || **t**G or **p**F || **p**G. In other words, Hamilton's insistence on 'quantification of the predicate' may possibly have certain advantages for a system of predicate logic, but it fails to translate naturally into human language.

If we forget about the quantification of the predicate and try to render the system in terms of quantifiers, we need the three quantifiers *all*, *some (but not all)* and *no*, corresponding with three sentence types **A** for *All F is G*, **I** for *Some but not all F is G* (equivalent with **I*** or *Some but not all F is not-G*), and **N** for *No F is G*. The three types **A**, **I** and **N** are mutually exclusive and thus form three pairs of contraries. They are defined as follows in Hamiltonian predicate logic ([[F]] and [[G]] are natural but not necessarily distinct sets):

A (*All F is G*)	true iff	[[F]] ⊆ [[G]]
I/I* (*Some F is (not-)G*)	true iff	[[F]] ⚭ [[G]] or [[G]] ⊂ [[F]]
N (*No F is G*)	true iff	[[F]] ○○ [[G]]

This is not a direct implementation of the set-theoretic restrictions formulated above. First, [[F]] and [[G]] are required to be natural sets (PNST-1), but they are not required to be distinct (PNST-2), as one sees from the definition of the truth condition for *All F is G*. Moreover, the reading of *some* as 'some but not all' only partially succeeds in making *some* a reflex of partial mutual intersection ((PNST-5). For that to be so, *Some F is G* must be assigned truth just in case [[F]] ⚭ [[G]] and is to be read as 'some but not all F is G and some but not all G is F'. In Hamiltonian logic it is read merely as 'some but not all F is G', without its converse. Consequently, in Hamilton's system $\mathbf{I} \equiv \mathbf{I}^*$ (*Some F is G* ≡ *Some F is not-G*) but the equivalence of **I** and **I!** (*Some G is F*; the converse of **I** is written as **I!**) does not hold.

A full implementation of the set-theoretic restrictions (1) to (6) yields a logic where the three sentence types are defined as follows ([[F]] and [[G]] are distinct natural sets):

A (*All F is G*)	true iff	$[[F]] \subset [[G]]$
I (*Some F is G*)	true iff	[[F]] ⚭ [[G]]
N (*No F is G*)	true iff	[[F]] ○○ [[G]]

Here, the sets involved must be both natural (PNST-1) and distinct (PNST-2). Then, for **A**, the condition is that [[F]] be *properly* included in [[G]], according to PNST-2, which says that two sets must be distinct. For **I**, the condition is *mutual* partial intersection (PNST-5). Now, *Some F is G* is to be read as "some but not all F is G and some but not all G is F". In such a logic, other than in the Hamilton system, $\mathbf{I} \not\equiv \mathbf{I}^*$, because when $[[G]] \subset [[F]]$, *Some F is G* is false but *Some F is not-G* is true. But the equivalence of *Some F is G* with its converse, *Some G is F* –that is, $\mathbf{I} \equiv \mathbf{I!}$ –holds in such a logic, but not in the Hamiltonian system, again owing to cases where $[[G]] \subset [[F]]$. In such cases, Hamilton makes *Some F is G* true but *Some G is F* false, while a logic that straightforwardly translates PNST–5 into *some* makes both false. A logic where both $\mathbf{I} \equiv \mathbf{I}^*$ and $\mathbf{I} \equiv \mathbf{I!}$ hold is inconsistent.

The fact that in Hamiltonian logic $\mathbf{I} \not\equiv \mathbf{I!}$ but $\mathbf{I} \equiv \mathbf{I}^*$ does not necessarily make it less natural, even though it goes against the hypothetical PNST–5. In fact, natural logical intuitions seem to support Hamilton in this respect. To test this, one has to find examples where it is clear from world knowledge or from linguistic meaning that $[[G]] \subset [[F]]$. Such examples are hard to get by, but here are a few:

(3) a. Some children are orphans.
b. Some people are Englishmen.
c. Some computers are laptops.
d. Some heavenly bodies are planets.

In all these cases, [[G]] is properly included in [[F]] and all four sentences are effortlessly taken to be true. This supports the Hamiltonian position that existentially quantified sentences where [[G]] ⊂ [[F]] are naturally considered true. Further confirmation for this position is found in the fact that the converses of these sentences make eyebrows go up:

(4) a. Some orphans are children.
b. Some Englishmen are people.
c. Some laptops are computers.
d. Some planets are heavenly bodies.

We have no choice but to accept the Hamiltonian position as regards *some*, taking both **I** and **I*** to be true when [[G]] ⊂ [[F]] (space 4) while giving up the intuition that **I** ≡ **I!**, *despite PNST–5*. Apparently, the symmetry of the existential quantifier is, though also supported by intuition,[10] less strongly so than the equivalence **I** ≡ **I*** valid in Hamilton's system. One may consider modifying basic-natural set theory by redefining intersection as requiring either mutual partial intersection or an asymmetrical proper inclusion, but one has to fear that this will not bear experimental testing. For the moment, therefore, we have to live with the fact that, in this respect, BNPC deviates from basic-natural set theory. Why or how BNPC has extended the condition for **I** with "or [[G]] ⊂ [[F]]", making *some* nonsymmetrical while the intuition that it is symmetrical still persists, is hard to say.

Thus, we take BNPC to be identical with Hamiltonian logic but for the general condition that in BNPC all sets involved are *distinct* natural sets and not just natural sets, as they are in the Hamilton system. **A** is thus considered true just in case [[F]] ⊂ [[G]] and both **I** and **I*** are considered true just in case either [[F]] ⚭ [[G]] or [[G]] ⊂ [[F]].

Since it makes sense to consider **A** to be equivalent with **N***, and correspondingly **A*** to be equivalent with **N** (the Modulo-*-Principle), we can set up the square notation and the VS-model for BNPC in fig. 8–a,b, respectively. The VS-model consists of twelve basic sentence types distributed over four spaces. Space 1 covers cases where [[F]] ⊂ [[G]]; space 2 is for cases where [[F]] ⚭ [[G]] or [[G]] ⊂ [[F]]; space 3 for cases where [[F]] ○○ [[G]];

10 As appears from the second chapter of Aristotle's *Prior Analytics*, where he posits that *All F is G* must be taken to entail *Some G is F* and that *Some F is G* is equivalent with *Some G is F*.

space 4, for cases where [[F]] = ∅, is left inoperative. The corresponding dodecagon is given in fig. 8–c. If space 4 is made operative and only **N** and **N*** are declared true in it, as in fig. 9, all that happens is that the logic is slightly weakened, but it remains sound. But note that in space 4 of fig. 9–b it is no longer required that [[F]] and [[G]] are distinct natural sets.

BNPC does not suffer from UEI because **¬A** and **¬A*** do not entail **I** or **I***. One notes the extraordinary multitude of logical relations in fig. 8–c, which forms a technically speaking complete dodecagon and is thus even richer and more powerful than the Square, while it does not suffer from the logical defect of UEI. It is worth observing that this insight is new, as neither BNPC nor the Hamiltonian system have ever been subjected to VS-modelling or the corresponding system of polygonal representation.

Does this now mean that this unexpectedly rich, powerful and sound predicate logic carries the day? Not quite yet, because there still are a few empirical problems, one of them consisting in the fact that BNPC fails to account for the strong natural intuition, observed by many authors (notably Jespersen 1917: 86–91), that **¬A** is equivalent with both **I** and **I***, which means that in the logical system we would like to see an equivalence relation between **¬A**, **I** and **I***. But BNPC fails to oblige as is easily checked in fig. 12.7–b,c. The entailments from **I** or **I*** to **¬A** hold, but natural intuition requires a stronger relation. In this respect, the Square fares somewhat better, since, in the Square, **¬A** and **I*** are equivalent, though **¬A** and **I** are mere subcontraries.

The pragmaticists tackle the problem by an appeal to the Gricean maxims, reinforced by Horn's theory of scalarity (Horn 1972, 1989), in virtue of which the negation, when applied to a quantifiable scale, cuts off only the higher part of the scale but leaves the remainder intact. Or, as Jespersen put it (1917: 86), "in negativing an *A* [i.e. *All F is G*; PAMS] it is the *absolute* element of *A* that is negatived". But this answer seems inadequate for a number of reasons. For example, it is not clear why the psychology of scalarity does not apply to the analogous propositional operators *and* and *or*, which can also be seen to form a quantifiable scale, in that *and* involves as many terms as there are members of the conjunction, while (exclusive) *or* involves just one. Yet *not-and* –that is, *Not(P and Q and R and ...)* –is not felt to entail *P or Q or R or ....* We must, therefore, look elsewhere for an explanation.

The explanation proposed here falls back on trivalent presuppositional logic, (section 8), in particular on the notion of topic-comment structure,

Figure 8: The Square Representation, VS-Model and Dodecagon of BNPC with Space 4 Idle

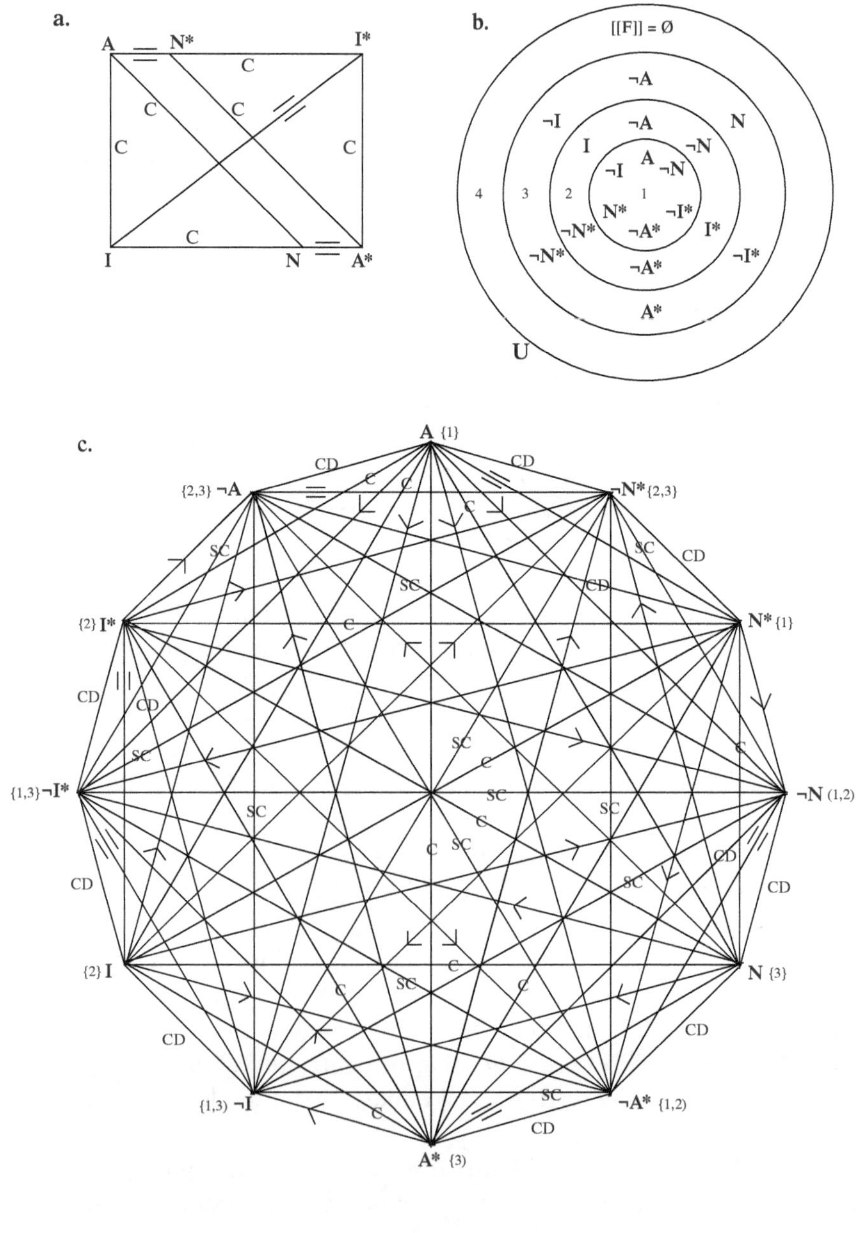

Figure 9: The Square, VS-Model and Dodecagon of BNPC with Space 4 Operative

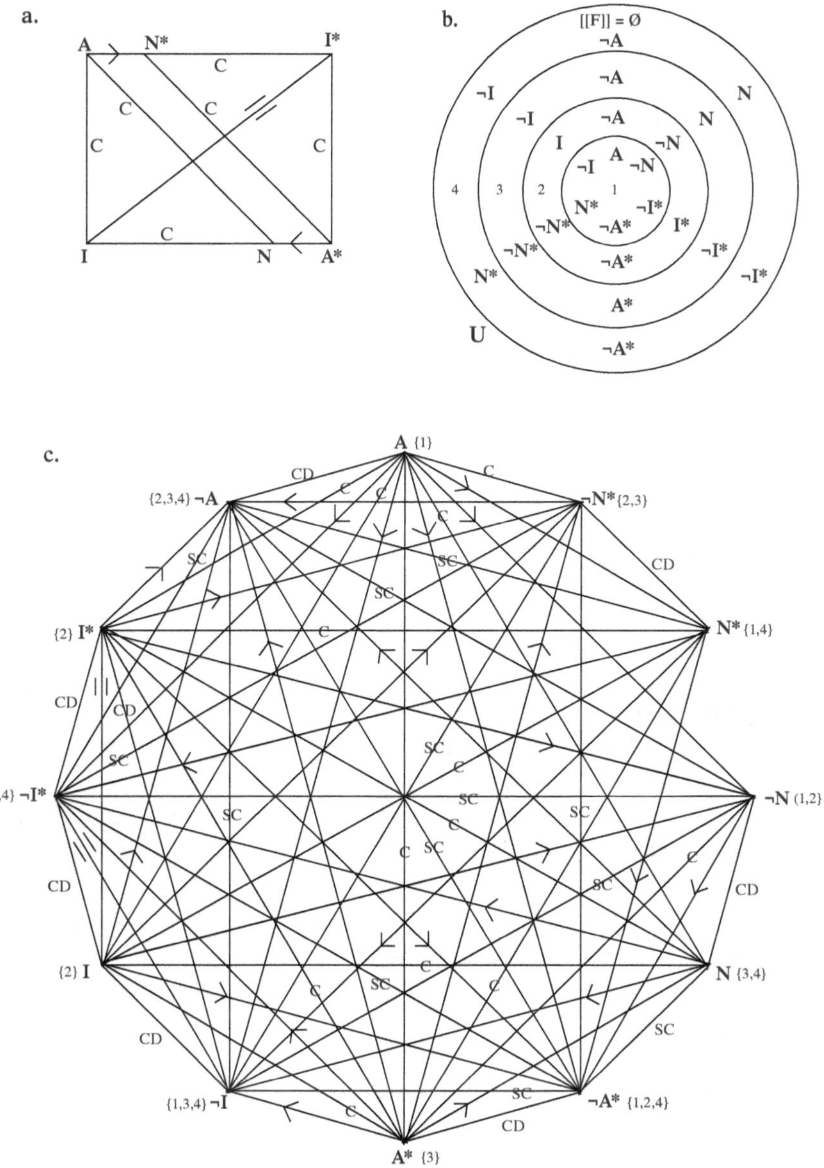

also often called 'information structure'. Topic-comment structure often comes with negative sentences. If we take it that topic-comment structure is formalizable as an underlying cleft structure and that *all* in a sentence like *Ben didn't eat all of his meal* has comment status, this sentence is then analysed as "what Ben ate of his meal was not all", which entails presuppositionally that Ben ate some of his meal, excluding the case that he ate nothing. This sentence thus *presupposes*, and hence *entails*, that Ben ate some of his meal and it *asserts* that he did not eat all of it. Similarly for a sentence like *Not all flags are green*, which, if analysed as "the flags that are green are not áll (flags)", excludes the case that there are no green flags. In sum, given the sentence form *The flags that are green are not all flags*, it is presupposed, and hence entailed, that some flags are green and that some flags are not green (given that **I** ≡ **I***). And because, in BNPC, both **I** and **I*** entail ¬**A**, it follows that ¬**A** both entails and is entailed by **I** and **I***, which makes the three equivalent. This solution requires, however, that ¬**A**-type sentences be given a topic-comment structure with ¬**A** as the comment and that BNPC be extended with a presuppositional component.

A further problem consists in the fact that, in cases where [[F]] = ∅, it seems natural to say that **N** is true. Given that this world has no mermaids, it is not hard to agree that *There are no mermaids living in London* must be considered true. But then the corresponding **N*** sentence *There are no mermaids (that are) not living in London* should also be true, because if there are no mermaids, there will be no mermaid among those entities that are living in London nor among those that are not. The disturbing fact is, however, that while *There are no mermaids living in London* is considered true, *There are no mermaids (that are) not living in London* is false for natural intuition and, in fact, equivalent with *All mermaids are living in London.* This latter equivalence is accounted for, since, as is shown in fig. 8-b, both **A** and **N*** are true in space 1 and false in the spaces 2 and 3, which makes them equivalent in a model where [[F]] ≠ ∅.

But in order to see if BNPC is tenable as a logical system, we must add a fourth space to fig. 8–b for cases where [[F]] = ∅. This has been done in fig. 9–b, where truth has been assigned to **N**, but then truth must also be assigned to **N*** on pain of making the semantics of the *no*-operator inconsistent. It thus follows that if we turn BNPC into a fully fledged logical system that also caters for cases where [[F]] = ∅, as in fig. 9, a gross unnaturalness appears, because truth for **N*** makes **N*** clash with natural intuitions. This is, however, not a problem for the logical system but only for the claim that BNPC reflects natural logical intuitions. Therefore, if this claim is to be upheld, it is essential that the fourth space for cases where $[[F]] = \varnothing$ should not be considered part of it –owing to PNST–1, which declares $\varnothing$ not to be a set.

Finally, there is still the fact that BNPC holds only without what we have called "further reflection". This leads to the following predicament. Suppose Joe is checking if all the 45 doors in the building are properly locked. He has come to number 15 and so far all has been well. We feel that, at this juncture, he ought to be able to say in truth that (at least) some doors are properly locked. But BNPC does not allow him to do so, because *Some doors are properly locked* entails that not all doors are. And Joe cannot vouch for that entailment. In fact, he must wait till he has checked all doors before he can say either that some doors are properly locked or that all are. All he can say after checking 15 doors is that it is not so that no door is properly locked, or *Not (no door is properly locked)*, which entails that at least some and perhaps all doors are properly locked. One might propose that natural-language *some* is ambiguous between "some perhaps all" (=*not no*) and "some but not all". But the rules of good methodology make one reluctant to do that. There is, thus, the basic intuition that *some* implies or entails "not all", but further reflection makes one see that *some* ought not to be taken that way. The Gricean maxim "be as informative as you can be" does not satisfy. So, what is the answer?

It would be presumptuous to claim that the answer is near at hand. But it seems reasonable at least to explore the possibility that we must distinguish, as we have done, between two levels of natural logic, a *basic-natural* and a *strict-natural* level. At the basic level, where no further reflection takes place, we would find BNPC. At the strict level we would expect to find either the Square or AAPC. SMPC would be considered to be way beyond the strict-natural level, it being the product of highly sophisticated abstract mathematical analysis, which takes place at the *constructed* level. Therefore, SMPC is not taken into account as a viable candidate.

Which is the better candidate for the position of strict-natural predicate logic: the Square or AAPC? AAPC is logically weaker than the Square, which is why we prefer to keep the Square in the basket, if at all possible, and see if it can be saved from logical error. In other words, can we have our logic cake and eat it too? The answer seems to be positive. Nature has, apparently, fitted out the human race with a logical possibility, the device of presupposition, that was never thought of by our mathematical friends, nor, for that matter, by logicians of past ages,[11] but is highly functional for linguistic interaction. Section 8 shows how this works.

7. How the Various Systems Score for Logical Intuitions

Let us now take provisional stock and see how the various systems score empirically, at this stage of the analysis, given their satisfaction conditions for *some*, *no* and *all*. Table 1 shows the satisfaction conditions for the quantifiers *some*, *no* and *all* in the four predicate-logic systems under discussion (the Hamiltonian system has been left out because it is identical to BNPC except that *All F is G* is true in BNPC just in case $[[F]] \subset [[G]]$ whereas in the Hamiltonian system it is true just in case $[[F]] \subseteq [[G]]$). Table 2 shows the empirical success scores of the various systems in terms of the following seven natural logical intuitions ("$\vdash$" and "$\equiv$" are to be read here as "*suggests* an entailment" and "*suggests* equivalence", respectively).

Natural logical intuitions:

1. Some F is G $\vdash$ Not-all F is G	**I $\vdash$ ¬A**
2. Some F is G $\equiv$ Some F is not-G	**I $\equiv$ I***
3. Some F is G $\equiv$ Some G is F	**I $\equiv$ I!**
4. All F is G $\vdash$ Some G is F	**A $\vdash$ I!**
5. All F is G $\vdash$ Some G is not-F	**A $\vdash$ I!***
6. No F is not-G $\equiv$ All F is G	**N* $\equiv$ A**
7. Not-all F is G $\equiv$ Some F is G $\equiv$ Some F is not-G	**¬A $\equiv$ I $\equiv$ I***

The first thing that strikes one is that none of the logical systems under discussion is able to account for intuition 7, which says that *Not-all F is G* is felt to be equivalent with both *Some F is G* and *Some F is not-G*, excluding *No F is G*. Apparently, intuition 7 is a special case. In section 8, which deals with presuppositional logic, it will become clear that intuition 7 is accounted for in BNPC once *Not-all F is G* is assigned a topic-comment

11 In Frege (1892), and also in some medieval works (see note 13), mention is made of presupposition, but no proposal was ever made to integrate presupposition into a logical system until Strawson (1950) and Dummett (1973: 421; see note 14).

Table 1: Satisfaction Conditions for Some, No **and** All **in the Systems Considered**

SOME:	satisfaction condition:
BNPC	$[[F]] \circ\!\!\circ [[G]]$ or $[[G]] \subset [[F]]$ ($[[F]]$, $[[G]]$ are distinct natural sets)
AAPC	$[[F]] \cap [[G]] \neq \varnothing$
the Square	$[[F]] \cap [[G]] \neq \varnothing$
SMPC	$[[F]] \cap [[G]] \neq \varnothing$
NO:	satisfaction condition:
BNPC	$[[F]] \circ\circ [[G]]$ ($[[F]]$, $[[G]]$ are natural sets)
AAPC	$[[F]] \cap [[G]] = \varnothing$
the Square	$[[F]] \cap [[G]] = \varnothing$
SMPC	$[[F]] \cap [[G]] = \varnothing$
ALL:	satisfaction condition:
BNPC	$[[F]] \subset [[G]]$ ($[[F]]$, $[[G]]$ are natural sets)
AAPC	$[[F]] \subseteq [[G]]$ and $[[F]] \neq \varnothing$
the Square	$[[F]] \subseteq [[G]]$ (applicable only if $[[F]] \neq \varnothing$)
SMPC	$[[F]] \subseteq [[G]]$

Table 2: Empirical Success Scores for Different Systems of Bivalent Predicate Logic

BNPC	1, 2, 4, 5, 6
AAPC	3, 4,
the Square	3, 4, 6
SMPC	3, 6

structure and BNPC has been extended with a presuppositional component, which, incidentally, makes the logic trivalent. Extending AAPC with a presuppositional component establishes the equivalence of *Not-all F is G* and *Some F is not-G*, lacking in nonpresuppositional AAPC. The equivalence felt to exist between *Not-all F is G* and *Some F is G* is, of course, not attainable in either AAPC or the Square, since these two systems have the subaltern entailment from **A** to **I**.

BNPC scores best, though it lacks an account for intuition 3 which makes the quantifier *some* symmetrical (**I** ≡ **I!**). We have here an instance of inconsistent intuitions: the intuition that **I** ≡ **I!** (symmetrical *some*) fits AAPC, the Square (and SMPC), while the intuition that **I** ≡ **I*** (exclusive *some*) fits BNPC. Apparently, the former has overridden the latter, but not without leaving a clearly felt trace.

AAPC and the Square successfully account for the intuitions 3 and 4 because in both systems *some* is symmetrical. Moreover, *All F is G* cannot be true, in these systems, unless *Some G is F* is also true, under the definitions for *some* and *all* valid in the systems at hand. One notes that the combination of the intuitions 3 and 4 makes for the intuition that the subaltern entailment (**A** ⊢ **I**) holds. The Square, in addition, accounts for intuition 6, because in the Square the Conversions hold, so that ¬**I*** ≡ **A**. In AAPC, as has been shown, all that can be said is that **A** ⊢ ¬**I*** but not vice versa. Finally, SMPC satisfies the intuitions 3 and 6. It satisfies intuition 3 because in SMPC **I** ≡ **I!** and it satisfies intuition 6 because in SMPC the Conversions hold.

Meanwhile it must be said, against BNPC, that natural-language *some* must sometimes be taken to mean 'at least some, perhaps all'. A speaker can vouch for the truth of *Some F is G* when only part of [[F]] has been checked and so far all elements in [[F]] satisfy the conditions of the predicate G. But BNPC does not allow a speaker to make an existential statement until complete knowledge of the situation at hand has been achieved. This conclusion is to some extent disconcerting, in that it amounts to saying that the basic-natural set-theoretic relations of mutual exclusion, mutual partial intersection and proper inclusion as expressed by the quantifiers *no*, *some* and *all*, respectively, make for a predicate logic (Hamilton or BNPC) that is applicable only when full knowledge of the situation at hand is available. If this were cognitive reality, it would put a brake on the extension of knowledge and thus on intellectual development as a whole. Aristotle made the world see that, *on reflection*, one must concede that **A**-sentences entail the corresponding **I**-sentences, thereby removing the main blemish in any predicate logic constrained by the principles of basic-natural set theory. Needless to say, this Aristotelian upgrading from basic to strict naturalness conquered the world, until it was replaced, a century ago, with the modern, highly constructed, system of predicate logic known as standard modern predicate calculus, which, ironically, banished again the subaltern entailment from **A** to **I** sentences so as to get rid of undue existential import.

In conclusion we say that SMPC, though currently considered inviolate, may be appropriate for mathematical reasoning but is a highly unlikely candidate for the position of logic of language. It is proposed here that the real predicate logic of language manifests itself in two guises: (a) basic-natural predicate calculus or BNPC, and (b) strict-natural predicate logic, which is identical with the Square, both extended with a presuppositional component.

8. The Device of Presupposition and Presuppositional Logic

What a presupposition is, is perhaps best illustrated with a few examples. Consider the following cases (">>" stands for "presupposes"):

(5) a.	Jack has come back.	>>	Jack has been been away
b.	Jack is divorced.	>>	Jack was married before
c.	I have forgotten your name.	>>	I knew your name before
d.	Only Jack left.	>>	Jack left
e.	It was Joe who said "damn"	>>	someone said "damn"
f.	The king is eating.	>>	there is a king
g.	She realized that she had lost.	>>	she had lost
h.	That she had lost surprised her.	>>	she had lost

(5a–c) are cases of so-called CATEGORIAL presupposition: the presupposition is clearly induced by the main lexical predicate (*come back*, *be divorced*, *have forgotten*), since one cannot come back if one hasn't been away first, one cannot be divorced if one hasn't been married first, and one cannot have forgotten something if one did not know it beforehand. The presupposition of (5d) is induced by the particle *only*, which turns the clause *Jack left* into a presupposition while it asserts that no-one (nothing) else deserves the predicate of that clause. (5e) is a presupposition induced by the CLEFT CONSTRUCTION. (The same type of presupposition is induced by contrastive accent.) (5f) instantiates a so-called EXISTENTIAL presupposition: for someone to be eating he or she must first exist. (5g,h) are cases of FACTIVE presupposition: the main predicate (*realize*, *be surprised*) is factive, which means that it induces the presupposition that what is said in the *that*-clause is true.

In all cases, presuppositions are entailments semantically fixed at type-level: the carrier sentence cannot be true without its presuppositions also being true, for strictly semantic reasons. Yet there is a difference with standard or classical entailments in that in a coherent text a presuppositional entailment naturally precedes its carrier sentence. The following bits of text are 'natural' and coherent:

(6) a. Jack has been been away and/but he has come back.
b. Jack was married before and/but he is divorced.
c. I knew your name before and/but I have forgotten it.
d. Jack left and he is the only one who did.
e. Someone said "damn" and it was Joe who did.
f. There is a king and he is eating.
g. She had lost and she realized it.
h. She had lost and that surprised her.

This observation has given rise to the now universally accepted view (first proposed in Seuren 1972) that each new sentence is added, or, in today's technical parlance, INCREMENTED, to a given discourse domain which acts as a memory store for the information carried by previous discourse. Presuppositions are thus restrictions on the usability of their carrier sentences in certain contexts: a sentence S presupposing P is usable only in a context where P has been incremented first.

Presuppositions are always recoverable from their carrier sentences and on many token presentations they actually have to be recovered for the proper interpretation of the utterance concerned. When a sentence S carrying a presupposition P is presented, it is not necessary to present P first, as P can be recovered from S, in principle on the basis of the linguistically codified lexical meaning of the main predicate of S –a process called ACCOMMODATION or POST HOC SUPPLETION. It is thus, in principle, superfluous to make the effort of actually pronouncing presuppositions explicitly. This makes presupposition a powerful energy-saving device cutting down on the effort of speaking, as the listener supplies the missing information on the basis of his or her knowledge of the language. The device of presupposition is thus tailor-made for quick and efficient linguistic interaction.[12]

12 Some (radical) pragmaticists maintain that presuppositions are phenomena at token level, not at type level. This view fails to account for the fact that presuppositions are

A further respect in which presuppositions differ from classical entailments is that they have a tendency to PROJECT upwards. That is, when the carrier sentence S is placed under a higher non-entailing operator, such as the verb *believe*, its presuppositions P are normally left no longer as full entailments but as default inferences (>). Sentence (7a) noncogently suggests that Jack has been away, but (7b), which instantiates classical entailment, is neutral as to whether Jack is dead or not.

(7) a. Joe believes that Jack has come back. > Jack has been away
b. Joe believes that Jack has been murdered. ≯ Jack is dead

The precise conditions of presupposition projection are still a matter of dispute. It has been suggested that the negation *not* functions like *believe* in that it allows the presuppositions of the carrier sentence to project upward as default inferences. But that cannot be true, because in some classes of cases, the presuppositions of a negated carrier sentence remain intact as full entailments, as in the following sentences:

(8) a. Not only Jack left. >> Jack left
b. It wasn't Jack who left. >> someone left
c. I haven't forgotten your name at all. >> I knew your name before
d. Not until 7 did she remember my name. >> she knew my name before
e. It wasn't Joe who said "damn". >> someone said "damn"
f. That she had lost did not surprise her. >> she had lost

In other negative sentences, the presuppositions are cancelled altogether, as in (9a–f), each containing a (boldface) POSITIVE POLARITY ITEM. The sentences in (9) are all naturally followed by the negation of the presupposition of the nonnegative carrier sentence: a sequence like *He did nót steal several books; there wére no books* is perfectly natural and coherent, especially with heavy accent on *not*. In all these cases, the negation calls for a correction of previous discourse, which causes a so-called "echo-effect".

systematically recoverable from their carrier sentences, regardless of any context. It thus also misses the point of the energy-saving function of presupposition – a point that should be dear to their hearts.

(9) a. He did nót **blow his top** about the fire. ≯ there was a fire
b. He did nót steal **several** books. ≯ there were books
c. He did nót steal **both** books. ≯ there were books
d. She has nót **already** come back. ≯ she has been away
e. He would nót **rather** leave. ≯ he could do something else
f. She does nót **still** live in Paris. ≯ she lived in Paris before

Note the stark contrast between (9f) and (10a,b), which presuppose and thus fully entail that she lived in Paris before:

(10) a. She does not live in Paris any more. >> she lived in Paris before
b. She no longer lives in Paris. >> she lived in Paris before

These observations strongly support the thesis that natural language uses sentential negation in at least two different ways, a default presupposition-preserving MINIMAL NEGATION ($\sim$) and a marked presupposition-cancelling RADICAL NEGATION ($\simeq$). The radical negation says, in effect, that the sentence in its scope suffers from presupposition failure. It is thus a meta-level comment on the sentence in its scope which requires a repair in the current discourse domain. (See Seuren 1985: 210–313; 1988; 1991 for ample comment.)

One should note that the radical negation can only be used as full sentence negation and it has to occur in construction with the main finite verb. Negations in other positions, as in (8a) or (8d), or in participials or infinitivals, or in constrruction with a negative polarity item, such as *at all* in (8c), are *per se* minimal negations and so are all internal negations in the logical structures considered in the present paper.

The literature on presupposition is relatively recent, as the phenomenon was not widely recognised until the 1950s.[13] The first author to present presupposition as a typical feature of human languages was the Oxford philosopher Peter Strawson (1919–2006) in Strawson (1950, 1952, 1954, 1964). Strawson focussed on the function of sentential negation in presupposition-carrying sentences. According to him, natural-language negation

13 In medieval philosophy, the concept of presupposition was known under the name of "praeiacens", but no presupposition theory was ever developed. See Seuren (1998: 426–441) for further comment.

(*not*) always preserves presuppositions. For Strawson, *Jack has not come back* still entails that Jack has been away. This made him conclude that natural language infringes upon the Aristotelian Principle of the Excluded Third (PET): a sentence is either true or false or, in cases of presupposition failure, without a truth value.

It was shown in Wilson (1975) that this is incorrect. Provided the negation is heavily accented, one can very well say, for example,

(11) Jack has NÓT come back. he has never been away!

This made her conclude that PET can be maintained. But Wilson failed to reckon with cases like those in (8), where presuppositions are fully preserved under negation, as well as with cases like those in (9), where presuppositions are eliminated altogether.

Unless and until a better proposal is put forward, I therefore conclude that natural language has two logically distinct negations, the presupposition-preserving MINIMAL NEGATION (∼) and the presupposition-cancelling RADICAL NEGATION (≃). Correspondingly, it has a trivalent logic with the three truth values TRUE, MINIMALLY FALSE (F1) and RADICALLY FALSE (F2).[14] The corresponding truth tables for the PRESUPPOSITIONAL PROPOSITIONAL CALCULUS (PPC) are as follows:

Figure 10: Three-Valued Presuppositional Propositional Calculus (PPC)

P	∼P	≃P	¬P
T	F1	F1	F1
F1	T	F1	T
F2	F2	T	T

∧ (P \ Q)	T	F1	F2
T	(1) T	(2) F1	(7) F2
F1	(3) F1	(4) F1	(8) F2
F2	(5) F2	(6) F2	(9) F2

∨ (P \ Q)	T	F1	F2
T	(1) T	(2) T	(7) T
F1	(3) T	(4) F1	(8) F1
F2	(5) T	(6) F1	(9) F2

The table for ¬P has been added for good measure, though it is, strictly speaking, superfluous, because, for any sentence (proposition) P, $\sim P \vee \simeq P \equiv \neg P$. That is, the disjunction of the minimal and the radical

14 The same proposal was made by the Oxford philosopher Michael Dummett (1973: 421), though for different, more philosophical, reasons. This proposal was never acknowledged in the literature.

negation equals classical bivalent negation. Moreover, $\simeq$P is equivalent with $\simeq\sim$P, because $\simeq$P is true only in cases of presupposition failure and when P suffers from that disability, so does $\sim$P, with the presupposition-preserving negation $\sim$.

The slots for conjunction ($\wedge$) and disjunction ($\vee$) have been numbered 1–4 for the cases where F2 is not at play and 5–9 for the cases where it is. This has been done to make it possible to state the various valuation spaces and hence the entailment relations holding in the system, while clearly distinguishing between the bivalent (T and F1) component of this logic and the extension with the third value F2. For example, the following VSs can now be listed:

(12) (1) $/P \wedge Q/ = \{1\}$
(2) $/P \vee Q/ = \{1,2,3,5,7\}$
(3) $/\sim(P \wedge Q)/ = \{2,3,4\}$
(4) $/\sim P \vee \sim Q/ = \{2,3,4,6,8\}$
(5) $/\sim(P \vee Q)/ = \{4,6,8\}$
(6) $/\sim P \wedge \sim Q/ = \{4\}$
(7) $/\simeq(P \wedge Q)/ = \{5,6,7,8,9\}$
(8) $/\simeq P \vee \simeq Q)/ = \{5,6,7,8,9\}$
(9) $/\simeq(P \vee Q)/ = \{9\}$
(10) $/\simeq P \wedge \simeq Q)/ = \{9\}$
(11) $/\neg(P \wedge Q)/ = \{2,3,4,5,6,7,8,9\}$
(12) $/\neg P \vee \neg Q)/ = \{2,3,4,5,6,7,8,9\}$
(13) $/\neg(P \vee Q)/ = \{4,6,8,9\}$
(14) $/\neg P \wedge \neg Q)/ = \{4,6,8,9\}$

One sees that standard propositional calculus is preserved when the calculus is restricted to the slots 1–4. When all slots are taken into account, $\sim(P \wedge Q) \vdash \sim P \vee \sim Q$ but not vice versa, while $\sim P \wedge \sim Q \vdash \sim(P \vee Q)$ but not vice versa. Also, $\sim P \vee \sim Q$ entails none of the other admissible conjunctions or disjunctions or their minimal or radical negations. De Morgan's laws thus do not hold for the minimal negation but have been reduced to one-way entailments.

It is tempting to see a connection with the fact, mentioned above, that (1a) is naturally reduced to (1b), while a reduction of (2a) to (2b) is highly unnatural and requires a good deal of reflection ((1) and (2) are repeated here for convenience):

(1) a. She doesn't like trains and she doesn't like planes.
b. She doesn't like trains or planes.

(2) a. She doesn't like trains or she doesn't like planes.
b. She doesn't like trains and planes.

Moreover, De Morgan's laws do hold for the radical negation $\simeq$, as is shown by (7)–(10) in (12), and also for the standard negation $\neg$, as is shown by (11)–(14). This means that PPC is isomorphic with standard bivalent propositional calculus for the negation operators $\simeq$ and $\neg$, but not for the minimal negation $\sim$.

As regards predicate logic, incorporating presuppositions, resulting in trivalent presuppositional predicate calculus, is a little more complex. First we take the Square as our point of departure for an extension with a presuppositional component, resulting in a presuppositional Square or P-Square, not only because the Square is most badly in need of such an extension owing to its UEI defect, but also because it looks as if it takes us straight to the natural logic of language.

I have adopted, in the spirit of the theory of generalized quantifiers, the logico-semantic template $\forall$x [Gx,Fx] and $\exists$x [Gx,Fx] for quantified sentences of the form *All F is G* and *Some F is G*, respectively. The predicate *Gx* is the SUBJECT or MATRIX term of the quantifying predicate $\forall$ or $\exists$; the predicate *Fx* is the OBJECT term, also called the RESTRICTOR term. The quantifiers $\forall$ (*all*) and $\exists$ (*some*) are thus higher-order predicates over pairs of predicate extensions (sets).

We use the following format for the definition of predicate extensions ("P^2" stands for a two-term or binary predicate):

(13) $[[P^2]] = \{<x,y> : \ldots$ (preconditions for x & y) $\ldots \mid \ldots$ (update conditions for x & y) $\ldots \}$
(the extension of the predicate P^2 is the set of pairs of objects x, y such that x and satisfy the preconditions and the update conditions defined for them)

The preconditions together with the update conditions form the SATISFACTION CONDITIONS of the predicate in question. Failure to satisfy the preconditions leads to F2, while failure to satisfy the update conditions leads to F1. For example, the binary predicate *kill* can be described as follows:

(14) $[[\text{Kill}]] = \{<x,y> : x, y$ exist; y is animate$\mid x$ causes y to die $\}$
(the extension of the predicate *kill* is the set of all pairs of objects x and y such that (preconditions) x and y actually exist and y is animate, and (update condition) x causes y to die)

The description given in (14) is likely to be empirically inadequate, but that is not the point. The point here is of a notational nature: the presupposition-generating preconditions are given first, after the colon, while the assertion-generating update conditions are given after the upright stroke.

The precondition of actual existence is a default for linguistic predicates and can thus be overridden. The predicate *worship*, for example, has a precondition of actual existence for the referent of its subject term but not for the referent of its object term, because one can very well worship a nonexistent entity, as is amply proved by both history and everyday life. This shows that the presupposition of existence that has played such a role in presupposition theory does not originate in definite determiners, as has been thought by all presupposition theorists so far, but in the precondition of actual existence defined for the terms of any given predicate. A sentence like *The Monster of Loch Ness has pneumonia* is thus radically false, assuming that this monster does not exist, because to have pneumonia one first has to enjoy actual existence. If it does exist (and if it has lungs), the sentence is either true or minimally false.

In accordance with the notation specified, we define the universal quantifier ∀ in the P-Square as follows:

(15) $[[\forall]] = \{<[[G]],[[F]]> :$ the elements in $[[F]]$ satisfy the preconditions of $G \mid [[F]] \subseteq [[G]]\}$
(the unversal quantifier ∀ denotes the set of all pairs of predicate extensions [[G]] and [[F]] such that (precondition) all elements of [[F]] satisfy the preconditions of the predicate *G*, and (update condition) [[F]] is included in [[G]])

As regards the existential quantifier ∃, there appears to be a choice. Either we don't do anything and leave its definition as it is in standard predicate logic, as in (16a), or we add the same precondition that has been added to the universal quantifier, as in (16b):

(16) a. $[[\exists]] = \{<[[G]],[[F]]> \mid [[G]] \cap [[F]] \neq \varnothing\}$
(the existential quantifier ∃ denotes the set of all pairs of predicate extensions [[G]] and [[F]] such that the intersection of [[G]] and [[F]] is nonnull)

b. $[[\exists]] = \{<[[G]],[[F]]> :$ the elements in [[F]] satisfy the preconditions of $G \mid [[G]] \cap [[F]] \neq \varnothing\}$
(the existential quantifier $\exists$ denotes the set of all pairs of predicate extensions [[G]] and [[F]] such that (precondition) all elements of [[F]] satisfy the preconditions of the predicate *G*, and (update condition) the intersection of [[G]] and [[F]] is nonnull)

The difference hinges on whether a sentence like *No mermaid has a bank account* or *There isn't a mermaid that has a bank account* is taken to be true or radically false owing to the lack of mermaids in this world. The answer "true" has a great deal going for it. But if that is accepted, *No mermaid does not have a bank account* or *There isn't a mermaid that hasn't a bank account* should likewise be true. But, as we have seen, natural intuitions go the other way, because it is clearly felt that the latter two are equivalent with *All mermaids have a bank account*, which is considered radically false in the absence of mermaids.

If (16b) is taken to be the more correct version for the Square, the VS-model of the P-Square is fig. 11-b, which assigns radical falsity (F2), rather than minimal falsity (F1), to *Some mermaids have a bank account*, which means that this sentence suffers from presupposition failure and calls for a context repair. Since, as has been said, $\simeq P \equiv \simeq{\sim}P$ for any sentence P, *No mermaid has a bank account* is then also radically, rather than minimally, false, just like the corresponding **~I*** type *There isn't a mermaid that hasn't a bank account* and the corresponding **A**-type *All mermaids have a bank account.* The unnatural element in this series of truth-value assignments is the assignment of radical falsity to **~I** (as well as to **I**) since a sentence like *No mermaid has a bank account* is naturally considered true in the absence of any mermaids. The equivalence of **A** and **~I*** is saved, but that advantage is bought at the cost of a great deal of unnaturalness elsewhere in the system.

If, however, one takes (16a) to be more correct, the VS-model is like fig. 11–a. Now **~I** (*No mermaid has a bank account*) counts as true, which is all right. But so does **~I*** (*There isn't a mermaid that hasn't a bank account*), which is not all right because we want this to be equivalent with *All mermaids have a bank account*, which, according to fig. 11–a, is radically false. I feel priority should be given to (16a) and thus to the VS-model in fig. 11–a, mainly because the desired equivalence holds within the confines of the restricted universe of situations $\mathbf{U}^{\mathbf{R}}$ in which all presuppositions are satisfied, which is what counts for natural logic. The situations outside U^R only count when the logic is tested for soundness.[15] This allows us to opt for the definition of the existential quantifier given in (16a) and hence for the VS-model in fig. 11–a, and still maintain the default equivalence of **A** and ~**I***.

The spaces 1, 2 and 3 have been subsumed under the name $\mathbf{U}^R$ because they jointly form the default class of situations where all presuppositions are fulfilled. $\mathbf{U}^R$ is the *default presuppositional subuniverse* of **U**. Space 4 in fig. 11–b contains only radically negated positive or negative sentences. In fig. 11–a, by contrast, the radical negations in space 4 are restricted to **A**- and **A***-sentences and their negations whereas **I**- and **I***-sentences are minimally negated, as a result of the fact that (16a) specifies no preconditions for ∃ so that **I**- and **I***-sentences cannot be radically false. In both cases, $\mathbf{U}^R$ comprises precisely those situations where all presuppositions are satisfied. The Square holds as long as the logic is restricted to $\mathbf{U}^R$. If this analysis is correct, we have saved the Square from logical disaster: UEI no longer spoils the system. And it does not correspond too badly with natural intuitions. We can thus eat our cake and have it.

One notes that the qualification "$[[F]] = \varnothing$" is no longer there in space 4 of fig. 11–a,b. This is because the boundary between **U** and $\mathbf{U}^R$ is now no longer defined by the condition that $[[F]] = \varnothing$ but by the condition of absence of presupposition failure in general, which affects almost all sentences in any language. The presuppositional variety of predicate logic is thus not only relevant for just the logical constants but has a direct and all-pervasive bearing on the logical analysis of language as a whole, since presuppositional preconditions adhere to almost every predicate in the lexicon of every language. Logic has thus become an integral part of the semantic description of languages, not as a modelling tool, as in model-theoretic formal semantics, but as part of the language machinery itself.

Let us now see what happens when a presuppositional component is added to the basic-natural predecessor of the Square, BNPC, resulting in presuppositional basic-natural predicate calculus or P-BNPC. As before,

15 It is interesting to note that ~**I***-type sentences allow for positive polarity items like *rather* to occur, as if the sentence were not negative but positive. This observation was made by the late Carl Leroy Baker in an old paper of 1970, where he gives the example:
(i) There is nobody here who wouldn't rather be in Montpelier.
This sentence carries no echo-effect and contains no radical negation, despite the occurrence of *rather*.

Figure 11: Two Candidates for the VS-Modelling of the P-Square

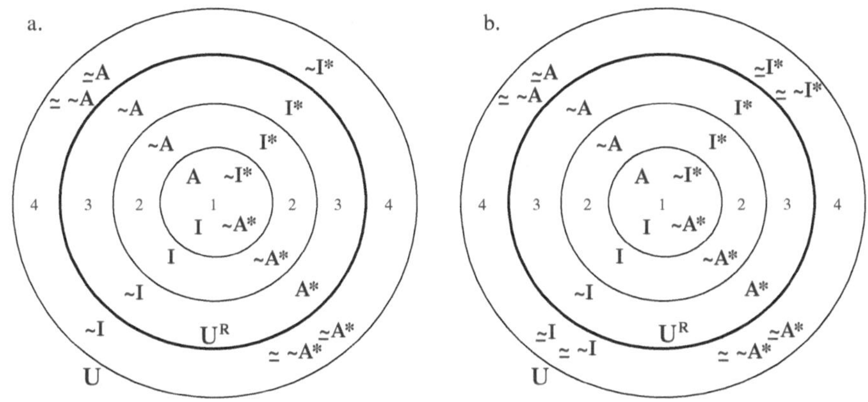

when the elements in [[F]] fail to satisfy the preconditions of the G-predicate, radical falsity is assigned to **A** and ~**A** but truth to ~**I** and ~**I***, and also to **N** and **N***. The satisfaction conditions for the quantifiers *all*, *some* and *no* in P-BNPC are thus as follows ([[F]] and [[G]] are distinct natural sets):

[[∀]] = {<[[G]],[[F]]> : the elements in [[F]] satisfy the preconditions of *G* | [[F]] ⊂ [[G]]}
(the unversal quantifier ∀ denotes the set of all pairs of predicate extensions [[G]] and [[F]] such that (precondition) all elements of [[F]] satisfy the preconditions of the predicate *G*, and (update condition) [[F]] is properly included in [[G]])

[[∃]] = {<[[G]],[[F]]> | [[F]] ⚭ [[G]] or [[G]] ⊂ [[F]]}
(the existential quantifier ∃ denotes the set of all pairs of predicate extensions [[G]] and [[F]] such that [[G]] and [[F]] are in mutual partial intersection or [[G]] is properly included in [[F]])

[[N]] = {<[[G]],[[F]]> | [[F]] ○○ [[G]]}
(the negative existential quantifier N denotes the set of all pairs of predicate extensions [[G]] and [[F]] such that [[G]] and [[F]] are in total mutual exclusion)

Figure 12–a shows the VS-model for P-BNPC, while fig. 12–b is repeated from fig. 9–b to enable the reader to compare the two. As with bivalent BNPC, the spaces 1 to 3 are defined for the following conditions: space 1 for [[F]] ⊂ [[G]], space 2 for [[F]] ⚭ [[G]] or [[G]] ⊂ [[F]], space 3 for [[F]] ○○ [[G]]. Space 4 in fig. 12–a is reserved for cases where the preconditions of the *G*-predicate are not satisfied. Space 4 in fig. 12–b is, of course, reserved for cases where [[F]] = ∅.

Figure 12: VS-Model for P-BNPC, Compared with BNPC Extended to a Full Logic

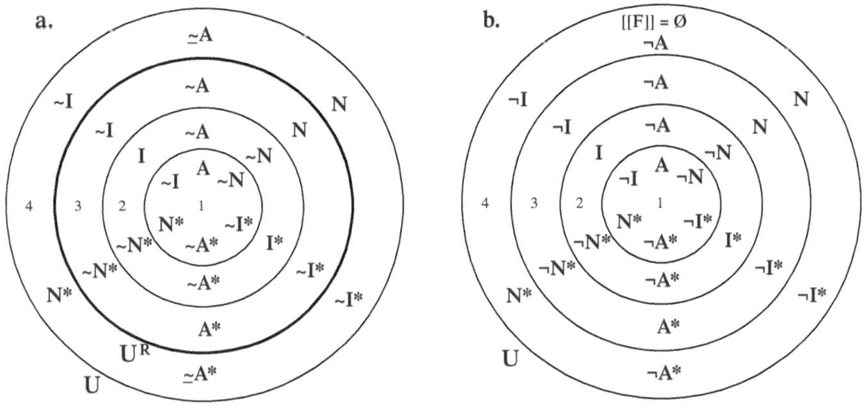

We can now revert to the problem discussed earlier of the intuitive equivalence of ~**A** with both **I** and **I*** (intuition 7). If it is assumed that, at least prototypically, *Not all F is G* is interpreted as the topic-comment (cleft) structure "the F that is G is not all F", then *Not all F is G* presupposes and thus entails *Some F is G* and also *Some F is not-G*, as the two are equivalent in the system at hand. This establishes the entailment from ~**A** to both **I** and **I***. Conversely, both **I** and **I*** entail ~**A** because that is how they are defined in the system of fig. 12–a: /**I**/ = /**I***/ = {2} while /~**A**/ = {2,3}, which means that **I**/**I*** ⊢ ~**A**. We thus have entailment both ways between ~**A** and **I** or **I***, and hence equivalence.

For the P-Square this result is not attainable. As one can read from fig. 11–a, the conjunction of ~**A** with its presupposition **I** gives the VS {2}, whereas /~**A**/ = {2,3}, which makes for a one-way entailment only.

Of course, ~**A** and **I*** are equivalent because the Conversions hold in the P-Square (with minimal negation). Nor is this result attainable for any presuppositional form of AAPC, as is easily checked.

The conclusion is that, provided one lets ~**A** presuppose **I** on account of topic-comment structure, we may add intuition 7 (*Not-all F is G ≡ Some F is G ≡ Some F is not-G*) to the score of BNPC as specified in section 7, which makes P-BNPC the clear overall winner of the systems considered. The only intuition still missing is intuition 3 (*Some F is G ≡ Some G is F* or **I** ≡ **I!**), but, as has been said, here we have no remedy, at least not for BNPC. This intuition falls in with the Square (and the P-Square) and cannot be considered to be part of the set of basic logical intuitions. Perhaps it is possible to test whether this particular intuition is weaker than the others, which would then presumably be due to the fact that cases where *Some F is G* is true while [[G]] ⊂ [[F]] are relatively rare and hence untypical.

9. An Intensional Natural Ontology Required

A final word is must be said regarding the ontology needed for presuppositional systems of predicate logic such as P-BNPC or the P-Square.

All nonpresuppositional predicate logic systems considered so far are defined as quantifying only over actually existing entities, leaving out of account nonexisting, virtual or intensional entities, which have no actual existence but are the creations of human imagination one way or another. In other words, modern logic is, on the whole, STRICTLY EXTENSIONAL. The American logician Richard Montague recognised the need for an intensional logic quantifying also over virtual entities (his 'entities in other possible worlds') and he accordingly devised a programme of 'extensionalization of intensions' in terms of an infinite number of 'possible worlds', each of a strictly extensional nature. This programme, however, foundered on the problem of substitutivity of expressions denoting impossible objects or impossible worlds (see Dowty et al. 1981: 175). In Montague's system, talking about a square round cupola is the same as talking about a man who is dead and alive at the same time. Likewise, the possible world account makes it mandatory to accept that sentences like (17a) and (17b) are equivalent, since the object term of the verb *believe* refers to the same set of possible worlds, namely the null set:

(17) a. Andrew believes that the square root of 36 is 7.
b. Andrew believes that all Parisians are both dead and alive.

It is obvious, however, that Andrew may well believe the one but not the other. It is now widely recognised that this is intuitively, and hence empirically, inadequate for the purposes of natural-language semantics. For that reason, it is now generally acknowledged that this makes the Montague account untenable.

Yet the fact remains that humans refer to and quantify over virtual entities as easily and as naturally as they do over actually existing ones, and virtual entities may produce truth and falsity as naturally as actually existing entities do. This is often forgotten. Yet we speak about all Greek gods or some Greek gods as naturally as we speak about all Greek statues or some Greek statues. The fact that Greek gods never existed whereas many Greek statues have survived till today, and that we know that, doesn't seem to bother the logical, semantic and interpretative machinery of language users in the least, though, of course, it makes a difference as regards the assignment of truth values. Thus, sentences like (18a,b) are unambiguously true despite the fact that neither Apollo nor any other Greek god has ever enjoyed actual existence:

(18) a. The god Apollo was worshipped in the island of Delos.
b. Some Greek god was worshipped in the island of Delos.

The fact that virtual entities may produce truth depends on the kind of predicate the terms denoting them take. As we have seen, the predicate *worship* is extensional with regard to its subject term but intensional with regard to its object term. This means that in order to worship something or somebody one needs actual existence but in order to be worshipped one does not, since virtual being suffices.

There are many such predicates, such as *love*, *think of*, *talk about*, *believe in*, *imaginary*, *afraid of*, etc. etc. They simply lack the otherwise default precondition of actual existence for the term with respect to which they are intensional. Philosophers, especially of the twentieth-century Anglo-Saxon variety, have argued strongly that virtual entities deserve no place in a sound ontology. But even if one takes this to be correct, they do deserve one in any sound theory of NATURAL ONTOLOGY, the ontology that humans operate with, or else there will be a horrific clash with intuitions and with widespread plain social facts. When I talk about Apollo I may be talking about something that does not actually exist, but I do not talk about nothing, as so many philosophers have it. And when I talk about Apollo, I do not talk about the monster of Loch Ness or any other virtual entity likewise lacking actual existence. Any sound theory of natural ontology must, therefore, allow for virtual entities.[16]

It is thus entirely wrong to attribute existential presuppositions to the occurrence of a definite determiner, as the vast majority of presupposition theorists have claimed. In principle (but there are exceptions, to do with predicates like *exist* or *imaginary*, where the actual existence or nonexistence of the subject-term referent is part of the update conditions, not of the preconditions), existential import is due to the extensional character, and hence to the preconditions, of predicates. The existential quantifier is thus deprived of its property of existential import. All it has left is an entailment of being, not of actual existence.

Apart from these general observations and considerations, however, we now see that the presupposition theory outlined above provides a further, specific, argument for a natural intensional ontology, for the simple reason that no truth value can come about unless the property expressed by the predicate is applied to an entity of some sort. Thus, (18a) cannot be true unless the property of being worshipped in the island of Delos is mentally assigned to something, in this case the god Apollo, which only has virtual being. Analogously, a sentence like *The present king of France is bald* can only have the truth value F2 if the property of being bald is mentally assigned to some object, in this case the nonexistent king of France. That (18a) is true whereas *The present king of France is bald* is radically false, though in both cases there is reference to a nonexistent, virtual object, results from the semantics of the predicates *worship* and *bald*, respectively: *be worshipped* is intensional but *bald* is extensional with regard to its subject term.

The tenability of the presupposition theory presented here thus depends not only on the soundness of the logic and the empirical adequacy of the

16 The necessity of the assumption of virtual entities for any theory of natural ontology is amply analysed and argued for in Chapter 2 of Seuren (2009).

analysis but also on the assumption that humans refer to and quantify over virtual entities besides really existing entities.

This analysis allows us to leave the definitions of the quantifiers given in (15) and (16a) unchanged. All that has to change is the definition of the notion of predicate extension, which now encompasses virtual entities as well for cases of predicates that are intensional with respect to one or more of their terms. The universal quantifier, as defined in (15) above, induces existential import with respect to a given term *t* only if the *G*-predicate is extensional with respect to *t*. When the *G*-predicate is intensional with respect to *t*, there is no existential import when *t* is universally quantified. Thus, *All mermaids are imaginary* carries no existential import with regard to mermaids because *imaginary* is intensional with regard to its subject term. Likewise, *Some mermaids are imaginary* carries no existential import, for the same reason. But *Some mermaids are not imaginary* does, because the complement of [[Imaginary]] consists of all those entities that have actual existence (or are reducible to such entities). It is thus clear that, provided predicates of existence and nonexistence are dealt with adequately, the P-Square no longer suffers from UEI, not because the universal quantifier is defined as requiring nonnullness for its *F*-term, but because it is defined as requiring presuppositional satisfaction by the members of the *F*-class of the preconditions of the *G*-predicate. In this sense, the analysis presented shows the relevance of logical analysis for a proper description of any natural language as a whole.

10. Conclusion

The least one can conclude from the above is that, between standard modern logic and pragmatics, the empirical question of the natural logic of language has been badly neglected over the past few decades. It is, therefore, legitimate to explore this field and see to what extent natural logic satisfies natural intuitions. This has the advantage of providing a 'digital' rather than an 'analogical' solution (in the words of my friend and colleague Dany Jaspers). It seems, moreover, that empirical adequacy cannot be achieved without the assumption of a presuppositionally defined discourse domain for the interpretation of utterances, which necessitates the assumption of two distinct kinds of falsity and hence of two distinct negation operators, default minimal falsity, corrected by the presupposition-preserving minimal negation, and metalinguistic radical falsity, corrected by the presupposition-cancelling radical negation.

References

Baker, C. L. (1970). Double negatives. *Linguistic Inquiry 1(2)*, 169–186.

Butterworth, B. (1999). *The mathematical brain*. London: Macmillan.

De Morgan, A. (1847). *Formal Logic: or, The Calculus of Inference, Necessary and Probable*. London: Taylor & Walton.

Dehaene, S. (1997). *The number sense*. New York: Oxford University Press.

Dehaene, S. (2005). Evolution of human cortical circuits for reading and arithmetic: the "neuronal recycling" hypothesis. In S. Dehaene, J.-R. Duhamel, M. D. Hauser, & G. Rizzolatti (Eds.), *From monkey brain to human brain. A Fyssen Foundation symposium* (pp. 133–157). Cambridge, Mass.: MIT Press.

Dehaene, S., Izard, V., Pica, P., & Spelke, E. (2006). Core knowledge of geometry in an Amazonian indigene group. *Science 311*, 381–384.

Dowty, D., Wall, R. E., & Peters, S.R. (1981). *Introduction to Montague Semantics*. Dordrecht: Reidel.

Dummett, M. (1973). *Frege. Philosophy of language*. London: Duckworth.

Frege, G. (1892). Ueber Sinn und Bedeutung. *Zeitschrift für Philosophie und philosophische Kritik 100*, 25–50.

Ginsburg, H. P., & Allardice, B. S. (1984). Children's difficulties with school mathematics. In B. Rogoff & J. Lave (Eds.), *Everyday cognition: its development in social context* (pp. 194–219). Cambridge, Mass.: Harvard University Press.

Grice, H. P. (1975). Logic and conversation. In P. Cole & J. L. Morgan (Eds.), *Speech Acts*. (= *Syntax and Semantics 3*, 41–58). New York-San Francisco-London: Academic Press.

Hamilton, W. (1866). *Lectures on Metaphysics and Logic*. Vol. IV (= *Lectures on Logic*. Vol. II, 2nd ed., revised). Eds. H. L. Mansel and John Veitch. Edinburgh and London: Blackwood & Sons.

Horn, L. R. (1972). *On the Semantic Properties of Logical Operators in English.* Distributed by Indiana University Linguistics Club 1976.
Horn, L. R. (1989). *A Natural History of Negation.* Chicago: The University of Chicago Press.
Jaspers, D. (2005). *Operators in the Lexicon. On the Negative Logic of Natural Language.* PhD thesis, Leiden University.
Jespersen, O. (1917). *Negation in English and Other Languages.* Det Kgl. Danske Videnskabernes Selskab, Historisk-filologiske Meddelelser I,5. Copenhagen: Andr. Fred. Høst & søn.
Kneale, W., & Kneale, M. (1962). *The Development of Logic.* Oxford: Clarendon Press.
Mostovski, A. (1957). On a generalization of quantifiers. *Fundamenta Mathematica 44*, 12–36.
Pica, P., Lemer, C., Izard, V., & Dehaene, S. (2004). Exact and approximate arithmetic in an Amazonian indigene group. *Science 306*, 499–503.
Quine, W. V. O. (1953). *From a Logical Point of View.* Cambridge, Mass.: Harvard University Press.
Seuren, P. A. M. (1972). Taaluniversalia in de transformationele grammatika. *Leuvense Bijdragen 61*(4), 311–370. (Partially republished in English in Seuren 2001: 293–311.)
Seuren, P. A. M. (1985). *Discourse Semantics.* Oxford: Blackwell.
Seuren, P. A. M. (1988). Presupposition and negation. *Journal of Semantics 6* (3/4), 175–226.
Seuren, P. A. M. (1991). Präsuppositionen. In A. von Stechow & D. Wunderlich (Eds.), *Semantik. Ein internationales Handbuch der zeitgenössischen Forschung / Semantics. An International Handbook of Contemporary Research* (pp. 286–318). Berlin-New York: Walter de Gruyter.
Seuren, P. A. M. (1998). *Western Linguistics. An Historical Introduction.* Oxford: Blackwell.
Seuren, P. A. M. (2001). *A View of Language.* Oxford: Oxford University Press.
Seuren, P. A. M. (2002). The logic of thinking. *Koninklijke Nederlandse Akademie van Wetenschappen, Mededelingen van de Afdeling Letterkunde, Nieuwe Reeks 65* (9), 5–35.
Seuren, P. A. M. (2006). The natural logic of language and cognition. *Pragmatics 16* (1), 103–138.
Seuren, P. A. M. (2009). *Language in Cognition.*(= Language From Within vol. 1). Oxford: Oxford University Press.

Seuren, P. A. M. (2010). *The Logic of Language*. (= Language From Within vol. 2). Oxford: Oxford University Press..
Strawson, P. F. (1950). 'On referring'. *Mind 59*, 320–344.
Strawson, P. F. (1952). *Introduction to Logical Theory*. London: Methuen.
Strawson, P. F. (1954). A reply to Mr Sellars. *The Philosophical Review 63* (2), 216–231.
Strawson, P. F. (1964). Identifying reference and truth-values. *Theoria 30* (2), 96–118.
Van Fraassen, B. (1971). *Formal Semantics and Logic*. New York-London: Macmillan.
Wilson, D. (1975). *Presuppositions and Non-Truth-Conditional Semantics*. London-New York-San Francisco: Academic Press.

Questions and Answers about Oppositions

FABIEN SCHANG

Abstract

A general characterization of logical opposition is given in the present paper, where oppositions are defined by specific answers in an algebraic question-answer game. It is shown that opposition is essentially a semantic relation of truth values between syntactic opposites, before generalizing the theory of opposition from the initial Apuleian square to a variety of alternative geometrical representations.

In the light of this generalization, the famous problem of existential import is traced back to an ambiguous interpretation of assertoric sentences in Aristotle's traditional logic. Following Abelard's distinction between two alternative readings of the **O**-vertex: *Non omnis* and *Quidam non*, a logical difference is made between negation and denial by means of a more fine-grained modal analysis.

A consistent treatment of assertoric oppositions is thus made possible by an underlying abstract theory of logical opposition, where the central concept is negation. A parallel is finally drawn between opposition and consequence, laying the ground for future works on an abstract operator of opposition that would characterize logical negation just as does Tarski's operator of consequence for logical truth.

Key words: affirmation, denial, existential import, negation, **NOT**, opposition, **QAS**

1. An Abstract Theory of Opposition

1.1. A Question-Answer Game for Oppositions

To define logical oppositions, we developed a question-answer game in which the logical value of a formula is not a truth-value but an answer to the corresponding question. We refer to this conceptual framework hereafter as Question-Answer Semantics (**QAS**).

1.1.1 What is a Logical Opposition?

While the concept of opposition occurs in various places and senses within Aristotle's works[1], the theory of logical opposition is the resulting group of four logical relations: contradiction, contrariety, subcontrariety, and subalternation. Let us symbolize by $\mathrm{Op}(\alpha, \beta)$ the 2-place relation between the two relata α and β, where Op is to be read as "– is opposed to …". If these are strictly concerned with the combination of truth and falsity, then only *propositions* can serve as values of α and β. Let us symbolize by T and F the logical values of truth and falsity, and let v be the interpretation function that assigns a logical value to propositions. Then $v(\alpha) = \mathrm{T} / \mathrm{F}$ means that the proposition α is true/false.

Recalling that the Aristotelian oppositions obey the Principle of Bivalence (**PBV**), namely: every proposition is either true or false but not both, every opposition $\mathrm{Op}(\alpha, \beta)$ could be seen as a 2-ary relation belonging to the set of ordered couples $\{\langle \mathrm{T}, \mathrm{T}\rangle, \langle \mathrm{T}, \mathrm{F}\rangle, \langle \mathrm{F}, \mathrm{T}\rangle, \langle \mathrm{F}, \mathrm{F}\rangle\}$. In other words, oppositions could be rendered by the perfect disjunctive normal forms (DNF) characterizing the relation $\mathrm{R}(\alpha, \beta)$: $(\alpha \wedge \beta) \vee (\alpha \wedge {\sim}\beta) \vee ({\sim}\alpha \wedge \beta) \vee ({\sim}\alpha \wedge {\sim}\beta)$, where every affirmative proposition (α or β) is read as T and every negative proposition (${\sim}\alpha$ or ${\sim}\beta$) is read as F.

1 See Aristotle (b), Δ10; and Aristotle (a), Chapter 10 for a general examination of the oppositions: by contrariety, by relation, by privation (and possession) and by affirmation (and negation).
While the "metaphysical" sense of opposition is stated as a set of relations between *concepts*, the logical sense only concerns relations between *propositions* and the subcase of opposition by affirmation and negation. As for the metaphysical oppositions of contrariety and contradiction, they are included into logical oppositions by reformulating their conceptual relata into propositions.

Definition 1. An *opposition* is a set of combined truth-values involving both truth and falsehood. Each opposition Op(α, β) is characterized by its specific DNF.

Definition 1.1. α and β are said to be *contradictory* to each other (Op: CD) if and only if (hereafter: iff) they cannot be T or F together.
$CD(\alpha, \beta) = \{\langle T, F\rangle, \langle F, T\rangle\}$
DNF: $(\alpha \wedge {\sim}\beta) \vee ({\sim}\alpha \wedge \beta)$

Definition 1.2. α and β are said to be *contrary* to each other (Op: CT) iff they cannot be T together but can be F together.
$CT(\alpha, \beta) = \{\langle T, F\rangle, \langle F, T\rangle, \langle F, F\rangle\}$
DNF: $(\alpha \wedge {\sim}\beta) \vee ({\sim}\alpha \wedge \beta) \vee ({\sim}\alpha \wedge {\sim}\beta)$

Subcontraries are also presented as "contradictories of contraries", that is: each component of the DNF for subcontraries is the negation of a corresponding component for contraries.

Definition 1.3. α and β are said to be *subcontrary* to each other (Op: SCT) iff they cannot be F together but can be T together.
$SCT(\alpha, \beta) = \{\langle T, T\rangle, \langle T, F\rangle, \langle F, T\rangle\}$
DNF: $(\alpha \wedge \beta) \vee (\alpha \wedge {\sim}\beta) \vee ({\sim}\alpha \wedge \beta)$

The last opposition of subalternation is more difficult to define for two main reasons: it is not a *symmetric* relation; the proper DNF of subalternation cannot be so easily obtained from one of the three preceding oppositions.

Definition 1.4. α is said to be *subalternate* to β (Op: SB) iff β cannot be F whenever α is T.
$SB(\alpha, \beta) = \{\langle T, T\rangle, \langle F, T\rangle, \langle F, F\rangle\}$
DNF: $(\alpha \wedge \beta) \vee ({\sim}\alpha \wedge \beta) \vee ({\sim}\alpha \wedge {\sim}\beta)$

Such definitions may strike as rather artificial, or at least incomplete: why not exhaust the set of possible DNFs? And why isn't SB a symmetric relation like the other oppositions? It can be observed that a complete list of possible DNFs does not consist in a set of 4 oppositions but, rather, in a set of 16 logical connectives for classical logic. A reformulation of oppositions in terms of a question-answer game enhances their understanding.

Definition 2. An *opposition* is a set of answers to questions about the compossible truth-values of α and β.

To the **Q** = 2 following questions $\mathbf{Q}_1$: "Possibly $v(\alpha) = v(\beta) = F$?" and $\mathbf{Q}_2$: "Possibly $v(\alpha) = v(\beta) = T$?", oppositions reply with an ordered set of two yes-no answers. Let **P** be symbolized by 1 for affirmative answers: "yes", and by 0 for negative answers: "no". Only one answer is to be given to each question, so that oppositions result in an ordered 2-tuple of answers: $\mathrm{Op}(\alpha,\beta) = \langle \mathbf{P}_1, \mathbf{P}_2 \rangle$.

The Aristotelian theory of quantified oppositions (see Aristotle (a), Ch.7) occasioned the famous Apuleian square **AEIO**, with a set of four sentential schemata $\{\alpha, \beta, \gamma, \delta\}$ and a set of four oppositions Op = {SB, CT, SCT, CD}. The two horizontal lines stand for CT (the line **AE**) and SCT (the line **IO**); the two diagonals correspond to CD (the lines **AO** and **EI**); as for the two vertical lines, the arrows indicate the asymmetric relation of SB from α to β (the lines **AI** and **EO**)[2].

Figure 1: Oppositions between Quantified Propositions

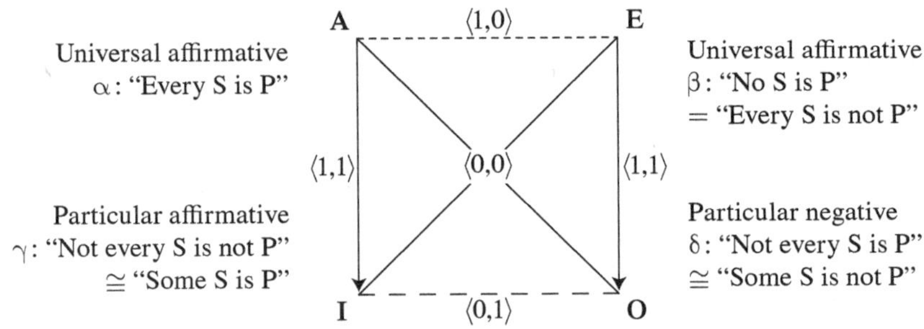

Definition 2.1. SB yields an affirmative $\mathbf{P}_1$ and an affirmative $\mathbf{P}_2$.
$(\alpha, \gamma) = (\beta, \delta) = \langle 1, 1 \rangle$

2 The vertices **A, E, I, O** are to be traced back to medieval times, with **AI** (**A**ff**I**rmo) for affirmative propositions and **EO** (n**E**g**O**) for negative propositions. It means that the capital letters have to do with the *quality* of the opposed formula, and not the *quantity*; consequently, they can and will be used hereafter for any applied oppositions beyond the single case of quantified oppositions.

Definition 2.2. CT yields an affirmative $\mathbf{P}_1$ and a negative $\mathbf{P}_2$.
$(\alpha, \beta) = \langle 1, 0 \rangle$

Definition 2.3. SCT yields a negative $\mathbf{P}_1$ and an affirmative $\mathbf{P}_2$.
$(\gamma, \delta) = \langle 0, 1 \rangle$

Definition 2.4. CD yields a negative $\mathbf{P}_1$ and a negative $\mathbf{P}_2$.
$(\alpha, \delta) = (\beta, \gamma) = \langle 0, 0 \rangle$

1.1.2 How Many Oppositions are There?

The reason why there are 4 oppositions in Aristotle's theory seems to be due to the cardinality of **Q** and **P**. Aristotle's theory includes only 2 truth-values T and F (by **PBV**). No further truth-value can be introduced to increase the number n of questions, and no question can be answered both affirmatively and negatively; hence there are $m = 2$ sorts of available answers to each question, so that the number of the Aristotelian oppositions is $m^n = 2^2 = 4$ in the light of **QAS**.

Now some objections could be raised against the meaning and the number of "genuine" oppositions. Against this algebraic characterization of **QAS**, Béziau (2003) argues that opposition is synonymous with "incompatibility": a propositional relation is an opposition only if its relata cannot be both true, in which case the appropriate question for characterizing Op is not Q_2 but only Q_1. And given that the answer to Q_1 must be affirmative accordingly, opposition is not defined any longer as a combinatorial game of questions-answers and should exclude both SCT and SB. Béziau however accepts SCT as a "verbal" opposition because of its derivation from CD and CT[3]; but he clearly excludes SB from his geometrical representations of opposition[4].

3 "Let us recall that Aristotle does not introduce explicitly the notion of "subcontraries", but refers to them only indirectly as "contradictories of contraries"; moreover he does not really consider them as opposed: '*Verbally* four kinds of opposition are possible, viz. universal affirmative to universal negative, universal affirmative to particular negative, particular affirmative to universal negative and particular negative to particular negative. But *really* there are only three: for the particular negative is only verbally opposed to the particular negative. Of the genuine opposites I call those which are universal contraries, e.g. "every science is good", "no science is good"; the others I call contradictories.' (Aristotle, *Prior Analytics*, 63b21–30)" (Béziau 2003; italics added.)

4 "I think that neither subalternation nor superalternation can be considered as relations of opposition. For example P is subaltern of $P \wedge Q$, and it does not really make sense to consider them as opposed." (Béziau 2003, 224)

Conversely, Sion (1996) increases the set of oppositions by including two further elements: implicance, and unconnectedness[5]. But he can do so only by modifying the content of the questions **Q**. Sion's view of oppositions proceeds from a set of $\mathbf{Q}=4$ alternative questions about whether the relata are possibly T or F, rather than compossibly. We thus have a new set of basic questions-answers (where $\mathbf{Q}=4$ and $\mathbf{P}=2$): $\mathbf{Q}_1$: "$v(\alpha)=T$, $v(\beta)=T$?", $\mathbf{Q}_2$: "$v(\alpha)=T$, $v(\beta)=F$?", $\mathbf{Q}_3$: "$v(\alpha)=F$, $v(\beta)=T$?", and $\mathbf{Q}_4$: "$v(\alpha)=F$, $v(\beta)=F$?", where to the additional oppositions of implicance and unconnectedness correspond the 4-tuples of answers $\langle 1,0,0,1\rangle$ and $\langle 1,1,1,1\rangle$.

But to do so is actually to characterize the meaning of a classical propositional connective. Should any such connective be considered as a proper opposition? It is well-known that each logical opposition corresponds to exactly one logical connective among the $m^n=2^4=16$ ordered answers $\langle \mathbf{P}_1,\mathbf{P}_2,\mathbf{P}_3,\mathbf{P}_4\rangle$. Thus CT amounts to incompatibility ($\uparrow(\alpha,\beta)=\langle 0,1,1,1\rangle$), CD to exclusive disjunction ($\underline{\vee}(\alpha,\beta)=\langle 0,1,1,0\rangle$), SCT to inclusive disjunction ($\vee(\alpha,\beta)=\langle 1,1,1,0\rangle$), and SB to conditional ($\rightarrow(\alpha,\beta)=\langle 1,0,1,1\rangle$). However, a geometrical representation of Sion's alleged oppositions does not result in the Apuleian square but a slightly different set of geometrical objects, namely: a logical *quatern*, depicted syntactically with a theory of quaternality by Gottschalk (1953) and semantically with a theory of reversibility (or INRC group) by Piaget (1949).

Sion's alleged oppositions of implicance and unconnectedness refer to the connectives of biconditional ($\leftrightarrow(\alpha,\beta)=\langle 1,0,0,1\rangle$) and tautology ($T(\alpha,\beta)=\langle 1,1,1,1\rangle$), respectively; but he stops the process there, instead of exhausting the combinatorial list of his answers. Thus there is a gap

We'll see later that Moretti (2009) presents a constructive argument against Béziau's present point: the geometry of oppositions cannot be expanded without using subalternation in its representations.

5 "By the 'opposition' of two propositions, is meant: the exact logical relation existing between them – whether the truth or falsehood of either affects, or not, the truth or falsehood of the other. In this context, note, the expression "opposition" is a technical term not necessarily connoting *conflict*. We commonly say of two statements that they are "opposite", in the sense of *incompatible*. But there, the meaning is *wider*; it refers to any mental confrontation, any logical face-off, between distinguishable propositions. In this sense, even forms which imply each other may be viewed as 'opposed' by virtue of their *contradistinction*, though to a much lesser degree than contradictories. Thus, the various relations of opposition make up a continuum." (Sion 1996, Ch.6, 1; italics added.)

Figure 2: The Logical Quatern

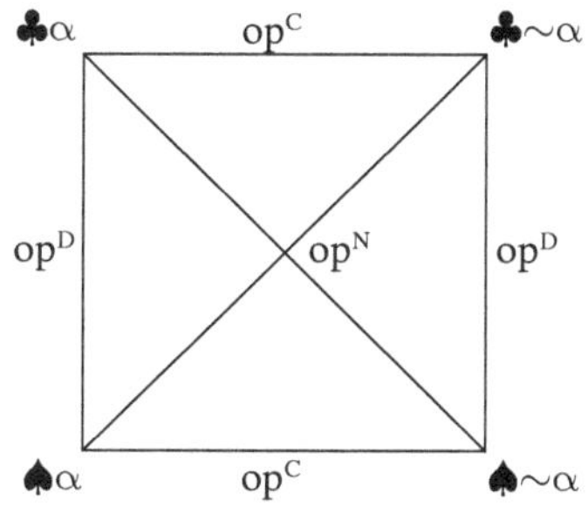

♣/♠: dual constants (♣α ⇔ ∼♠∼α)
α/∼α: affirmative sentences/negative sentences

op^C: contraduality
op^N: negation
op^D: duality

between the compossible truth-values of our own characterization of Op and the possible truth-values of Sion's characterization.

Given that Sion's view of opposition is not only unjustifiably incomplete but also gives rise to a geometrical object that is different from Aristotle's square, we argue that the meaning of Op eventually relies on a combination of compossible truth-values. Roughly speaking, our own characterization in **QAS** constitutes a trade-off between two views of opposition: *intensive* (according to its meaning by **Q**), and *extensive* opposition (according to its extensions by **P**). As a minimal precondition, **Q** must be about the compossibility of truth-values in order to follow the traditional theory of opposition. Such a precondition can make room for a generalization of the initial square by changing the parameters **Q** and **P**.

1.2 Generalization of Oppositions: Syntactic and Semantic Views

1.2.1 Syntactic Generalization

A syntactic generalization of the theory of opposition relies upon a structural extension of the relata in Op(Φ, Ψ), without any semantic reference to truth-values. As pointed out in the Figure 2, dual constants can be applied to propositions to give rise to a range of syntactic transformations.

Blanché (1953) noted this structural point, arguing that Aristotle's theory of opposition still prevails beyond the sole case of quantified propositions (see Figure 1) and includes modal oppositions. Furthermore, the formulas Φ and Ψ serve as structured formulas $\Phi = \clubsuit\alpha$ and $\Psi = \spadesuit\alpha$, where the modal functions $\{\clubsuit, \spadesuit\}$ can be variously interpreted: as null modality $\{\varnothing\}$, quantifiers $\{\forall / \exists\}$, alethic modalities of necessity and possibility $\{\Box / \Diamond\}$, deontic modalities of obligation and permission {op / P}; and so on[6]. An additional, purely syntactic definition of opposites op (where op $\neq$ Op) can thus be given as a set of construction rules for quaterns, where op^X means "operates as a X".

Definition 3. A logical *opposite* op is a 1-ary function on one formula Φ, where op is an operation from Φ to Ψ. To the two initial questions Q_1: "Is the dual constant interchanged?" and Q_2: "Is the sentence interchanged?", each sort of opposition op^X is a 1-ary operation that yields a 2-tuple of yes-no answers $\langle P_1, P_2 \rangle$.

Definition 3.1. Duality (X = D) yields an affirmative P_1 and a negative P_2.
$op^N = \langle 1, 1 \rangle$

Definition 3.2. Negation (X = N) yields an affirmative P_1 and an affirmative P_2.
$op^D = \langle 1, 0 \rangle$

Definition 3.3. Contraduality (X = C) yields a negative P_1 and an affirmative P_2.
$op^C = \langle 0, 1 \rangle$

Definition 3.4. Identity (X = I) yields a negative P_1 and a negative P_2.
$op^I = \langle 0, 0 \rangle$

6 Φ and Ψ stand for 1-ary functions in (Blanché 1953, von Wright 1951) (where $\clubsuit = \Box$ and $\spadesuit = \Diamond$); but they can also stand for 2-ary functions, as in (Piaget (1949), Gottschalk (1953)) (where $\clubsuit = \wedge$ and $\spadesuit = \vee$).
Unlike von Wright (1951), who strictly referred to 'modalities' as the modes in which a sentence is said to be true or false (necessarily, obligatorily, and so on) and distinguished them from the truth-operators (truth and falsity), we use "modalities" to refer only to a 1-ary sentential function throughout the present paper.

Since Ψ results from Φ by applying one of the four preceding operations to Φ to get op(Φ) = Ψ, any opposition relation Op(Φ, Ψ) amounts to the opposition of Φ and one of its opposites op(Φ).

Definition 4. For any opposition relation Op(Φ, Ψ) and any oppositional operation op^X on Φ (where X∈{N, D,C, I}), we have: Op(Φ, Ψ) = Op(Φ, op^X(Φ)).

A famous expansion of the Apuleian square is the hexagon of alethic modalities in (Blanché 1953, Sesmat 1951) where ♣ = □. The difference between logical squares and hexagons does not reside in their relation Op, which is the same for the two; but rather in their relata: they are related by the same sorts of opposition. Thus Blanché's hexagon results from the union of two triangles of contrariety and subcontrariety by their subaltern opposites.

Figure 3: The Hexagon of Modal (Alethic) Oppositions

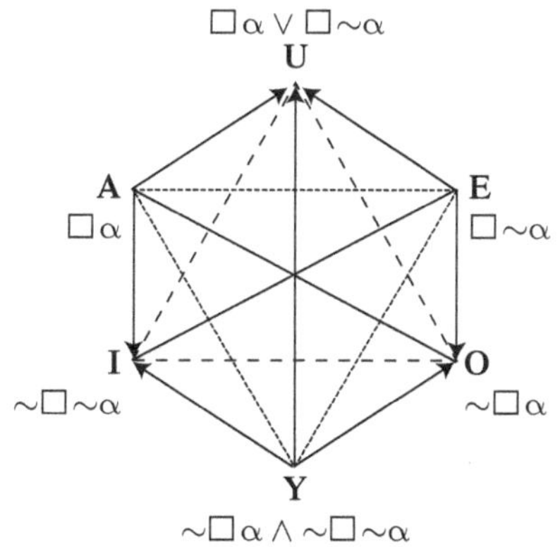

Y amounts to the non-basic modality of *contingency* (two-sided possibility), and its contradictory **U** is introduced in order to respect the central symmetry for any paired opposites in the geometrical representation. The above transition from 4 to 6 vertices has been generalized by Moretti (2009) with his theory of *n*-opposition (**NOT**): a logical square is a bi-segment resulting from the subalternate connection of 2 segments or "simplexes" (viz. contrariety and subcontrariety); a logical hexagon amounts to a bi-triangle resulting from the subalternate connection of 2 triangles; and so on, where each expanded logical polygon goes from one to an indefinite number of simplexes.

Aristotle's square is thus only one geometrical opposition within a series of increasing cases, and the syntactic structure of opposed terms accounts for the complexity of logical polygons in **NOT**.

1.2.2 A Syntactic Key Notion for Oppositions: Negation

A distinction has just been made between an opposition Op and an opposite op: opposition is a relation, whereas opposites are relata. The previous figures have shown that Aristotle's square was a geometrical representation of opposition to be characterized by *semantic* questions about truth-values, while Gottschalk's and Piaget's quatern was a geometrical representation of opposites encoded by *syntactic* questions about structured formulas.

An intermediary position between semantic oppositions and syntactic opposites would be a syntax of oppositions, all the more that a central topic for **QAS** is the syntactic concept of *negation*. In contradistinction to Moretti's semantic approach, opposition has been thought in Béziau (2006) as a theory about classical and non-classical negations in the controversy against Slater's objection to paraconsistent negation (see Slater (1995)). For this very reason, oppositions can be viewed as a relation between any formula and one of its negated counterparts.

Definition 5. A logical *opposite* is a set of answers to questions about negating the components of an initial formula $\clubsuit\alpha$.
To the $\mathbf{Q}=2$ following questions $\mathbf{Q}_1$: "Is $\clubsuit$ negated?" and $\mathbf{Q}_2$: "Is α negated?", oppositions reply by an ordered set of $\mathbf{P}=2$ yes-no answers (where the 2-tuple $\langle 0,0\rangle$ is the trivial case of self-identity).

Definition 5.1. A subaltern opposite (op^D) yields an affirmative $\mathbf{P}_1$ and an affirmative $\mathbf{P}_2$.
$op^D(\clubsuit\alpha)=(\sim\clubsuit\sim\alpha)=\langle 1,1\rangle$

Definition 5.2. A contradictory opposite (op^N) yields an affirmative $\mathbf{P}_1$ and a negative $\mathbf{P}_2$.
$op^N(\clubsuit\alpha)=(\sim\clubsuit\alpha)=\langle 1,0\rangle$

Definition 5.3. A (sub-)contrary term (op^C) yields a negative $\mathbf{P}_1$ and an affirmative $\mathbf{P}_2$.
$op^C(\clubsuit\alpha) = (\clubsuit{\sim}\alpha) = \langle 0, 1 \rangle$

Figure 4: The Logical Square

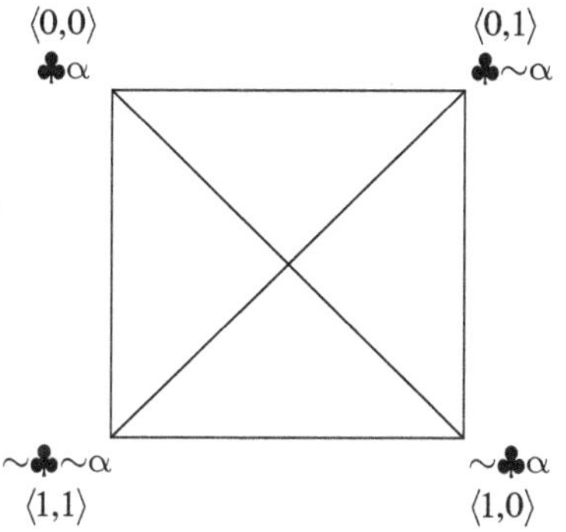

The advantage of such questioning is that it characterizes quaterns without any reference to duals, and defines $\spadesuit\alpha$ syntactically as ${\sim}\clubsuit{\sim}\alpha$. It also explains why subalternation is seen as a contradictory of contraries, given that the subaltern ${\sim}\clubsuit{\sim}\alpha$ results from the successive application of op^C and op^N to the initial formula. Finally, it provides a relevant explanation for why subalternation is not a proper opposition: if any opposition includes an opposite-forming operator that proceeds by negating one or several components in a formula, then subalternation amounts to a double negation and, thus, yields an affirmation from a classical perspective.

The disadvantage is, again, that a *sub*contrary opposite cannot be gathered from any unnegated formula: the basic $\clubsuit\alpha$ has no subcontrary opposite, but only a contrary one $\clubsuit{\sim}\alpha$. Likewise, subcontrariety cannot be gathered by reading its expression 'contradictory *of* contraries' as a function of function: to say that SCT is the relation $Op(\clubsuit\alpha,(op^N(op^C(\clubsuit\alpha))$ is not a correct translation, because if $\clubsuit\alpha = \langle 0, 0 \rangle$ then $op^N(op^C(\clubsuit\alpha)) = op^N(\clubsuit{\sim}\alpha)$ $= ({\sim}\clubsuit{\sim}\alpha) = op^D(\clubsuit\alpha)$. Hence $Op(\clubsuit\alpha, op^D(\clubsuit\alpha) = SB$, and not SCT. Moreover, this syntactic characterization cannot be successfully applied to structurally more complex formulas Φ such as non-basic modal sentences: the **Y**- and **U**-vertices of a logical hexagon cannot be specified by means of

Q_1 and Q_2 above, which means that a semantic characterization eventually prevails for a complete account of oppositions.

1.2.3 Semantic Generalization

Another way to generalize the geometry of oppositions is to turn what Moretti (2009) calls "logical bi-simplexes" into logical p-simplexes, including $p > 2$: a syntactic generalization has to do with the structure of opposed formulas, whereas a semantic generalization concerns the number of their compossible truth-values. It is clear that Aristotle's theory of opposition is entirely based on the **PBV**. But there is no reason not to introduce $p > 2$ truth-values in a **QAS** abstract characterization, thereby turning the classical bi-simplexes into non-classical p-simplexes. By doing so, **NOT** sheds a new light on the controversial meaning of negational principles such as the excluded middle (**PEM**) or the non-contradiction (**PNC**).

The device should sound appealing for any champion of many-valued logics, especially in the variety of philosophical logics where "standing propositions" are superseded by mere "variable sentences". Furthermore, Aristotle himself seemed to contest the universal applicability of **PBV** as regards the case of future contingents[7]. Despite this controversy about bivalence, no attempt was made to expand the theory of opposition to embrace $p > 2$ truth-values before Moretti (2009), who overcomes Slater's objection to paraconsistent negation in the light of a non-classical reading of **PNC**[8].

From our abstract perspective of **QAS**, a semantic generalization of oppositions entails a modification in the content of **Q** or **P**: either the questions are to be asked about the compossibility of alternative truth-values beyond truth and falsity, so that $n > 2$; or the answers may be extended beyond "yes" and "no", so that $m > 2$. Let us take Łukasiewicz's 3-valued logic of indeterminacy $Ł_3$ as an example, where $v(\alpha) = ½$ means that α is

7 See Aristotle (a), Chapter 9. The customary interpretation is that Aristotle renounced to the universal application of **PBV** while maintaining **PEM**. Some other commentaries blamed this view for confusing on *sentence* and a *proposition*, however. Such differences will be returned to in the next section about illocutionary modalities.

8 The question is whether **PNC** is about contradiction in particular, or negation in general. It is argued in Béziau (2006) that a negation needn't be a contradictory-forming operator: not every negation op^X verifies $Op(\Phi, op^X(\Phi)) = CD$. But since **PNC** is about CD, it is hardly relevant to say that $Op(\Phi, op^R(\Phi)) = SCT$ when the point is op^N and not op^C. Consequently, a further challenge would be to find a non-classical version of op^N for which $Op(\Phi, op^N(\Phi)) = CD$ is verified while departing from **PNC**.

indeterminate. Indeterminacy can be interpreted in two different ways in **QAS**. Either it is added as a further truth-value besides truth and falsity, in which case a new question $\mathbf{Q}_3$: "Possibly $v(\alpha)=v(\beta)=\frac{1}{2}$?" is introduced; or it is interpreted as an indefinite answer beyond affirmation ("yes") and denial ("no"), so that a new answer $\mathbf{P}_3$: "maybe" is introduced. The resulting set of questions-answers will amount to either $m^n=2^3=8$ oppositions, or $m^n=3^2=9$ oppositions.

These non-classical oppositions give rise to three problems: how does one account for the difference in the number of oppositions when $m=3$ or n $=3$, since the results in the two cases are supposed to have the same meaning? How should one interpret the non-classical oppositions? Isn't it possible to reduce the non-classical oppositions to classical ones? It is out of the scope of the presented paper to examine these questions.

2. Application of the Theory of Opposition

After describing the abstract properties of opposition, we now apply the logical analysis presented to settle several problems from philosophical logic. Among these are the traditional problem of *existential import*, and the intermediary case of epistemic oppositions.

It is argued in the sequel that epistemic modalities help to settle the problem of existential import by making a distinction between two senses of the underlying concept of *truth*, namely: an ontological sense of "being true", and an epistemological sense of "being held to be true". Opting for the latter allows the construction of the so-called "illocutionary modalities", a linguistic trade-off between epistemic modalities and truth-claims. The results of Moretti (2009) will be used as a way to represent our solution to existential import in an alternative geometry of oppositions.

2.1 The Core Problem of Opposition: Existential Import

It has been generally claimed that Aristotle's theory of opposition was flawed unless one implicitly assumed the non-emptiness of its component terms. The reason for this claim is that Aristotle did not address the specific problem of the alleged existential import for negative propositions **E-O**. Our answer to the initial question of knowing whether negative propositions are existentially committed is clearly no. But before developing our own diagnosis, let us first consider the issue.

Hypothesis 1. Universal affirmatives have existential import.
Let **A** be: "Every griffin is an animal", and let v be an interpretation function for every proposition in **AEIO**. Granted there are no griffins at all, it is naturally assumed that $v(\mathbf{A}) = \mathrm{F}$: nothing can be truly predicated about whatever does not exist, which implies that $v(\mathbf{A}) \neq \mathrm{T}$; now by **PBV**, $v(\mathbf{A}) \neq \mathrm{T}$ entails $v(\mathbf{A}) = \mathrm{F}$.

O is the contradictory of **A**, hence $v(\mathbf{O}) = \mathrm{T}$. **A** is a universal affirmative, hence **O** is a particular negative stating that there is at least one griffin that is not an animal. Hence there is at least one griffin and there are no griffins at all (by assumption), which is self-contradictory. Such a consequence is absurd, by **PNC** and so is its premise, which in turn must be refuted: $v(\mathbf{A}) \neq \mathrm{F}$, so $v(\mathbf{A}) = \mathrm{T}$.

The initial assumption that griffins don't exist is compatible with $v(\mathbf{A}) = \mathrm{T}$, provided that universal propositions don't presuppose the existence of their subject terms. Modern logic meets this proviso, by encoding a universal proposition in the form of a conditional whose antecedent needn't be true. If so, then **A** means: "If x is a griffin, then x is an animal". Given that the universal affirmative is the negation of the corresponding particular negative: "Every A is B" = "Not-(Some A is not B)", $v(\mathbf{A}) = \mathrm{T}$ iff there is no griffin which is not an animal. Now there is no griffin at all. Hence $v(\mathbf{A}) = \mathrm{T}$.

Hypothesis 2. Universal affirmatives don't have existential import.
That $v(\mathbf{A}) = \mathrm{T}$ entails that $v(\mathbf{E}) = \mathrm{F}$. If $v(\mathbf{E}) = \mathrm{F}$, then its contradictory is T, *i.e.* $v(\mathbf{I}) = \mathrm{T}$. **I** is a particular affirmative, thus meaning that there is at least one griffin that is an animal. Now the initial assumption claims that there are no griffins, whatever the interpretation of the universal may be (whether existential or not). Hence $v(\mathbf{I}) = \mathrm{F}$ and $v(\mathbf{I}) = \mathrm{T}$. This is self-contradictory, by **PNC**. Hence the initial assumption must be rejected again.

Conclusion.
Both hypotheses $v(\mathbf{A}) = \mathrm{T}$ and $v(\mathbf{A}) = \mathrm{F}$ result in an inconsistent theory of oppositions when applied to propositions with empty terms.

In the light of the foregoing *reductio ad absurdum*, two general reactions are conceivable.

In order to avoid inconsistency in the theory of opposition, one either assumes the non-emptiness of subject terms as an *ad hoc* hypothesis. This was Russell's stance on the semantic paradox, when he proposed his ramified type theory as a way out.

Or one sustains that propositions with empty terms (without referent) are as much meaningful as any other, thus opening the way for improving Aristotle's theory of opposition without begging the point. We will endorse this second attitude in the sequel.

There is a rich literature on existential import. Horn (1989, p. 24) lists four different positions on the subject as follows:

There are, as it happens, at least four distinct ways of answering such questions:

(i) Existential import is determined by the *quality* of the proposition; affirmative (**A** and **I**) propositions entail existence, while negative ones (**E** and **O**) do not.

(ii) Existential import is determined by the *quantity* of the proposition: universals (**A** and **E**) have no existential import, while particulars (**I** and **O**) do.

(iii) Existential import corresponds to a *presupposition* associated with **A**, **E**, **I** and **O** propositions.

(iv) The question of existential import is entirely absent from the Square of Opposition.

The present paper rejects (ii) and (iv) while endorsing (i) and (iii): **E** and **O** entail (don't entail) existence whenever the latter is presupposed (is not presupposed), and such a presupposition depends upon the speaker's intention. The problem is not of a syntactic order but has to do with the ambiguous interpretation of **O**, in the sense that not every contradictory of a universal affirmative amounts to a negative *particular*. The difficulty thus lies in the natural translation of the **O**-vertex, which is customarily rendered as a particular negative in the Apuleian square. Our approach squares with Abelard's position, as described by (Horn 1989, p. 26):

> For Aristotle, *Not every man is white* was indeed taken to be the canonical contradictory of *Every man is white* (*De Int.*, 24b6), but there is no suggestion that it is not considered to be equivalent to *Some man is not white*; for Apuleius and Boethius, these

> two forms were explicitly taken to be notational variants. Abelard's results, despite the consistency of his argumentation, were apparently too counterintuitive to be taken seriously; later medieval (and modern) logicians almost without exception rejected this distinction between *non omnis* and *quidam non*[9].

Such a distinction will be taken seriously and made intuitive in the sequel, on the basis of a twofold reading of Aristotle's assertoric propositions: an *ontological* reading, which concerns the truth and falsity of propositions; an *epistemological* reading, which concerns the attitude of speakers toward such propositions. The source of the problem seems to reside in a deep confusion between the "affirmation" and "negation" of *assertoric* oppositions; the suggested way out is a more fine-grained analysis of negation, especially the **O**-vertex of the initial logical square.

2.2 *Epistemic Modalities*

An epistemological reading of assertoric oppositions refers to epistemic modalities, where a propositional attitude is attached to a sentential content. The most familiar epistemic concepts are knowledge and belief, symbolized by two modal operators K and B. Now three main difficulties arise if we want to construct a polygon of epistemic oppositions; these concern the logical relations between knowledge and belief, knowledge and truth, and the assumption of mere consideration.

2.2.1 Knowledge and Belief

Firstly, knowledge and belief don't seem to be included in the same polygon of opposition. A tentative theory of epistemic oppositions was proposed by Engel (2007), with the square of Figure 5.

Although the affirmative subalternation **AI** matches with the "entailment thesis": $K\alpha \Rightarrow B\alpha$, it is easily seen that it leads to some counterintuitive results: **IE** states that whatever is ignored is not believed (and conversely), while **EO** claims that no ignored proposition can be believed. Furthermore, ignorance is currently viewed as the negation of knowledge ($\sim$K) and should thus be opposed to knowledge contradictorily rather than contrarily. As for

9 About Apuleius' conflation of *non omnis* and *quidam non*, see also Monteil (2003) and Parsons (2006). Expansion from the Apuleian square to a new logical polygon, as proposed in the sequel, is actually aimed at making Abelard's results more intuitive.

disbelief, the ambiguous meaning between B∼α and ∼Bα cannot be made sufficiently clear in a polygon of opposition with only 4 vertices. The crucial point is that K and B are equally strong modalities paralleling necessity ♣ = □, so that both of them should be located at one and the same vertex **A**. There are two alternative ways to settle the problem.

A first move is to turn Engel's logical square into a logical hexagon of (strictly) epistemic and doxastic modalities, where K and B occur in two separate polygons and are contrary to a common epistemic counterpart of contingency, namely: doubt (the **Y**-vertice). The following figure depicts similar oppositions for ♣ = K or B, while recalling the apparent impossibility to combine these into one and the same polygon (Figure 6).

Figure 5: A Square for Epistemic Oppositions

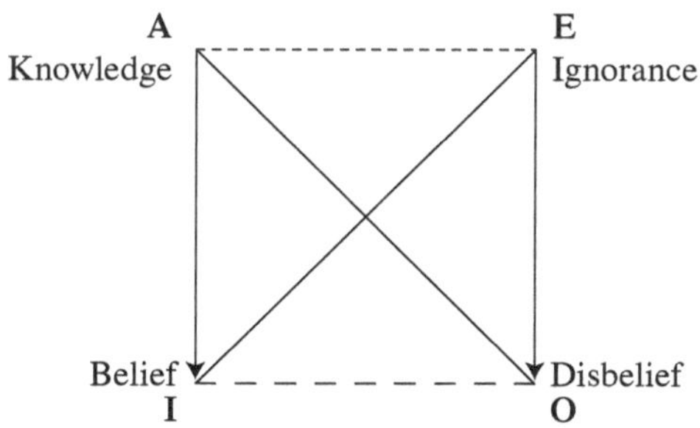

Figure 6: A Hexagon for Epistemic Oppositions

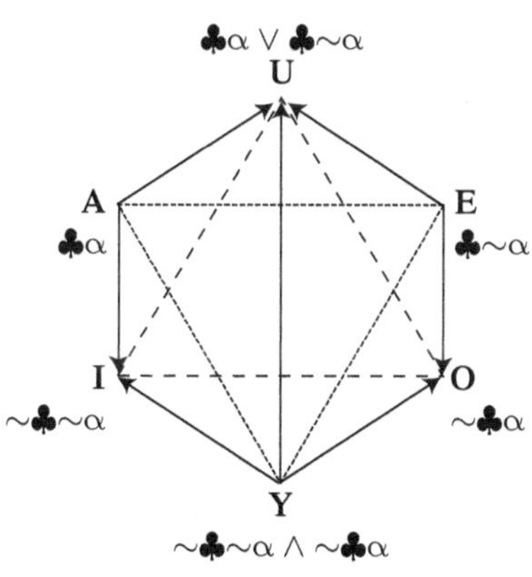

2.2.2 Knowledge and Truth

A second move is to merge knowledge and belief into one and the same modality, in order to produce a common modality of strong belief or subjective certainty.

Although some logics of knowledge as strong belief already occurred in the literature of epistemic logic, the mainstream view of epistemic modalities follows Plato's definition of knowledge as justified true belief (*Theaetetus*, 200d-e). The concept of truth makes the substantial difference between belief as a psychological attitude and knowledge as an objective state: such an "ontological" difference means that every proper knowledge is a belief that is made true by a state of affairs. In accordance with this philosophical tradition, the logical framework of Kripke's relational semantics gave rise to a series of modal systems with specific accessibility relations between "possible worlds"; the conspicuous domination of Kripke structures thus requires an examination of several modal systems for characterizing K and B.

Epistemic modalities are currently characterized as **KT4**(**5**) or **KD45** modalities, depending on whether the modality at hand is knowledge or belief. This means that the following axiom schemata variously hold for $\clubsuit = \mathrm{K}$ or B: (**K**): $(\clubsuit\alpha \wedge (\alpha \Rightarrow \beta)) \Rightarrow \clubsuit\beta$, (**T**) $\clubsuit\alpha \Rightarrow \alpha$, (**D**) $\clubsuit\alpha \Rightarrow {\sim}\clubsuit{\sim}\alpha$, (4) $\clubsuit\alpha \Rightarrow \clubsuit\clubsuit\alpha$, (5) ${\sim}\clubsuit\alpha \Rightarrow \clubsuit{\sim}\clubsuit\alpha$. The reduction laws of modalities for the iterative systems **S4** and **S5** help to introduce corresponding polygons of opposition, as illustrated by Moretti (2009).

But it can also be noted that such a characterization of epistemic modalities can be made intuitive by an alternative, illocutionary interpretation of epistemic modalities where "truth" is only synonymous with "truth-claim"[10]. So is the case with the *Moore's Paradox*, according to which it is absurd but not inconsistent to say "It rains, but I don't believe it" $(\alpha \wedge {\sim}\mathrm{B}\alpha)$. The prominent diagnosis argues that its *statement* departs from the sincerity condition for every uttered statement: any speaker is supposed to believe what it states. In other words, a correct formalization of the Moorean statement is not $(\alpha \wedge {\sim}\mathrm{B}\alpha)$ but its iterated version $\mathrm{B}(\alpha \wedge {\sim}\mathrm{B}\alpha)$. The distributivity of epistemic operators entails that the Moorean sentence amounts to a statement of the logical form $(\mathrm{B}\alpha \wedge \mathrm{B}{\sim}\mathrm{B}\alpha)$, thus violating the axiom schema (**4**) and making it explicitly inconsistent. More generally, Fitch (1963) has shown that $\clubsuit(\alpha \wedge {\sim}\clubsuit\alpha)$ is inconsistent (logically false) for any

10 See Schang (2007) for a detailed account of epistemic paradoxes and the motivation of their illocutionary treatment.

value concept ♣, whether it be instantiated by belief, knowledge, or even truth. The same inconsistency occurs in the subsequent *Fitch's Paradox of Knowability*, where any epistemic modality ♣ turns its sentential content α into the content of a statement ♣α with illocutionary clauses.

In short, an illocutionary reading of epistemic modalities consists in merging knowledge and belief into one and the same attitude of *truth-claim* and replacing the criterion of justification for K by its degree of force: knowledge is then synonymous with certainty, or strong truth-claim, in such a logic of statements where no realistically-minded difference is maintained any longer between the concepts of knowledge and truth.

Whether the reader favours a realist or anti-realist stance about knowledge, the point is that such a reduction of truth (of a sentential content) to truth-claim (of its corresponding statement) will be endorsed in the next section about existential import; this will be done after considering another difficulty with epistemic oppositions and their formulation.

2.2.3 Knowledge and Mere Consideration

The statement of doubt (in symbols: D) was characterized above as an epistemic counterpart of contingency. That is, whoever has doubts about α also disbelieves both α and $\sim\alpha$: $D\alpha =_{df} \sim B\alpha \wedge \sim B\sim\alpha$, where disbelief is rendered as the negation of belief.

But there is a difference between non-belief, as a mere absence of belief, and disbelief. As noted in Englebretsen (1969) and Hart (1980), not every sentential content α can be doubted by anyone. It is so, because disbelief assumes a self-conscious activity by a subject who considers, *i.e.* thinks about the sentential content. Thus Socrates couldn't have been said to doubt (or to have doubts about) Obama's victory during the US elections 2008, because he wouldn't have been able to think about such a referent. This results in a more complex picture of epistemic oppositions, where every such attitude towards a sentential content α requires its consideration as a minimal precondition.

In order to construct a modal logic of doubt, Hart (1980) claims in that the latter is to be defined as a believed non-belief, so that we obtain the compound characterization $D\alpha =_{df} B(\sim B\alpha \wedge \sim B\sim\alpha)$. A **S5**-modal logic of belief would help to include the non-basic modality of doubt by reducing $B\sim B$ to $\sim B$; but such a characterization would conflate whatever is not believed self-consciously or not. In a nutshell, non-beliefs are to be opposed to disbeliefs; the problem is thus to find the right vertex in the right polygon of epistemic oppositions supplemented with doubt.

For this purpose, Englebretsen (1969) takes the basic notion of consideration into account by adding a new scope of negation to the epistemic oppositions. It is to be noticed that Englebretsen proceeds by returning to the sources of Aristotle's tradition or "term logic", where negation could be applied either to the copula or to the predicate term.

Figure 7: The Four in Term Logic

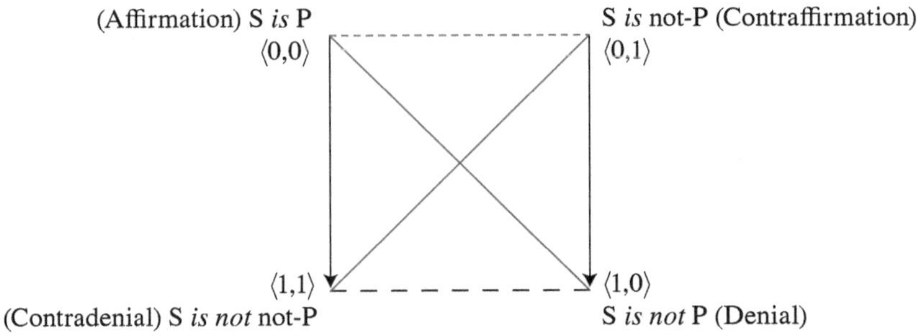

By analogy with the abstract structure of quaterns (see Figure 2), we fall here again on a compound structure ♣α where ♣ = S and α = P. A relevant difference should be made between a *predicate* negation "not" and a *predicate term* negation "not-": the former is attached to the whole predication "S is P" and proceeds like sentential negation; the latter uses a hyphen that restricts the scope of negation to the predicate term P. The first results in contraffirmation (♣∼α) and the second in mere denial (∼♣α). No wonder the opposites are as numerous in term logic (Figure 6) as in quantified propositions (Figure 1): their structured formulas result from the same question-answer game with $m^n = 2^2 = 4$, Q_1: "Is ♣ negated?", and Q_2: "Is α negated?".

In addition to the foregoing more fine-grained scope of negation, the introduction of the component of consideration turns the structure of epistemic modalities from Φ = ♣α into the more complex structure @(Φ) = @♣α, where @ is a consideration-operator that applies to the whole epistemic statement ♣α and means that α is considered. Accordingly, Englebretsen's conflation of term and epistemic logic results in a

Figure 8: Epistemic Multi-Modal Oppositions

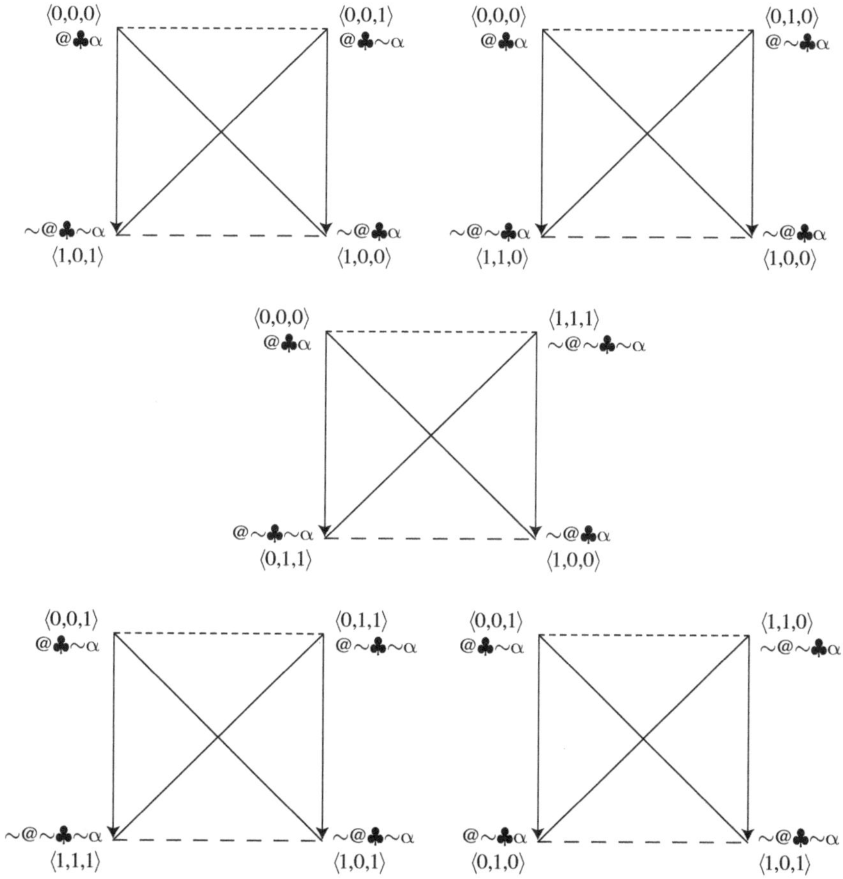

multi-modal logic with $n = \mathbf{P}^{\mathrm{Q}} = 2^3 = 8$ available opposites, given $\mathbf{Q}_1$ = "Is @ negated?", $\mathbf{Q}_2$ = "Is ♣ negated?", and $\mathbf{Q}_3$ = "Is α negated?". The 8 subsequent answers are described in Englebretsen (1969) by a group of 5 different squares, where ♣ = K but could also stand for B in our merged perspective of epistemic modalities.

The **NOT** framework should shed theoretical light on the construction of a complete polygon including all of these epistemic oppositions to replace Englebretsen's five separate squares. Pending such a construction for multi-modalities, the following semantics for assertoric oppositions suggests an alternative reading of epistemic mono-modalities where consideration is vacuously included.

2.3 Illocutionary Modalities

2.3.1 Assertoric Statements

The *illocutionary* import of epistemic attitudes stems from Austin's seminal distinction between a locutionary, illocutionary and perlocutionary feature of speech-acts (symbols: Fα): the locutionary aspect refers to the sentential content α and its semantic properties; the illocutionary aspect refers to the intended aim of its utterance by a speaker; the perlocutionary aspect refers to its expected effect upon the hearer.

A formal pragmatics has been proposed accordingly by Searle & Vanderveken (1985), where every speech-act Fα essentially includes an "illocutionary force" F with six main components among five classes of speech acts. Assertive acts are one case in point: a statement such as "Obama won the US elections 2008" is an assertion, where the speaker intends to tell the truth and succeeds in doing so only if its sentential content α describes how things really are.

Without going into further details on the framework of illocutionary logic, we need two of its main components in what follows, namely: the sincerity condition of an assertive act and its degree of strength. Indeed, the so-called "direction of fit" between a sentential content and the state of affairs expressed by it assumes a correspondence theory of truth that couldn't be endorsed from an antirealist perspective, so that assertive acts occur hereafter as truth-claims that don't really depart from belief. Moreover, the logical status of denial is not made clear by Searle & Vanderveken: it is not clearly argued whether denial belongs to the specific class of assertive acts or serves as a general negation for any illocutionary force. While such an *illocutionary negation* is compared with intuitionistic negation[11], it will be argued below that the latter is just a subcase of denial.

11 "Illocutionary negation is like intuitionistic negation from this perspective since, in intuitionistic logic, it is equally invalid that $\alpha \vee \sim\alpha$ and $\sim\sim\alpha \rightarrow \alpha$." (Searle & Vanderveken, p. 154; also quoted in Vernant (2003, footnote 22 p. 5).

2.3.2 An Assertoric Logic of Affirmation and Denial

In order to give a more fine-grained analysis of denial (illocutionary negation), the technical use of relational semantics by Vanderveken's formal pragmatics is replaced here by an algebraic semantics for affirmation and denial. Here is the characterization of assertive or assertoric statements $F = \{A, R\}$, where each speech-act $\clubsuit = F$ is the expression of a corresponding epistemic attitude.

Definition 6. An *assertoric statement* $S\alpha$ (with $F = S$) is a value-claim or logical judgment about the truth-value of a sentence α, where a truth-value $\{T, F\}$ is predicated of the sentential content.
To the $\mathbf{Q} = 2$ initial questions about a sentential content α: "Do you hold $v(\alpha) = T$?" and "Do you hold $v(\alpha) = F$?", assertoric modalities reply by an ordered set of $\mathbf{P} = 2$ yes-no answers.

Definition 6.1. An *affirmation* (or acceptance) (A) yields an affirmative $\mathbf{P}_1$, irrespective of $\mathbf{P}_2$.
$(\alpha) = \langle 1,- \rangle$
The epistemic counterpart of affirmation is belief: $B\alpha$.
There are two subspecies of acceptance.

Definition 6.1.1. A *conjecture* (A^-) yields an affirmative $\mathbf{P}_1$ and a affirmative $\mathbf{P}_2$.
$(\alpha) = \langle 1, 1 \rangle$
The epistemic counterpart of conjecture is weak belief: $B\alpha \wedge B{\sim}\alpha$.

Definition 6.1.2. An *assertion* (A^+) yields an affirmative $\mathbf{P}_1$ and a negative $\mathbf{P}_2$.
$(\alpha) = \langle 1, 0 \rangle$
The epistemic counterpart of assertion is strong belief: $B\alpha \wedge {\sim}B{\sim}\alpha$.

Definition 6.2. A *denial* (or rejection) (R) yields a negative $\mathbf{P}_1$, irrespective of $\mathbf{P}_2$.
$(\alpha) = \langle 0,- \rangle$
The epistemic counterpart of denial is disbelief: ${\sim}B\alpha$.
There are two subspecies of rejection.

Definition 6.2.1. A *neg-assertion* (R^-) yields an affirmative $\mathbf{P}_1$ and a negative $\mathbf{P}_2$.
$(\alpha) = \langle 0, 1 \rangle$
The epistemic counterpart of neg-assertion is strong disbelief: $\sim B\alpha \wedge B{\sim}\alpha$.

Definition 6.2.2. A *doubt* (R^+) yields an affirmative $\mathbf{P}_1$ and a negative $\mathbf{P}_2$.
$(\alpha) = \langle 0, 0 \rangle$
The epistemic counterpart of doubt is weak disbelief: $\sim B\alpha \wedge {\sim}B{\sim}\alpha$.

A formal semantics for assertoric modalities can be characterized as follows, where these are taken to be 1-ary statement-forming operators upon sentences (in the vein of Bochvar's external operators or von Wright's truth logics). In addition to truth-values for sentences, truth-claim values are introduced for statements such that $v(A^+\alpha) = 1$ iff $v(\alpha) = \langle 1, 0 \rangle$ and $v(A^+\alpha) = 0$ otherwise.

Definition 7. A statement $\clubsuit\alpha$ is *right* (1) or *wrong* (0) if it expresses the speaker's epistemic attitude about α.

Definition 7.1. $v(A^-\alpha) = 1$ iff $v(\alpha) = \langle 1, 1 \rangle$, and $v(A^-\alpha) = 0$ otherwise.

Definition 7.2. $v(A^+\alpha) = 1$ iff $v(\alpha) = \langle 1, 0 \rangle$, and $v(A^+\alpha) = 0$ otherwise.

Definition 7.3. $v(R^-\alpha) = 1$ iff $v(\alpha) = \langle 0, 1 \rangle$, and $v(R^-\alpha) = 0$ otherwise.

Definition 7.4. $v(A^+\alpha) = 1$ iff $v(\alpha) = \langle 0, 0 \rangle$, and $v(A^+\alpha) = 0$ otherwise.

Illocutionary oppositions are reformulated correspondingly in terms of compossible truth-claim values rather than truth-values.

Definition 8. An *opposition* is a set of answers to questions about compossible truth-claims for the statement $\clubsuit\alpha$ and its opposite $op^X(\clubsuit\alpha)$ (where $\clubsuit \in \{A, R\}$ and $X \in \{I, N, C, D\}$).

To the $\mathbf{Q} = 2$ following questions $\mathbf{Q}_1$: "Possibly $v(\clubsuit\alpha) = v(op(\clubsuit\alpha)) = F$?" and $\mathbf{Q}_2$: "Possibly $v(\clubsuit\alpha) = v(op(\clubsuit\alpha)) = T$?", oppositions reply by an ordered 2-tuple of answers: $Op(\clubsuit\alpha, op(\clubsuit\alpha)) = \langle \mathbf{P}_1, \mathbf{P}_2 \rangle$.

The same ordered 2-tuples hold here as in Definition 2, with $Op = \{SB, CT, SCT, CD\}$.

It is easily seen that such a logic of statements departs from classical logic, in the light of the answers $v(\alpha) = \langle 1, 1 \rangle$ and $v(\alpha) = \langle 0, 0 \rangle$: the truth of a sentence classically entails the falsehood of its negation, but not here. This is due to the formulation of **Q**, in our corresponding question-answer semantics: the point about statements is not whether their sentential content *is* true or false, but whether it *is held to be* so.

Each of the four answers refers to a special assertoric force, whose classification depends both upon its quality (affirmative or negative) and its degree of force (X^+ or X^-).

On the one hand, it is to be noted that "affirmation" and "negation" refer here to **P** (and not α): an affirmative answer may be given about whether a negative sentence is true, (e.g., $v(\sim\alpha) = \langle 1, - \rangle$), and conversely (e.g., $v(\alpha) = \langle 0, - \rangle$). Let us recall that *locutionary* affirmation and negation concern the quality of a sentential content α, whereas *illocutionary* affirmation or "yes"-answer ($A = \{A^+, A^-\}$) and negation or "no"-answer ($R = \{R^+, R^-\}$) concern the quality of an answer **P**[12].

On the other hand, assertion shouldn't be assimilated to affirmation in our non-classical semantics: it is the strongest assertoric force, because it is a one-sided affirmation that is closer to truth-claim than any of its counterparts in $\{A, R\}$. In other words, the degree of force of an assertoric statement is determined by a linear ordering relation $>$ between definite and indefinite answers **P**: $\langle 1, 0 \rangle > \langle 1, 1 \rangle > \langle 0, 0 \rangle > \langle 0, 1 \rangle$. The previous modality of consideration @ is vacuously assumed in every aforementioned speech act, in the sense that any answer **P** is self-consciously given by the speaker.

As regards the existential import, an important point about assertoric statements is that only two of them are mentioned as usual speech acts: assertion A^+ and denial R, whose utterances are rendered by such ambiguous statements as "S is P" and "S is not P" respectively. The reason is that assertoric statements ordinarily aim at telling the truth and nothing but the truth, whether it concerns affirmative sentences or their negation. Hence the unique judgment stroke $\vdash$ in Frege's *Begrisschrift*, which means that the

12 A way to argue that denial (illocutionary negation) is prior to sentential (locutionary) negation is that the former essentially occurs as a *definiens* of the latter. Firstly, no "$\sim$" occurs in the formulation of **Q** and **P** but only the expressions of falsity (F) or denial (0). Secondly, locutionary negation cannot occur without a corresponding illocutionary negation, but the converse does not hold: not every "no"-answer to $\mathbf{Q}_1$ entails a "yes"-answer to $\mathbf{Q}_2$, as witnessed by $\mathbf{P} = \langle 0, 0 \rangle$. In other words, every negative statement proceeds from an illocutionary denial before being rendered by a locutionary negation.

following judgeable content α is held to be true by the speaker and holds as an axiom or derived theorem. But while Frege (1952) equated the opposite act of denial with an assertion of the negative content: $\vdash\sim\alpha$, denial may mean a mere rejection of α without any corresponding affirmation of $\sim\alpha$. Now any mere refusal doesn't stand for a particular speech act but a negative psychological attitude, and the indefinite statements of conjecture or doubt are just hinted by the tone of voice or qualifying clauses like "It seems that" (for conjecture) or "I don't say that" (for doubt). Hence any utterance of α is naturally understood as the expression of $\langle 1,0\rangle$ or $\langle 0,1\rangle$, and the problem of existential import is symptomatic of such a narrow understanding of affirmative and negative statements.

2.3.3 Back to the Existential Import

Recalling Aristotle's term logic, the modality of neg-assertion amounts to a contraffirmation and differs from denial. Thus Figure 6 can be partly reconstructed by means of assertoric modalities with the same structure $\clubsuit\alpha$, where $\clubsuit = $ A or R.

Figure 9: The Four in Assertoric Logic

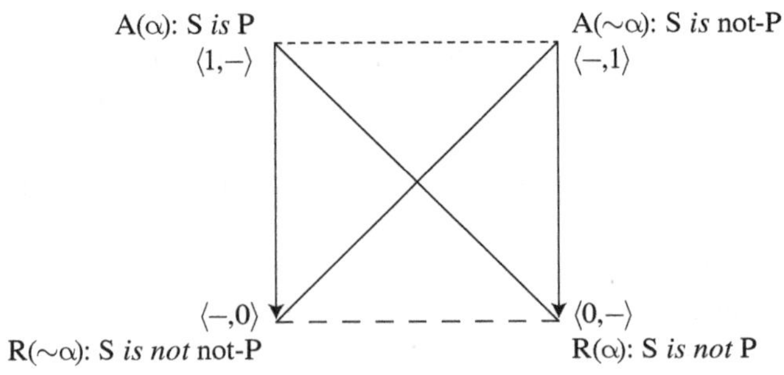

It clearly appears that not every assertoric statement is represented in the above figure: A and R are expressed by means of corresponding affirmative or negative statements, but nothing is said about their degree of force. This is so because Aristotle's traditional logic was uniquely composed of categorical assertions, where **PBV** restricts the range of **P** to $\langle 1,0\rangle$ and

$\langle 0, 1\rangle$. Now the case of empty subject terms created a difficulty in Aristotle's semantics: as argued by Horn (1989), Aristotle generally accepted that $v(\sim\alpha) = \mathrm{T}$ whenever the subject term does not exist; this entails that any denial of such a sentential content was held to be true by Aristotle[13]. Thus $v(\alpha) = \langle 0, 1\rangle$ seems to be the right answer for any statement whose sentential content includes an empty subject term.

But this is not the case, in the light of the relevant distinction between contraffirmation (neg-assertion) and denial: a sentence can be denied without its sentential negation being thereby affirmed, and an appropriate translation for Aristotle's attitude about empty terms should be rather $v(\alpha) = \langle 0, 0\rangle$. After all, Aristotle was said to accept the truth of denials such as "Not every griffin is an animal" while denying the truth of "Some griffin is not an animal". This means that only negative predications were accepted by the Stagirite in such cases, but not negated predicate terms; and given that only the first case corresponds to an illocutionary negation in our logic of assertoric statements, it follows from this term distinction that what was accepted by Aristotle is not the truth of a negative sentence $\sim\alpha$ but the denying attitude of any speaker about α: $\mathrm{R}(\alpha)$ is right, not $\mathrm{A}(\sim\alpha)$.

Does this illocutionary interpretation of negation give a consistent picture of the theory of opposition embracing the case of empty terms? On the one hand, the preceding figure shows that the logical square is insufficient for a geometrical representation of illocutionary oppositions; on the other hand, the syntactic generalization due to Moretti (2009) yields a group of polygonal oppositions for a logic of assertoric statements that takes account of their degree of strength (see Figure 10).

There are not $\mathrm{CT} = 2$, as in Aristotle's square, but $\mathrm{CT} = 4$ sorts of contrary statements: any two assertoric modalities exclude from each other (e.g., whatever is asserted is not conjectured), and the previous definition of subcontraries as contradictories of contraries gives rise to an identical number for subcontrary terms.

We thus obtain a gathering of 3-dimensional bisimplexes, according to the terminology of Moretti. An exhaustive geometrical representation of these illocutionary oppositions is composed of two tetrahedra of contrariety

13 In connection with footnote 13, (Horn 1989, p. 24) says that the "qualitative approach (i) has its roots in Aristotle (…) The existence of Socrates, as we have seen, is a necessary condition for the truth of any singular proposition concerning him (e.g., *Socrates is ill*), while his nonexistence is a sufficient condition for the truth of the corresponding (contradictory) negation (*Socrates is not ill*)".

Figure 10: Two Tetrahedra of Simplexes and One Cube of Subalternation

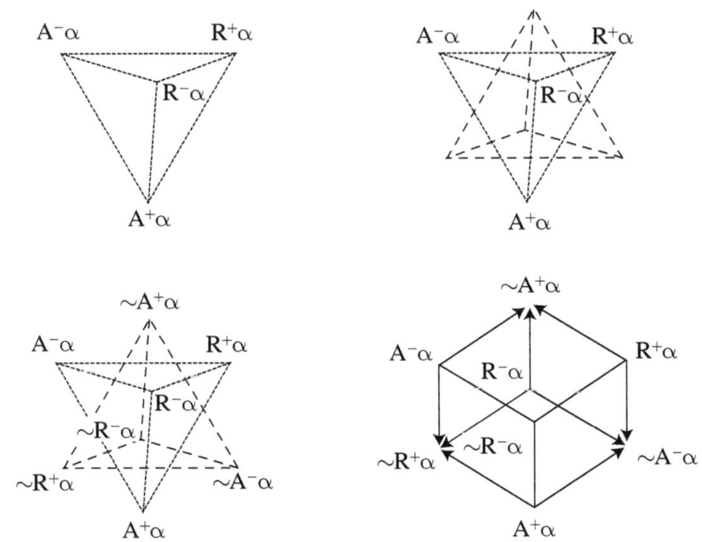

Figure 11: A Tetraicosahedron of Assertoric Oppositions

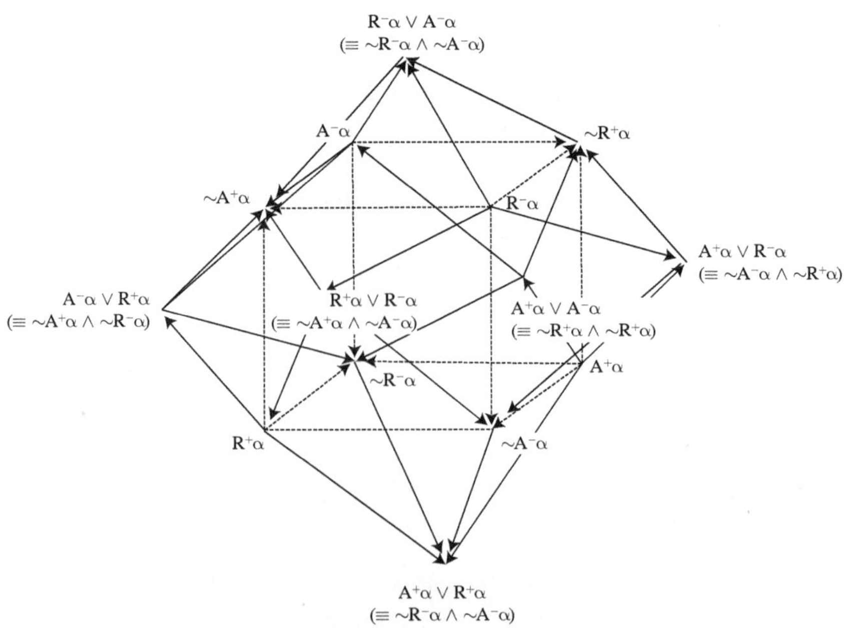

and subcontrariety (connected by contradictories) together with one cube of subalternation, where the four vertices of Aristotle's square (see Figure 8) occur as partial cases (see Figure 11).

The difference between the neg-assertion A^- and the denial R was thus rendered using Abelard's medieval distinction between the two alternative expressions of **O**: "Not every S is P", and "Some S is not P". We stick here to the first interpretation of **O** as an *illocutionary* negation of **A**, without following Apuleius' and Boethius' translation as a particular negative: "S is not-P". In other words, nothing is to be affirmed about non-existing entities like griffins: neither $v(\alpha) = T$, nor $v(\alpha) = F$ (*i.e.* $v({\sim}\alpha) = T$), so that the proper formalization of **O** is not $A^-({\sim}\alpha)$ but $R(\alpha)$. To be more precise, the initial sentence α: "Every griffin is an animal" (or, equivalently: "Griffins are animals") is as much denied as its locutionary negation: "Some griffin is not an animal", in the sense that no affirmation (" – *is* …") could be rightly uttered about a non-existing entity. If so, then α and ${\sim}\alpha$ are equally denied and the corresponding statement is not neg-assertion but doubt: $R^+(\alpha)$, or $v(\alpha) = \langle 0, 0 \rangle$.

In the light of the preceding figure highlighting assertoric oppositions, Aristotle's **A**- and **O**-vertices turn out to be contradictory to each other in a restrictive sense of negation: "Not every S is P" appears as the *illocutionary* negation of "Every S is P", in accordance to the equivalence ${\sim}A(\alpha) \Leftrightarrow R(\alpha)$ (whatever is not affirmed is denied, and conversely). By contradistinction, any affirmation about griffins will be *contrary* to a neg-assertion about them: both are incorrect, *i.e.* $v(A^+(\alpha)) = v(A^+({\sim}\alpha)) = v(R^-(\alpha)) = 0$. It amounts to saying that **O** doesn't mean the illocutionary affirmation of a negative sentential content (viz. a particular negative) but rather the illocutionary negation of an affirmative sentential content (viz. a universal affirmative). In other words: while the Abelardian "Not every S is P" attaches the negation to the universal quantifier, its (Apuleian or Boethian) variant "Some S is not P" is misleading in the light of our modal structuration. For the contradictory of $\clubsuit\alpha$ is not $\clubsuit{\sim}\alpha$ but ${\sim}\clubsuit\alpha$, which appears clearly only if the illocutionary aspect of an utterance is made explicit by $\clubsuit$.

3. Conclusion: the Future of Logical Opposition

Logical opposition has been considered throughout this paper in its main properties and its various applications. It has mainly been argued that:

(1) A question-answer game **QAS** properly used supports a complete theory of opposition, beyond Aristotle's initially quantified theory; a semantic generalization helps to characterize oppositions as 2-ary relations between any propositions, while a syntactic generalization helps to define logical opposites as its relata.

(2) The theory of opposition can be made totally consistent and settle the case of existential import by adopting an illocutionary view of oppositions that relates statements rather than mere sentences and distinguishes two senses of negation.

The theory of logical opposition may thus be completely overhauled, in the light of **NOT** and its formulation within the conceptual framework of **QAS**. Why not think about logic as a theory of opposition, rather than as a theory of consequence?

Is it because of the imprecise meaning of an opposition as compared to the precise characterization of logical consequence as truth-preservation? This paper has attempted to advance the perfectible theory of opposition by placing the emphasis on the concept of negation rather than on truth.

The incompleteness of Aristotle's theory of opposition has left a crucial gap between traditional and modern logic.

Our solution to existential import is an attempt to show that the theory of opposition is beyond traditional logic but can adequately be dealt with in purely modern logic terms such as relations, negation, and truth-values.

Given the prominent role of logical consequence in modern logic, the author will present opposition Op (and its opposite-forming operator op) as a counterpart of abstract consequence Cn (and its associated conclusion-forming operator) in a future work.

In a nutshell, opposition does not deserve to be considered as a mere tourist curiosity with old-fashioned properties. Given a proper facelifting, logical opposition can clearly emerge as a thick-skinned concept. This motivates our interest in **NOT** and **QAS**. The ageless concept of logical opposition needs to be clarified if the concern of logic is not to be just about consequence and truth but also about opposition and negation. The ultimate goal of our future works is to show that a theory of opposition comes to a logical theory of negation just as a theory of consequence comes to a logical theory of truth:

$$\frac{\text{Consequence}}{\text{Truth}} = \frac{\text{Opposition}}{\text{Negation}}$$

References

Aristotle (a). *De Interpretatione.*

Aristotle (b). *Metaphysics.*

Béziau, J.-Y. (2003). New light on the square of opposition and its nameless corner, *Logical Investigations, 10,* 218–233.

Béziau, J.-Y. (2006). Paraconsistent logics! A reply to Slater, *Sorites, 17,* 17–25.

Blanché, R. (1953). Sur l'opposition des concepts, *Theoria, 19,* 89–130.

Engel, P. (2007). *Can there be an epistemic square of opposition?* Talk presented during the 1st International Congress about the Square of Opposition (June 1–3).

Englebretsen, G. (1969). Knowledge, negation, and incompatibility, *The Journal of Philosophy, 66,* 580–585.

Englebretsen, G. (1981). *Logical negation*, Van Gorcum, Assen.

Fitch, F. (1963). A logical analysis of some value concepts, *Journal of Symbolic Logic, 28,* 135–142.

Frege, G. (1952). Negation. In *Translations from the Philosophical Writings of Gottlob Frege.* P.T. Geach and M. Black (Eds.).

Gottschalk, W.H. (1953). The Theory of Quaternality, *The Journal of Symbolic Logic, 18,* 193–196.

Hart, A.H. (1980). Toward a logic of doubt, *International Logic Review, 21,* 31–41.

Hintikka, J. (1962). *Knowledge and Belief*, Ithaca Press, New York.
Horn, L. (1989). *The Natural History of Negation*, Univ. of Chicago Press (2nd ed. 2001).
Monteil, J. F. (2003). Du nouveau sur Aristote. Remarques sur deux traductions arabes du De Interpretatione, *L'Enseignement Philosophique 53*, 3–13.
Moretti, A. (2009). *The Geometry of Logical Opposition*, PhD Thesis, University of Neuchâtel (Switzerland).
Parsons, T. (2006). The traditional Square of Opposition. In *Stanford Encyclopedia of Philosophy*. http://plato.stanford.edu/entries/square/index.html.
Piaget, J. (1949). *Traité de logique* (*Essai de logistique opératoire*), A. Colin (2nd ed. 1972).
Schang, F. (2007). *Philosophie des modalités épistémiques* (*La logique assertorique revisitée*). PhD Thesis, University of Nancy 2 (France).
Searle, J., & Vanderveken, D. (1985). *The Foundations of Illocutionary Logic*, Cambridge University Press.
Sesmat, A. (1951). *Logique II. Les Raisonnements. La syllogistique*. Paris: Hermann.
Slater, B. H. (1995). Paraconsistent logics? *Journal of Philosophical Logic*, *24*, 451–454.
Sion, A. (1996). *Future Logic*, especially Chapter 6: Oppositions, Geneva, http://www.thelogician.net/2_future_logic/2_chapter_06.htm.
Vernant, D. (2003). Pour une logique dialogique de la dénégation. In F. Armengaud, M. D. Popelard and D. Vernant (Eds.), *Du dialogue au texte. Autour de Francis Jacques*. Paris: Kimé. http://web.upmf-grenoble.fr/SH/PersoPhilo/DenisVernant/Denegation.pdf.
von Wright, G.H. (1951). *An Essay in Modal Logic*, Amsterdam: North-Holland, .

Part 4: Mathematical Logic

Definitional Reasoning in Proof-Theoretic Semantics and the Square of Opposition

PETER SCHROEDER-HEISTER

Abstract

Within a framework of clausal definitions of atoms, four forms of judgement are distinguished: Direct and indirect assertion, and direct and indirect denial. Whereas direct assertion and direct denial are established by directly applying a definitional clause ("definitional closure"), indirect assertion and indirect denial result from showing that all possible premisses of the opposite judgement can be refuted ("definitional reflection"). The deductive relationships between these four forms of judgement correspond to those represented in the square of opposition, if direct assertion and direct denial are placed in the *A* and *E* corners, and indirect assertion and indirect denial in the *I* and *O* corners of the square.

Keywords: Proof-theoretic semantics, definitional reasoning, denial, definitional reflection, square of opposition

1. Introduction

The square of opposition can be looked at from various points of view. This paper will only be concerned with the deductive relationships between its nodes. In our inferential framework we shall define four forms of judgement to be placed in the *A*, *E*, *I* and *O* corners of the square of oppositon, as well as a denial operation between judgements, such that

- the judgements in the *A* and *E* corners oppose each other without necessarily being denials of one another – *contrariety*,

- the judgements in the A and O corners, and those in the E and I corners are denials of one another – *contradictoriness*,
- the judgements in the I and O corners can be inferred from those in the A and E corners, respectively, but not necessarily vice versa – *subalternation*,
- the denials of the judgements in the I and O corners oppose each other, without necessarily being denials of one another – *subcontrariety*.

Proof-theoretic semantics denotes an approach to reasoning according to which the semantics of propositions is given in terms of proofs, and in which proofs themselves are philosophically primary (see Schroeder-Heister 2011). This may be viewed as an interpretation of the slogan "meaning is use", if proofs are related to the use of linguistic expressions. There are several variants of proof-theoretic semantics. The "standard" one, which is due to Prawitz and Dummett, is based on a proof-theoretic notion of validity and is described in detail in Schroeder-Heister (2006). Here we are concerned with a somewhat different, and more general, approach developed jointly with Lars Hallnäs, originally in the context of logic programming (Hallnäs & Schroeder-Heister 1990/91, Schroeder-Heister 1993). It is called "definitional reasoning" and investigates rule-based reasoning in a broad sense, not just logical reasoning, i.e. reasoning with logical constants.

2. Clausal Definitions

Unlike other approaches, which deal with logically complex formulas, ours deals with atomic formulas or *atoms*, for which a clausal definition is given. For the sake of simplicity, and in order to make the basic points as clear as possible, we only consider the very simple propositional case without any term structure and without individual variables. Therefore atoms are the same as propositional veriables, which will be denoted by lower case letters. Upper case letters denote (possibly empty) lists of atoms. A clausal definition, in short *definition*, has the form

$$\mathbb{D}\begin{cases} a_1 & \Leftarrow & B_{11} \\ & \vdots & \\ a_1 & \Leftarrow & B_{1m_1} \\ & \vdots & \\ a_n & \Leftarrow & B_{n1} \\ & \vdots & \\ a_n & \Leftarrow & B_{nm_n} \end{cases}$$

As usual in logic programming, the heads of the clauses are put in front, and clauses are read as:

$$\begin{array}{ccc} a_1 & \text{if} & B_{11} \\ a_1 & \text{if} & B_{12} \\ & \vdots & \end{array}$$

The reader who is not familiar with this notation, may use the left-to-right notation

$$\mathbb{D}\begin{cases} B_{11} & \Rightarrow & a_1 \\ & \vdots & \\ B_{1m_1} & \Rightarrow & a_1 \\ & \vdots & \\ B_{n1} & \Rightarrow & a_n \\ & \vdots & \\ B_{nm_n} & \Rightarrow & a_n \end{cases}$$

to be read as

$$\mathbb{D}\begin{cases} \text{if } B_{11} & \text{then} & a_1 \\ \text{if } B_{12} & \text{then} & a_1 \\ & \vdots & \end{cases}$$

However, it is important to be aware that these clauses are not understood as (material) implications, but rather as production rules. So

$$a \Leftarrow B$$

is intended to mean that from B one may produce a. In order to avoid any confusion with inference rules, which later on will be defined *with respect to* such entities, but otherwise are different from them, we avoid the term "rule" and speak of "clauses" instead, again following the terminology of logic programming. However, here another *caveat* must be put in place: We do not mean "clause" in the particular sense of resolution theory where it is often understood as a disjunction of literals. Clauses in our sense do not presuppose logic. Understanding them as elementary productions, i.e. devices that generate their heads from their bodies, is sufficient. The "real" meaning of clauses is explained by the inference rules motivated in the next two sections.

Bodies of clauses are also called *conditions*. If a clause has the form

$$a \Leftarrow B$$

then B is called a condition of a. As bodies may be empty, a clause has either the form

$$a \Leftarrow b_1, \ldots, b_n$$

or the form

$$a \Leftarrow$$

As there may be more than one clause for a, a may have more than one condition according to a given definition. In this sense definitions are not necessarily deterministic. For example

$$\begin{cases} a & \Leftarrow & b,c \\ a & \Leftarrow & d \\ a & \Leftarrow & e,f,g,h \end{cases}$$

is a definition of a in terms of three conditions, using three clauses for a. Thus definitions are not to be confounded with explicit definitions in the usual sense, which can be read bidirectionally. Definitions in our sense correspond to *inductive definitions* in the mathematical sense. In fact, in a general setting, inductive definitions are best understood as systems of clauses in our sense, by means of which a domain of objects can be generated (see Aczel 1977). Clauses themselves are logic-free, except that they are based on some *structural* logic in the sense that the comma within conditions is understood conjunctively.

3. Putting Definitions into Action: Definitional Closure and Definitional Reflection

The way clauses are understood is determined by *inference rules* which handle them, put them into action or make their meaning explicit (to use three different phrases for what is intended). The simplest inference rule says that if we can assert the defining conditions of a, then we can assert a itself. If

$$a \Leftarrow b_1, \ldots, b_n$$

is a defining clause for a, this clause can be applied using the inference rule

$$\frac{b_1 \ \ldots \ b_n}{a}$$

which, as a limiting case, covers

$$\frac{}{a}$$

This rule is an introduction rule for a, as it allows us to introduce a from one of its conditions. As a general schema it is called the rule of *definitional closure*, as it enables us to close a set of given atoms under the clauses of a definition. It allows us to build proof trees from given definitions. For example, given the definition

$$\mathbb{D}_1 \begin{cases} a & \Leftarrow \; b,c \\ b & \Leftarrow \\ c & \Leftarrow \\ d & \Leftarrow \; a,b \\ e & \Leftarrow \; a,f,h \\ h & \Leftarrow \; g \end{cases}$$

we can derive a in $\mathbb{D}_1$ as follows:

$$\dfrac{\dfrac{\dfrac{}{b}\quad\dfrac{}{c}}{a}\qquad\dfrac{}{b}}{d}$$

As usual we write

$$\vdash_{\mathbb{D}_1} d$$

to express the corresponding provability statement. An example of a proof of an atom from assumptions would be the following

$$\dfrac{\dfrac{\dfrac{}{b}\quad\dfrac{}{c}}{a}\qquad f\quad\dfrac{g}{h}}{e}$$

establishing

$$f,g \vdash_{\mathbb{D}_1} e$$

Graphically, an assumption occurs as an atom without an inference bar on top of it.

This idea of reasoning with definitions can be extended by also giving elimination rules for the atoms defined. Whereas the introduction rules are very elementary – they just interpret clauses as productions –, generalizing the idea of elimination rules from logic to clausal systems makes definitional reasoning particularly powerful. This idea was called "definitional reflection" by Hallnäs and is described in Hallnäs (1991, 2006), Hallnäs &

Schroeder-Heister (1990/91), Schroeder-Heister (1993, 2004, 2007). Here we discuss instead the idea of *denial rules* which is based on definitional reflection as well.

Suppose with every atom a there is associated its denial $-a$ in a sense that makes $-a$ the contradictory opposite to a. More precisely, suppose there is a metalinguistic denial operation $\widehat{\ }$ such that

$$\begin{aligned} \widehat{a} &:= -a \\ \widehat{-a} &:= a \end{aligned}$$

Thus we assume that with each atom a there is associated its (objectlinguistic) denial $-a$, such that a and $-a$ are metalinguistic denials of one another. To avoid the (object- vs. meta-linguistic) ambiguity in the term "denial", we sometimes speak of the *complement* $\hat{\ell}$ of a *literal* ℓ, where literals are atoms a and their denials $-a$. As our framework is quite general, we do not presuppose any particular understanding of "denial", as long as a complementation operation with the properties mentioned is available. One way of understanding it (but not the only, and definitely not the most central one) is to read it as classical Boolean negation. Note that in the simplified framework dealt with here, we do not draw a distinction between a pragmatic denial operation (which cannot be iterated) and the negation connective (which may occur nested inside a formula), as the only formulae we are dealing with are literals, i.e. atoms and their denials. We are interested in the way denials of atoms give rise to new inference rules without changing the clausal definition (in which, for the time being, no denial is allowed to occur).

Our motivation is that, if the atom a is defined by the clauses

$$\begin{cases} a & \Leftarrow & B_1 \\ & \vdots & \\ a & \Leftarrow & B_n \end{cases}$$

then, if asserting *one* of a's defining conditions suffices to assert a (a *introduction*), denying *all* defining conditions of a suffices to deny a, i.e. to establish $-a$ ($-a$ introduction). This leads to a rule of the form

$$\frac{\widehat{B_1} \ \dots \ \widehat{B_n}}{-a}(-)$$

For this rule expression to make sense, we have to define the denial (or complement) $\widehat{B}$ of a condition B. As a condition is a list $b_1,\dots,b_n$ of atoms, it is natural to consider it to be rejected, if one of its elements is rejected. So we let $\widehat{b_1,\dots,b_n}$ stand indefinitely for the denial $\widehat{b_i}$ of anyone of its members, so that every proof of $-b_i$ for some element b_i counts as a proof of $\widehat{b_1,\dots,b_n}$. Since we do not want to introduce $\widehat{b_1,\dots,b_n}$ as a formal expression into our object-language, we adopt the following convention: When we state

$$\frac{\widehat{B}}{a}$$

as an inference rule, where B is $b_1,\dots,b_n$, then this stands for the list of inference rules

$$\frac{-b_1}{a} \ \dots \ \frac{-b_n}{a}$$

and the notation for the schema of $-a$ introduction

$$\frac{\widehat{B_1} \ \dots \ \widehat{B_n}}{-a}(-)$$

stands for the list of all possible rules of the form

$$\frac{-b_1' \ \dots \ -b_n'}{-a}$$

where b_i' is an arbitrarily selected element of B_i for every i $(1 \leq i \leq n)$. Related expressions occurring later in this paper are understood accordingly. For example, if a is defined by the clauses

$$\begin{cases} a & \Leftarrow & b,c \\ a & \Leftarrow & d \\ a & \Leftarrow & e,f \end{cases}$$

then the schema of $-a$ introduction $(-)$ stands for the following inference rules:

$$\frac{-b \quad -d \quad -e}{-a} \qquad \frac{-b \quad -d \quad -f}{-a} \qquad \frac{-c \quad -d \quad -e}{-a} \qquad \frac{-c \quad -d \quad -f}{-a}$$

In the formulation of $(-)$ it is important that $B_1,\ldots,B_n$ *exhaust* the conditions of a. If there were a further condition B_{n+1} of a, based on the additional clause

$$a \Leftarrow B_{n+1}$$

we would have to extend this rule to

$$\frac{\widehat{B_1} \ \ldots \ \widehat{B_n} \ \widehat{B_{n+1}}}{-a}\ (-)$$

Thus the rule of denial introduction can be considered as an interpretation of what is sometimes called the "extremal clause" in inductive defintions: "Nothing else is a defining condition of a". This sort of inference is called "definitional reflection", as, when carrying it out, we reflect on the definition as a whole. More precisely, we should call it *definitional reflection with respect to denial*, since the original application of this principle is with respect to elimination inferences, which we do not consider here. Obviously, definitional reflection makes derivability non-monotonic with respect to the given definition, i.e., extending the definition may reduce the set of literals derivable.

The denial operation $-$, for which we have just formulated the introduction rule $(-)$, will be called *indirect denial*. This is because its governing rule $(-)$ is *derived* from denial-free clauses for atoms and in this way "parasitic" on them. Indirect denial will be distinguished from direct denial, which is established by individual clauses in (an extended notion of) definition. So "indirect" is another way to express what "definitional reflection" means. It precisely corresponds to the distinction between direct and indirect means of establishing a proposition in Dummett's theory of meaning (see Dummett 1991). The direct means are those *encoded in* the definition, whereas the indirect ones are those obtained by *reflection on* the definition. We give two examples for the application of definitional reflection with respect to denial: Given the definition

$$\mathbb{D}_2 \begin{cases} a & \Leftarrow & b \\ a & \Leftarrow & c \end{cases}$$

we obtain

$$-b, -c \vdash_{\mathbb{D}_2} -a$$

and given the definition

$$\mathbb{D}_3 \begin{cases} a & \Leftarrow & b \\ a & \Leftarrow & c, d \end{cases}$$

we obtain

$$-b, -c \vdash_{\mathbb{D}_2} -a$$

as well as

$$-b, -d \vdash_{\mathbb{D}_2} -a$$

So far there are no closed (i.e., assumption-free) proofs using (–), as (–) has only denials as premisses, and we have no way of establishing them (except by denial introduction). This again reflects the indirect character of the denial operation considered. If a proof of the denial of *a* is called a *refutation* of *a*, we can say that so far, there may be categorical (= closed = assumption-free) proofs of atoms, but only hypothetical (= open = assumption-dependent) refutations of atoms.

To be fully precise, we should define what definitional reflection means in the case of premiss-free clauses and in the case of atoms for which no defining clause at all is given. In the first case there is simply no rule of definitional reflection defined, i.e. no rule (–) is available. In the second case, $-a$ can be inferred vacuously, as no defining condition (not even the empty one) is available for *a*. However, these limiting cases will not be discussed here in detail.

4. Denials in Definitional Clauses

Suppose we now extend our framework by allowing literals and not only atoms to occur in the bodies of clauses, i.e., we consider clauses of the form

$$a \Leftarrow \ell_1,\ldots,\ell_n$$

for literals $\ell_1,\ldots,\ell_n$. Let us call such a clause an *assertion clause* (i.e., a clause for the assertion of an atom, i.e., a clause with an atom as its head). Then we can formulate a definition such as

$$\mathbb{D}_4 \begin{cases} a & \Leftarrow & b \\ a & \Leftarrow & -c,d \end{cases}$$

which yields

$$-b,c \vdash_{\mathbb{D}_4} -a \text{ and } -b,-d \vdash_{\mathbb{D}_2} -a$$

as both

$$\frac{-b \quad c}{-a} \text{ and } \frac{-b \quad -d}{-a}$$

are instances of (–). This makes it possible in principle to derive a denial without assumptions. Suppose, e.g., the definition

$$\mathbb{D}_5 \begin{cases} a & \Leftarrow & -b \\ b & \Leftarrow & \end{cases}$$

is given. Then the proof

$$\dfrac{\overline{b}}{-a}\,(-)$$

demonstrates that

$$\vdash_{\mathbb{D}_5} -a$$

holds, i.e., a is refuted outright (= categorically = without assumption).

Therefore, by using the inference rules of definitional closure (introduction of atoms = proof of atoms) and of definitional reflection (introduction of denials of atoms = rejection of atoms), we obtain a calculus for atomic reasoning, in which atoms can be proved or rejected on the basis of assertion clauses.

5. Clauses Defining Denials

The natural next step in extending our framework of definitional reasoning is to admit *denial clauses*, i.e. clauses whose head is a denial. If by that we mean admitting clauses of the form

$$-a \Leftarrow \ell_1,\ldots,\ell_n$$

then, in general, by considering clauses of the form

$$\ell \Leftarrow \ell_1,\ldots,\ell_n$$

we can dualize the whole framework developed so far. Suppose the literal ℓ is defined by the clauses

$$\begin{cases} \ell & \Leftarrow & B_1 \\ & \vdots & \\ \ell & \Leftarrow & B_n \end{cases}$$

Then we obtain as inferences based on definitional closure:

$$\frac{B_1}{\ell} \quad \cdots \quad \frac{B_n}{\ell}$$

whereas for definitional reflection we obtain

$$\frac{\widehat{B_1} \ \ldots \ \widehat{B_n}}{\hat{\ell}}$$

While the schema of definitional closure now covers introductions of denials, the schema of definitional reflection now covers introductions of assertions (in case ℓ is of the form $-a$. For example, given the definition

$$\mathbb{D}_6 \begin{cases} -a & \Leftarrow & b \\ -a & \Leftarrow & c, -d \\ -b & \Leftarrow & e \\ -c & \Leftarrow & e \\ d & \Leftarrow & e, -c \\ e & \Leftarrow & \end{cases}$$

and labelling steps of definitional closure with **c** and those of definitional reflection with **r**, the following proof demonstrates (the assertion of) a, with definitional reflection now leading to an assertion claim:

$$\dfrac{\dfrac{\dfrac{}{e}\,\mathbf{c}}{-b}\,\mathbf{c} \qquad \dfrac{\dfrac{}{e}\,\mathbf{c} \quad \dfrac{\dfrac{}{e}}{-c}\,\mathbf{c}}{d}\,\mathbf{c}}{a}\,\mathbf{r}$$

which establishes

$$\vdash_{\mathbb{D}_6} a$$

With this proposal we could finish our discussion of assertion and denial in definitional reasoning. We have obtained completely symmetric assertion and denial inferences for atoms, based either on definitional closure or on definitional reflection.

However, this would not only mean to miss out on the square of opposition (which is not a value in itself, of course), but we would also have to retract certain intuitions already developed. One of them was that the rules of definitional closure in a sense represent the *direct* way of establishing a proposition, whereas the rules of definitional reflection are *indirect* ways of doing so. So we distinguished between definitional knowledge stated explicitly in the definition (as the head of a definitional rule) and knowledge only *obtained* from the definition by reflecting on its structure. Putting this in a more traditional terminology, it is the difference between analytic and non-analytic knowledge. Analytic knowledge would be what can be extracted immediately from the definition; it is, so to speak, already there as the head of a clause. In contradistinction to that, non-analytic knowledge results by *interpreting* the definition. It has to be extracted from it in a way that goes beyond just forward reasoning along the definitional clauses. It is not like the "beam in the house" but like the "plant in the seed" to use Frege's (1984, § 88) picture. If we want to keep this fundamental meaning-theoretical distinction, we must terminologically distinguish between a denial which is directly established by means of definitional closure as the head of a definitional clause, and the denial which is established indirectly by definitional reflection, i.e., by means of a non-monotonic inference rule which requires taking the definition as a whole into consideration. In the following we shall write "$\sim a$" for the direct denial of a, in contradistinction to "$-a$" for the indirect denial established by definitional reflection. Denial clauses would then be clauses of the form

$$\sim a \Leftarrow B$$

rather than

$$-a \Leftarrow B$$

Thus in general, a definitional clause has the form

$$\langle\sim\rangle a \Leftarrow \langle-\rangle b_1,\ldots,\langle-\rangle b_n$$

where the angle brackets indicate that denial operators may occur at the places indicated.

Now, the operators $\sim$ and $-$, being both denial operations, should not remain unrelated to one another. It is natural to assume that the indirect denial is weaker, or at least not stronger, than the direct denial. So if we have rejected a directly, by inferring the denial of a by definitional closure, this should have at least the power of an indirect rejection, so we should be able to proceed as if with $\sim a$ we also had established $-a$. This leads to assuming a rule of subalternation with respect to denial:

$$\frac{\sim a}{-a}\,(Sub)$$

For example, given the following definition

$$\mathbb{D}_7 \begin{cases} \sim a & \Leftarrow \quad b \\ b & \Leftarrow \\ c & \Leftarrow \quad -a, -d \\ d & \Leftarrow \quad -b \\ d & \Leftarrow \quad a, c \end{cases}$$

we can establish

$$\vdash_{\mathbb{D}_7} c$$

by means of the following proof:

$$\dfrac{\dfrac{\dfrac{\dfrac{}{b}}{\sim a}}{-a}\,(Sub) \qquad \dfrac{\dfrac{}{b} \quad \dfrac{\dfrac{\dfrac{}{b}}{\sim a}}{-a}\,(Sub)}{-d}}{c}$$

In this proof, at two places, by use of subalternation the direct denial $\sim a$ of a is turned into the indirect denial $-a$ of a to obtain the premiss $-a$ of the closure rule for c (left application of (Sub)) and of the reflection rule for $-d$ (right application of (Sub)).

As already indicated at the beginning of this section, there is no reason to restrict definitional reflection to obtain only denial inferences. We can argue that we obtain an indirect sort of assertion from the denial of all possible conditions of directly denying an atom. For that we introduce an additional form of judgment $+a$ to express the *indirect assertion* of a, $+a$ being the contradictory opposite (complement) of $\sim a$:

$$\begin{aligned}\widehat{a} &:= -a\\ \widehat{-a} &:= a\\ \widehat{\sim a} &:= +a\\ \widehat{+a} &:= \sim a\end{aligned}$$

This differs from the conception without the direct/indirect distinction, where we had only a single pair of opposites ($a\,/\,-a$). Definitional reflection with respect to assertion now leads to the indirect assertion $+a$ of an atom a: Given

$$\begin{cases}\sim a & \Leftarrow\quad B_1\\ & \vdots\\ \sim a & \Leftarrow\quad B_n\end{cases}$$

as (direct) denial clauses for a, the rule of definitional reflection w.r.t. assertion is

$$\frac{\widehat{B_1}\ \ldots\ \widehat{B_n}}{+a}$$

For example, the definition

$$\mathbb{D}_8\begin{cases}a & \Leftarrow\quad b\\ a & \Leftarrow\quad c\\ \sim a & \Leftarrow\quad d,-e\\ \sim a & \Leftarrow\quad f\end{cases}$$

yields

$$-d,-f \vdash_{\mathbb{D}_8} +a \quad\text{and}\quad e,-f \vdash_{\mathbb{D}_8} +a$$

in addition to

$$b \vdash_{\mathbb{D}_8} a \quad\text{and}\quad c \vdash_{\mathbb{D}_8} a$$

In the bodies of clauses we have admitted assertions of the form a and denials of the form $-a$. From the point of view reached so far, this means that we have admitted direct assertions and indirect denials, which is an obvious asymmetry with respect to assertion and denial. This has, for example, the effect that the indirect $+a$ has no function in proofs, except as an end formula. It can neither occur as a premiss of definitional closure (as it is not admitted to occur in the body of a clause), nor can it occur as a premiss of definitional reflection (as its denial [complement] $\sim a$ cannot occur in the body of a clause either). We remove this asymmetry by allowing for both the direct and the indirect assertions and direct and indirect denials to occur in bodies of clauses. This makes it possible in principle to use any of the possible four forms of judgement as a premiss in an inference of definitional closure or definitional reflection. Furthermore, in analogy with subalternation for denial, we postulate the corresponding rule of subalternation for assertion:

$$\frac{a}{+a}(Sub).$$

Even though in the bodies of clauses any form of judgement is allowed, the heads of clauses remain to be either direct assertions or direct denials. Our framework hinges on the distinction between *direct* and *indirect*, and this distinction is drawn by saying that "direct" means "being the head of a program clause".

This leads to the following definition. Let a *direct literal* be either of the form a or of the form $\sim a$ for an atom a. Let an *indirect literal* be either of the form $+a$ or $-a$. Let a *literal* be either a direct or an indirect literal. Then a definitional clause has the form

$$\ell \Leftarrow \ell_1,\ldots,\ell_n$$

where ℓ is a direct literal, and $\ell_1,\ldots,\ell_n$ are literals. A definition $\mathbb{D}$ is a list of definitional clauses. Every definitional clause gives rise to a rule of *definitional closure*

$$\frac{\ell_1 \ \ldots \ \ell_n}{\ell}$$

whereas for every atom a occurring in the definition, the definition $\mathbb{D}$ as a whole gives rise to a negative and a positive schema of *definitional reflection*

$$\frac{\widehat{B_1} \;\dots\; \widehat{B_m}}{-a} \qquad \frac{\widehat{C_1} \;\dots\; \widehat{C_n}}{+a}$$

provided $B_1,\dots,B_m$ are the conditions of a and $C_1,\dots,C_n$ are the conditions of $\sim a$ with respect to the definition $\mathbb{D}$, i.e., the clauses for a and $\sim a$ in $\mathbb{D}$ are as follows:

$$\begin{cases} a & \Leftarrow & B_1 \\ & \vdots & \\ a & \Leftarrow & B_m \end{cases} \qquad \begin{cases} \sim a & \Leftarrow & C_1 \\ & \vdots & \\ \sim a & \Leftarrow & C_n \end{cases}$$

Furthermore, indirect assertions and denials are subalternates to the corresponding direct ones:

$$\frac{a}{+a}(Sub) \qquad \frac{\sim a}{-a}(Sub)$$

The following is an example showing the interaction of all rules. Note that only the subalternation rules have been labelled as such. In all other cases it is uniquely determined which rule is being applied: Rules of definitional closure only lead to direct literals, rules of definitional reflection only lead to indirect literals.

$$\mathbb{D}_9 \begin{cases} a & \Leftarrow & c,e \\ a & \Leftarrow & \sim d \\ \sim b & \Leftarrow & d \\ c & \Leftarrow & b \\ c & \Leftarrow & \sim d \\ e & \Leftarrow & -a,+d \end{cases}$$

The following proof demonstrates that $d \vdash_{\mathbb{D}_9} e$.

$$\dfrac{\dfrac{\dfrac{\dfrac{\dfrac{d}{\sim b}}{-b}(Sub) \quad \dfrac{d}{+d}(Sub)}{-c} \quad \dfrac{d}{+d}(Sub)}{-a} \quad \dfrac{d}{+d}(Sub)}{e}$$

6. The Square of Opposition

We have now obtained a structure which fits into the square of opposition (see Figure 1). The direct forms of judgement a and $\sim a$ are placed in the A and E nodes, and the indirect forms of judgement $+a$ and $-a$ in the I and O nodes, respectively. Due to our subalternation rules (Sub) and the complementation operation $\widehat{\ }$ with the properties

$$\begin{aligned} \hat{a} &:= -a \\ \widehat{-a} &:= a \\ \widehat{\sim a} &:= +a \\ \widehat{+a} &:= \sim a \end{aligned}$$

we have already interpreted the right and left sides (subalternation) and the diagonals (contradictoriness) of the square. It remains to give a meaning to contrariety and subcontrariety.

Contrariety is the opposition between direct judgements a and $\sim a$, i.e. between direct assertion and direct denial. If we want to relate it to the square, we have to say what it means in our framework that contrary judgements cannot both be true (but may both be false). In our proof-theoretic context, the most natural way is to interpret truth as provability and falsity as the provability of the denial (complement). Now it is easy to present a definition in which both a and $\sim a$ are provable, the most trivial one being

$$\mathbb{D}_{10}\left\{\begin{array}{rl} a & \Leftarrow \\ \sim a & \Leftarrow \end{array}\right.$$

So there is no way of claiming that contrary opposites *cannot* be proved.

Figure 1

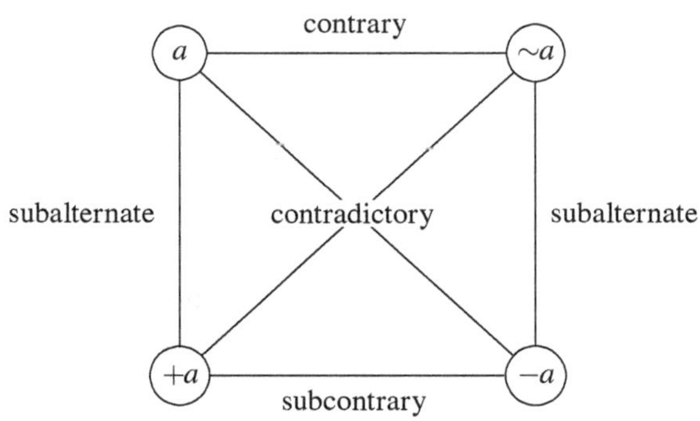

Therefore we propose a change of perspective. Rather than claiming that the square depicts deductive relationships naturally enforced by our inference rules, we use it as a structure to classify definitions. Some definitions may obey principles represented in the square, others may not. We might, e.g., say that a definition *respects* contrariety, because it is consistent, and so on. In this sense, the square is read as specifying certain deductive features of a given definition. We do not argue here that these specifications have to be fulfilled.

Thus we propose the following (metalinguistic) definition:

Definition 1. A definition $\mathbb{D}$ *respects*

(1) *contrariety*, if for no atom a, both $\vdash_{\mathbb{D}} a$ and $\vdash_{\mathbb{D}} \sim a$,
(2) *subcontrariety*, if for no atom a, both $\vdash_{\mathbb{D}} \widehat{+a}$ and $\vdash_{\mathbb{D}} \widehat{-a}$
(3) *weak contradictoriness*, if for no atom a, both elements of a complementary pair $(a/-a)$ or $(\sim a/+a)$ are provable

(4) *strong contradictoriness*, if for every atom a, exactly one element of each complementary pair $(a/{-a})$ and $({\sim a}/{+a})$ is provable.

$\mathbb{D}$ *weakly* respects the *square of opposition*, if it respects (1)–(3). It *fully* respects the *square of opposition*, if it respects (1)–(4).

Obviously, respecting contrariety and respecting subcontrariety means the same. This is something we are used to from the traditional square of opposition. To say that *SaP* and *SeP* are not both true is the same as saying that their negations $\widehat{SaP}$ and $\widehat{SeP}$, i.e., *SoP* and *SiP*, are not both false. This is nothing but the duality of true and false with respect to negation (assuming, of course, that *SoP* and *SiP* are indeed the negations of *SaP* and *SeP*, respectively).

The distinction between weak and strong contradictoriness is due to the fact that in our proof-theoretic framework the provability of a literal ℓ and the provability of its complement $\hat{\ell}$ do not necessarily exhaust all possibilities, as it may happen that neither ℓ nor $\hat{\ell}$ is provable in the given definition. Respecting strong contradictoriness corresponds to some sort of completeness and is therefore a very strong condition.

An example demonstrating that in (1) both a and ${\sim a}$ can be false, which in our context is expressed by $\vdash_{\mathbb{D}}\hat{a}$ and $\vdash_{\mathbb{D}}\widehat{\sim a}$, would be given by an a which is not defined in $\mathbb{D}$, i.e., for which there is no clause in $\mathbb{D}$ with a or ${\sim a}$ as its head. In this case, $\vdash_{\mathbb{D}}{-a}$, which is the same as $\vdash_{\mathbb{D}}\hat{a}$, and $\vdash_{\mathbb{D}}{+a}$, which is the same as $\vdash_{\mathbb{D}}{\sim a}$, hold vacuously. At the same time, this is an example demonstrating that in (2) both $+a$ and $-a$ can be true, which in our context is expressed by $\vdash_{\mathbb{D}}{+a}$ and $\vdash_{\mathbb{D}}{-a}$.

In the previous section, we took the position that indirect judgements should be intrinsically related to their direct forms by means of the rules of subalternation. However, from the specification perspective taken now, one might consider giving up postulating (*Sub*) as built-in rules. One might instead argue that the rules of definitional closure and of definitional reflection are the only genuine rules putting a definition into action, and in this sense are sufficient to explicate the meaning of direct and indirect assertion and denial. Subalternation would then be a non-primitive inference rule that may (but need not) be admissible on the basis of the closure and reflection rules. This means that the admissibility of subalternation would not always be trivial, but a matter of (mathematical) fact. If we choose this option, we must add a clause to the above definition yielding the following definition.

Definition 2. A definition $\mathbb{D}$ respects

(1) *contrariety*, if for no atom a, both $\vdash_{\mathbb{D}} a$ and $\vdash_{\mathbb{D}} \sim a$,
(2) *subcontrariety*, if for no atom a, both $\vdash_{\mathbb{D}} \widehat{+a}$ and $\vdash_{\mathbb{D}} \widehat{-a}$
(3) *subalternation*, if the rules (*Sub*) are admissible with respect to $\mathbb{D}$
(4) *weak contradictoriness*, if for no atom a, both elements of a complementary pair $(a/-a)$ or $(\sim a/+a)$ are provable
(5) *strong contradictoriness*, if for every atom a, exactly one element of each complementary pair $(a/-a)$ and $(\sim a/+a)$ is provable.

$\mathbb{D}$ *weakly* respects the *square of opposition*, if it respects (1)–(4). It *fully* respects the *square of opposition*, if it respects (1)–(5).

Whether to consider this option or not must be left open here. In the framework without subalternation as a primitive rule, we can show the following:

Theorem 1. In the framework without subalternation as a primitive rule, every definition $\mathbb{D}$ respects weak contradictoriness.

Demonstration. We proceed by induction on the length of proofs. Suppose a and $-a$ have both been proved. Then the last step of the proof of a has the form

$$\frac{\ell_1 \ \dots \ \ell_n}{a}$$

where

$$a \Leftarrow \ell_1, \dots, \ell_n$$

is a definitional clause for a, whereas the last step of the proof of $-a$ has the form

$$\frac{\dots \ \widehat{\ell_i} \ \dots}{-a}$$

for some i. This means that there are proofs of ℓ_i and $\widehat{\ell_i}$ of shorter length.

This result is an argument for the framework without subalternation, as it guarantees a property one should expect to hold in any case. With subalternation as a primitive rule, we might be able to prove both a and $-a$ (e.g., in the case of $\mathbb{D}_{10}$) and thus would destroy the idea of the complementary relationship expressed by the operation $\frown$.

As a corollary of this theorem we immediately obtain:

Corollary. If $\mathbb{D}$ *respects subalternation, it respects contrariety.*

This result is quite significant, too. It relates the consistency of a definition to the relation between direct and indirect assertion/denial. Only in a consistent definition we can infer the indirect from the direct forms.

If subalternation is presupposed as a primitive rule, we only have the following: result.

Theorem 2. In the framework with subalternation as a primitive rule, every definition $\mathbb{D}$ *which respects contrariety respects weak contradictoriness.*

Demonstration. Given proofs of a and of $-a$, the second proof cannot be obtained by subalternation (since otherwise we would have proofs of a and of $\sim a$). This means that we can argue as in the demonstration of Theorem 1.

The following is an example of a definition, which fully respects the square of opposition.

$$\mathbb{D}_{11}\begin{cases} a & \Leftarrow & b,c \\ \sim a & \Leftarrow & -b \\ \sim a & \Leftarrow & -c \\ b & \Leftarrow & \\ \sim b & \Leftarrow & -b \\ \sim c & \Leftarrow & \\ c & \Leftarrow & +c \end{cases}$$

As can easily be checked, the literals $\sim a, -a, b, +b, \sim c, -c$ are provable, whereas $a, +a, \sim b, -b, c, +c$ are unprovable in $\mathbb{D}_{11}$.

7. Discussion

We presented a framework for definitional reasoning, in which two denials and two assertions are distinguished. In contradistinction to approaches that introduce some strong form of negation into logical reasoning such as Nelson (1949), we also extended the notion of assertion and received full symmetry between assertion and denial. This symmetry led to our square of opposition. As there is a weak ("indirect") denial opposed to strong ("direct") assertion, there is a weak ("indirect") assertion opposed to strong ("direct") denial. The contradictory opposite to the strong concept is the opposite weak concept. The strong concepts mean that we can establish the proposition in question by means of the given clauses directly, applying in the last step an inference following the *direction* of the clause (thus "direct"). The weak concepts mean that we can establish a proposition by showing that the premisses of its opposite can be rejected. Here "rejection" does not mean *failure to prove*, but *proving the contradictory opposite* of at least one premiss. This reasoning is *indirect* as we have to use definitional reflection, i.e., we have to reflect on the definition as a whole. This represents a novel sense of duality which is different from treatments of weak and strong negation.

The notions of "direct" and "indirect" assertion and denial appeal to intuitions which are related to the idea of direct vs. indirect proofs in Dummett-Prawitz-style proof-theoretic semantics. However, there they are used to describe the relationship between introduction and elimination rules and not between assertion and denial. Moreover, we are here dealing with reasoning on the basis of arbitrary atomic definitions, i.e., with definitional reasoning in general. Nonetheless, in spite of these differences, there is some underlying general duality at work which manifests itself in various special forms like "assertion" vs. "assumption", "assumption" vs. "denial", "conditions" vs. "consequences", "introduction" vs. "elimination", "right" vs. "left" [sequent calculus] – a duality, which needs to be further elaborated (see Hallnäs 1991, 2006).

In order to gain a clear picture, we have exclusively dealt with the relation between assertion and denial, keeping aside elimination inferences as well as proofs from assumptions. So what we have sketched here, was only a first step. Our approach can be extended in various directions.

(1) We could combine the duality of assertions and assumptions (or of introduction and elimination rules) with the duality of assertions and denials (or of direct vs. indirect introduction rules), and investigate how these notions interact. This opens a new field of interesting observations with new harmony principles for the classification of definitions. As elimination rules bring with them the "standard" (normally intuitionistic) notion of negation, we may examine how direct denial relates to (and perhaps coincides with) standard negation.

(2) We might investigate what happens if we allow for implications to occur in the bodies of clauses. So far, we have only considered literals divided by commas as bodies. Including implication is not trivial at all, as one enters the field of non-monotonic inductive definitions. Furthermore, there are several options of denying implication, where

$$\widehat{a \to b} := a, \hat{b}$$

is only one possibility among others. Pursuing this idea would have to take ideas from connexive logic into account (see Wansing 2010).

(3) For real applications we must leave the area of propositional reasoning behind and consider clauses with first-order atoms containing individual variables. In that case defining definitional reflection for denial and assertion will require considerations quite analogous to inversion principles in standard proof-theoretic semantics. There are several options to formulate them, all of which require a careful handling of variables and substitutions (see Schroeder-Heister 2007).

(4) One might develop logic programming paradigms, in which logic programs have the form of definitions with assertions and denials, and define a way to compute bindings (substitutions). This would lead into the area of extended logic programming where one considers negated heads in program clauses (see Damásio & Pereira 1998).

Conclusion

It is fascinating to see how such an elementary figure as the square of opposition can exhibit a structure relevant to an area that at first sight looks quite remote to it. I should like to thank Jean-Yves Béziau for asking me to contribute a paper to the First World Congress on the Square of Oppostionand for encouraging me to think about how to relate certain aspects of my work to the square. Special thanks are due to Wagner de Campos Sanz for numerous helpful comments on a draft of this paper, and for many stimulating discussions on its topic. I also thank two anonymous reviewers for various substantial comments and suggestions. – The completion of this work was supported by the ESF EUROCORES programme "LogICCC – Modelling Intelligent Interaction" (DFG grant Schr 275/15-1) and by the joint French/German ANR/DFG project "Hypothetical Reasoning: Logical and Semantical Perspectives" (DFG grant Schr 275/16-1).

References

Aczel, P. (1977). An introduction to inductive definitions. In J. Barwise (Ed.), *Handbook of Mathematical Logic* (pp. 739–782). Amsterdam: North Holland.

Damásio, C.V., & Pereira, L.M. (1998). A survey of paraconsistent semantics for logic programs. In D.M. Gabbay & P. Smets (Eds.), *Handbook of Defeasible Reasoning and Uncertainty Management Systems, Vol. 2* (pp. 241–320). Dordrecht: Kluwer.

Dummett, M. (1991). *The Logical Basis of Metaphysics*, London: Duckworth.

Frege, G. (1984). *Die Grundlagen der Arithmetik: Eine logisch mathematische Untersuchung über den Begriff der Zahl.* Breslau: Koebner. Centenary edition by C. Thiel, Hamburg: Meiner 1986.

Hallnäs, L. (1991). Partial Inductive Definitions. *Theoretical Computer Science*, *87*, 115–142.

Hallnäs, L. (2006). On the proof-theoretic foundation of general definition theory. In R. Kahle & P. Schroeder-Heister (Eds.), *Proof-Theoretic Semantics*. Special issue of *Synthese, 148,* 589–602.

Hallnäs L., & Schroeder-Heister, P. (1990/91). A proof-theoretic approach to logic programming. I. Clauses as rules. *Journal of Logic and Computation, 1,* 261–283; II. Programs as definitions, *ibid.,* 635–660.

Nelson, D. (1949). Constructible falsity. *Journal of Symbolic Logic, 14,* 16–26.

Schroeder-Heister, P. (1993). Rules of definitional reflection. In: *8th Annual IEEE Symposium on Logic in Computer Science (Montreal 1993)* (pp. 222–232). Los Alamitos: IEEE Computer Society Press.

Schroeder-Heister, P. (2004). On the notion of *assumption* in logical systems. In: R. Bluhm & C. Nimtz (Eds.), *Selected Papers Contributed to the Sections of GAP5, Fifth International Congress of the Society for Analytical Philosophy, Bielefeld, 22–26 September 2003* (pp. 27–48). Paderborn: mentis.

Schroeder-Heister, P. (2006). Validity concepts in proof-theoretic semantics. In R. Kahle & P. Schroeder-Heister (Eds.), *Proof-Theoretic Semantics. Synthese, 148* (Special issue), 525–571.

Schroeder-Heister, P. (2007). Generalized definitional reflection and the inversion principle. *Logica Universalis, 1,* 355–376.

Schroeder-Heister, P. (2011). Proof-theoretic semantics. In E. Zalta (Ed.), *Stanford Encyclopedia of Philosophy.* http://plato.stanford.edu.

Wansing, H. (2010). Connexive logic. In E. Zalta (Ed.), *Stanford Encyclopedia of Philosophy (Fall 2010 edition).* http://plato.stanford.edu.

2-Opposition and the Topological Hexagon

RÉGIS ANGOT-PELLISSIER

Abstract

From Aristotle's square, Blanché has developed a famous hexagon. After him, more logical hexagons have been discovered, among which Béziau's ones endowed with paracomplete and parconsistent negations, and even a logical cube has been exhibited. This sequence of geometrical figures, which are in fact n-bisimplexes, gives birth to Moretti-Pellissier n-opposition theory (for $n \geq 3$) where hexagons represent 3-oppositions and the logical cube 4-opposition. Two kinds of n-opposition exist: the weak kind and the strong kind. In this paper, we shall give a general theorem of geometrical characterization of the weak and the strong n-opposition for all $n > 0$. The most astonishing result of this theorem is that weak 2-opposition is in fact Aristotle's square, starting point of the whole n-opposition theory. Furthermore, its two horizontal segments can be proved to be one of them in intuitionistic opposition and the other one in paraconsistent opposition. From the tight link between these logics and Heyting and co-Heyting algebras, we deduce a topological square embedded in a more remarkable topological hexagon, whose vertices are open or closed subsets and which contains three kinds of opposition (classical, intuitionistic, paraconsistent).

Keywords: n-bisimplex, co-Heyting algebra, Heyting algebra, close negation, open negation, strong n-opposition, weak n-opposition, logical square, topological hexagon, intuitionistic opposition, paraconsistent opposition

1. Introduction

The theory of n-opposition, based on the notion of "n-bisimplex", was developed by A. Moretti (2004) in order to generalize the completions of Aristotle's square into 3-oppositional hexagons discovered by Blanché, Béziau, Moretti and Smessaert. To give a mathematical definition of the notion of n-opposition, we have in "Setting n-opposition" ([Pel]) translated Moretti's modal graphs theory (a given modal system's "modal graph" shows which arrows hold, if any, between this system's "basic modalities") into set-theory by identifying the letters of the modal graph with subsets of a given set. In doing so, we have exhibited a translation of the logic generated by the modal graphs into the Boolean logic of the subsets of a set. Therefore, we succeeded in reducing n-opposition to the simpler notion of "n-partition of the true".

But we also discovered in the same article a "weak" kind of n-opposition: this weak n-opposition forms the heart of the structure containing all n-oppositional figures. For instance, the weak hexagons form the logical cube, the heart of the "logical tetraicosahedron" exhibited in [Pel], which is the geometrical solid including all 3-oppositional hexagons. Therefore in [Pel], we have split the unique notion of n-opposition into the two notions of strong and weak n-opposition. Strong n-opposition is expressed as an n-partition of the true or of the total set. And weak n-opposition is an n-partition of a subset of the total set. (For a synthetic view of n-opposition theory, see [Mor2].) But we have defined this distinction only for $n>2$ and, for $n=2$, we have let the 2-opposition, understood as strong 2-opposition, be defined as the contradiction.

The fact that the question of weak 2-opposition was pertinent emerged from discussions with Moretti about our mathematical definition of weak and strong n-opposition. In order to discover what weak 2-opposition means, we shall examine in the first part of this article the notion of weak n-opposition and give a general theorem of geometrical characterization of the weak and of the strong n-opposition for all $n>0$, generalizing the fact that strong hexagons are planar and weak hexagons are non planar. By application of this theorem to the case $n=2$ in the second part of our article, we shall demonstrate that weak 2-opposition is a logical square and then discuss its properties. In the same part, to have an exhaustive view of n-opposition, we shall examine strong and weak 1-opposition.

This astonishing singularity of weak 2-opposition, whose figure is formally non connex, has forced us to examine the signification of the two horizontal segments from which the square is constructed. But our set translation of the *n*-opposition gives the clue: one segment is an intuitionistic opposition and the other a paraconsistent one. In order to illustrate these notions, we shall recall in a third part the notions of topological spaces with their open and closed subsets and the notions of Heyting and co-Heyting algebras. The main results recalled in this part are that the open subsets form an intuitionistic propositional calculus and that the closed subsets form a paraconsistent propositional calculus.

With these notions in hand, we shall exhibit in the fourth part the topological hexagon, the main discovery of this article, whose particularity is that it contains the three types of negation: the classical set-theoretical one, the open intuitionistic one and the closed paraconsistent one, the first one being its internal strong 2-opposition and the last two forming a topological square which is actually one of its weak 2-oppositions. We shall conclude by illustrating the topological hexagon with simpleinstantiations and by seeing what work remains in front of us.

2. Strong *n*-Opposition versus Weak *n*-Opposition

2.1 A Brief Historical Introduction to n-Opposition Theory

Let us begin by recalling the history of *n*-opposition theory. The starting point of the whole story is Aristotle's logical square:

Figure 1

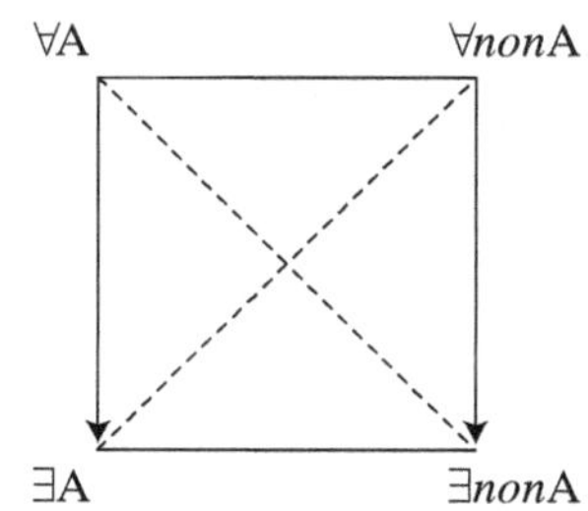

This logical square has four vertices and four types of binary relations between the vertices. The diagonal of the square represents the contradiction relation: two terms are in contradiction if and only if their conjunction is false and their disjunction is true. These two properties are the basic axioms of Aristotle's classical logic and are known respectively under the names of the non contradiction axiom and the excluded middle axiom. The former means that one logical proposition and its negation cannot be true in the same time and the latter means that reality can only be divided between one proposition and its negation without a third term or, in other terms, that it is not possible that a proposition and its negation are false at the same time. Today we know of non classical logics, among which intuitionistic logics, which do not verify the axiom of the excluded middle, and paraconsistent logics, which do not verify the axiom of non contradiction while the logic doesn't have every sentence as a theorem.

The two vertical arrows of the square name the subalternation relation; they are implications from the general to the particular. More interesting are the two other relations. The top edge of the square is the contrariety relation where two contrary terms have a false conjunction but their disjunction is not always true. The bottom edge named subcontrariety has the dual property: two subcontrary terms have a true disjunction but their conjunction is not always false. We can already see that the contrariety relation verifies the non contradiction axiom but not the excluded middle axiom. Dually, the subcontrariety relation verifies the excluded middle axiom but not the non contradiction one. Therefore, in Aristotle's square we have three kinds of negations: a classical one, the contradiction, and two others which have the flavour of, respectively, intuitionistic and paraconsistent negation. We shall develop this topic later in this article when, after studying the weak 2-opposition, we shall exhibit the topological square extracted from the topological hexagon.

Twenty five centuries afterward, Sesmat (1951) and Blanché (1953) discovered independently that Aristotle's logical square is not complete and so they proposed (in [Ses] and [Bla]) the completion of the square into the following logical hexagon:

Figure 2

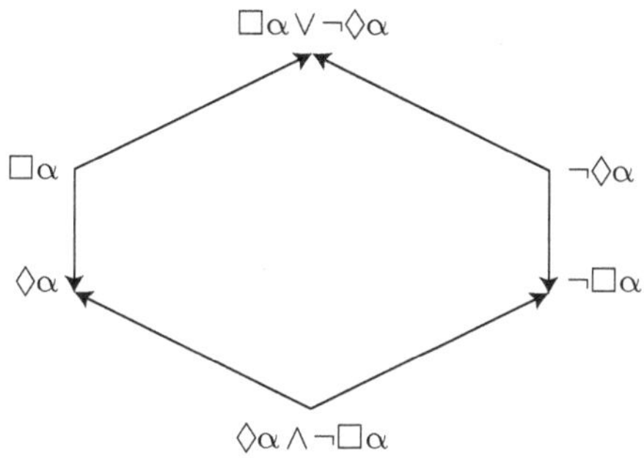

This hexagon contains the four relations forming Aristotle's square: the contradiction relation represented by central symmetry around the hexagon's center, the subalternation relation or implication in the border of the hexagon, the contrariety relation and its dual, the subcontrariety relation, which become here ternary relations represented by the following triangles:

- the contrariety triangle:

Figure 3

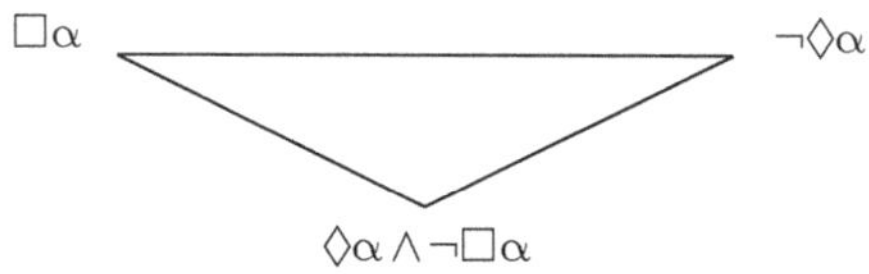

- the subcontrariety triangle (central symmetric of the contrariety triangle):

Figure 4

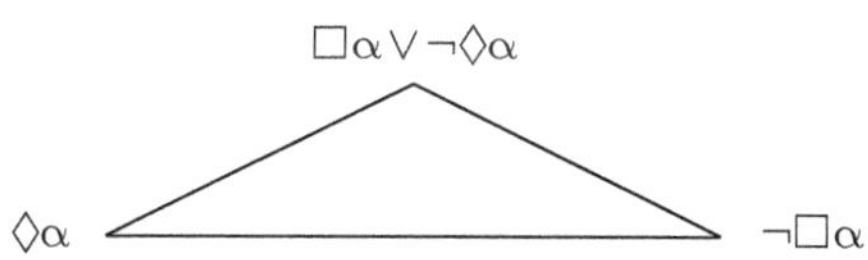

This hexagon is the perfect example of strong 3-opposition: it is a 2-bisimplex (an hexagon) built from a 2-simplex (a triangle) of contrariety and its symmetric by central symmetry, the subcontrariety triangle, by connecting non-contradictory vertices of the first triangle into the second. In the following figure, we can see the fitting of the two symmetrical triangles into the hexagon.

Figure 5

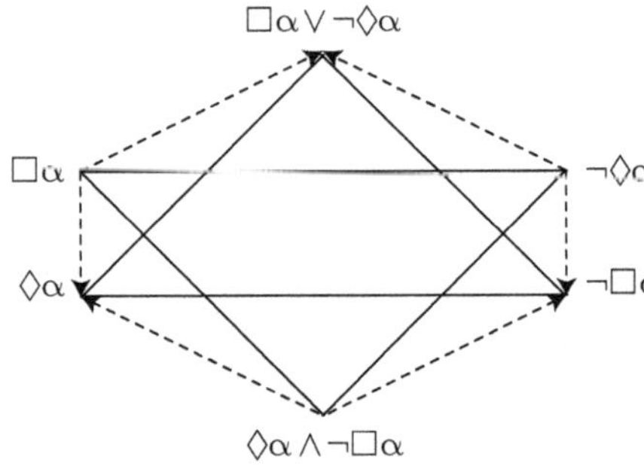

Furthermore we can express logically strong 3-opposition by following our set translation of the *n*-opposition (see [Pel]): strong 3-opposition is entirely determined by its triangle of contrariety which is nothing more than a 3-partition of the true. More explicitly, the three propositions in contrariety is expressed as three subsets of a set with the following properties: their set-theoretical union is the total set and their pair wise intersections are the empty set. We translate in logical terms these properties as follows: their disjunction is true and their pair wise conjunctions are false. Symmetrically, the three propositions in subcontrariety have the dual properties: their intersection is the empty set and their pair wise unions are the total set (or in logical terms, their conjunction is false and their pair wise disjunctions are true), so that we can say that the subcontrary propositions form a 3-co-partition of the false.

Béziau in [Bez], Moretti in [Mor] and Smessaert in [Sme] have discovered five other logical hexagons in Blanché style, so that we have all in all six logical hexagons. In order to understand the relative positions of these hexagons between them, to construct a geometrical solid included all logical hexagons and to generalize the ternary contrariety relation to a *n*-ary opposition relation, Moretti in [Mor] has developed the theory of

n-opposition and created the useful tool of modal graphs in order to exhibit ("decorate") the n-oppositional figures. Therefore Moretti's early definition of n-opposition follows the above remark about the construction of Blanché square and it amounts to saying that a n-oppositional figure (for any $n \geq 3$) is the fitting together of two symmetrical $(n-1)$-simplexes, representing n-ary contrariety and subcontrariety, whose vertices are connected as in the Blanché hexagon.

2.2 The Geometrically Perplexing Weak 3-Opposition Case

But, while undertaking to give a mathematical framework to Moretti's n-opposition theory, we have discovered that this type of hexagon is not the only type of 3-opposition. There exist non planar hexagons (or skew hexagons, whose vertices do not lie in a plane) which follow the rules of Blanché's hexagon except for the so-called "head" and "tail". In Blanché's, as in Béziau's, Moretti's and Smessaert's hexagons, the proposition on the top vertex, called the head, is the disjunction of its two neighbours and dually the tail or bottom proposition is the conjunction of its two neighbours. But this property tying the head and tail with the other four propositions is forced upon the structure of the logical hexagon. And in fact we have discovered a lot of logical hexagons which do not verify the rule of tail and head (see [Pel]). Therefore we have called weak 3-opposition this free type of 3-opposition because such hexagons are non planar, as the following one extracted from [Pel] demonstrates:

Figure 6

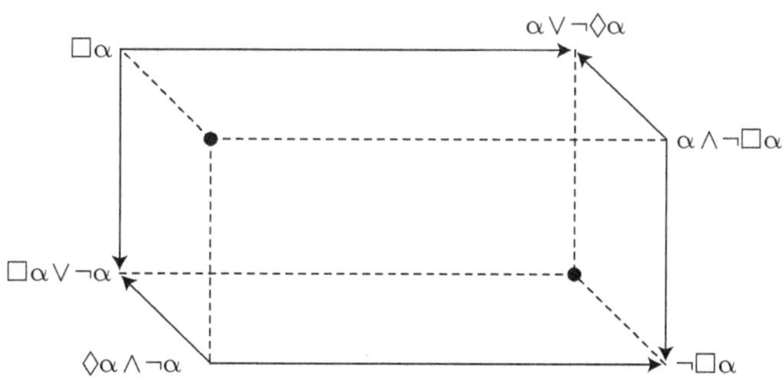

Geometrically, this weak hexagon is built as Blanché's hexagon, which is a 2-bisimplex, from a 2-simplex (a triangle) of contrariety and a symmetric one by central symmetry, the subcontrariety triangle, by connecting non-contradictory vertices of the first triangle into the second. But here the two symmetric triangles are in two parallel planes and therefore their fitting is not planar as it is in the case of the strong hexagon. Why? Because in logical terms, deduced from our set translation of Moretti's theory, weak 3-opposition is formed by a triangle of contrariety representing this time a 3-partition of a neither true nor false proposition and by its central symmetric triangle, the triangle of subcontrariety, which is the 3-co-partition of the classical negation of the above proposition. In the following figure, we can see the two triangles living in parallel planes from which the non planar hexagon is constructed.

Figure 7

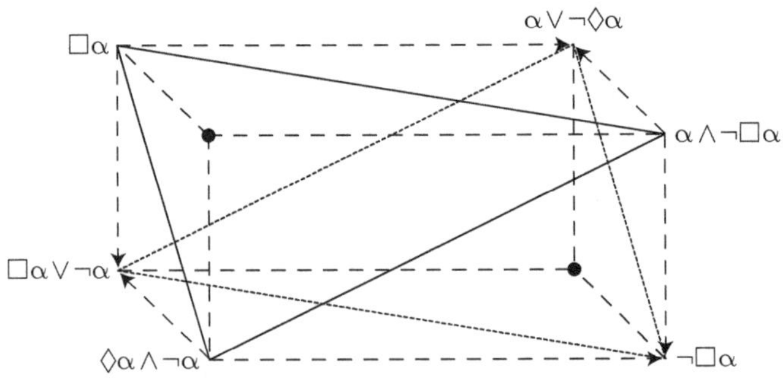

Therefore we can better see the difference between the two types of 3-opposi–tion: the strong one is a 3-partition of the true and its dual 3-co-partition of the false, whereas the weak one a 3-partition of some neither true nor false proposition and its dual 3-co-partition of the negation of this proposition. To be in a triangle of contrariety or subcontrariety means to be in equal distance of its gravity center. In the first case, these gravity centers are respectively the true and the false. And, as the central symmetry is the contradiction, its center is the true as well as the false, i.e. true and false are identified in the center of contradictory symmetry. That is why in the strong case the two gravity centers of the contrariety and subcontrariety triangles are identified with the center of the central symmetry of contradiction. Therefore the triangles, which are naturally in two parallel planes, are in this case coplanar, rendering this impression of two symmetrical triangles fitting together into the star-shaped hexagon. But in the weak case, the two gravity centers of the triangles are two mutually contradictory propositions and neither one nor the other being true or false can be identified with the center of the central symmetry of contradiction, so that the two symmetrical triangles remain in two parallel planes.

2.3 *Definitions of the Main n-Oppositional Notions*

As this reasoning can be generalized to the case of any integer n, after recalling the definitions of strong and weak n-opposition given in [Pel], we can establish a theorem about the geometrical properties of their respective geometrical figures. To begin with, we shall define the notion of n-bisimplex on which n-opposition is constructed.

Definition 1. A *n-bisimplex*, for $n \geq 1$, is formed by a n-simplex, called the contrariety simplex, and by its symmetric by central symmetry, called the subcontrariety simplex, each vertex of the first simplex being orientedly connected to each non-symmetric vertex of the second, by the subalternation relations. The symmetric vertices are naturally contradictory.

As in [Pel], we must remark that these n-bisimplexes are called $a(n+1)$-structures in [Mor] and that these $a(n+1)$-structures can be true, false or tautological in accordance with the truth value of subalternation relations. In this paper as in [Pel], we shall only consider n-bisimplexes whose subalternation relations are true and non tautological, i.e. implications are always true implications but never equivalences.

We can now give the definitions of the two types of n-opposition.

Definition 2. A *strong n-opposition*, for $n \geq 1$, is a $(n-1)$-bisimplex whose $(n-1)$-simplex of contrariety has the property that pair wise conjunctions of its vertices are false and whose disjunction of all its vertices is true.

In this case, we shall say that the n vertices of the $(n-1)$-simplex of contrariety form a *n-partition of the true*. And dually, we shall say that the

n vertices of the $(n-1)$-simplex of subcontrariety form a *n-copartition of the false* to mean that their pair wise disjunctions are true and their global conjunction is false.

Definition 3. A *weak n-opposition*, for $n \geq 1$, is a $(n-1)$-bisimplex whose $(n-1)$-simplex of contrariety has the property that pair wise conjunctions of its vertices are false but whose disjunction of all its vertices is not true.

In this case, the n vertices of the $(n-1)$-simplex of contrariety do not form a n-partition of the true but only a n-partition of a neither true nor false proposition and, dually, the symmetrical n vertices of the $(n-1)$-simplex of subcontrariety form a n-copartition of the classical negation of this proposition.

2.4 Theorem of Geometrical Characterization of Weak and of Strong n-Opposition

After this recall of the main definitions of n-opposition theory, we can now establish, using the following theorem, the geometrical distinction between the two types of n-opposition.

Theorem 1. In a weak n-opposition, for $n \geq 1$, the two $(n-1)$-simplexes of contrariety and subcontrariety are in parallel hyperplanes.

In a strong n-opposition, for $n \geq 1$, the two $(n-1)$-simplexes of contrariety and subcontrariety are in the same hyperplane and so fit together.

Proof: First of all, recall that the n-bisimplex lives in a n-dimensional affine space. Furthermore each $(n-1)$-simplex generates an affine hyperplane, or affine space of dimension $(n-1)$, of which it is a barycentric reper. As the two $(n-1)$-simplexes of contrariety and subcontrariety are symmetrical by a central symmetry whose center is the origin of the affine reper, their generated hyperplanes are parallel. This demonstrates the first part of the theorem.

To prove the second part of the theorem, we must remark that in the strong case we have equations between the vertices of the $(n-1)$-simplexes of contrariety and subcontrariety: the first equation is given by the n-partition of the true, the second one by the n-copartition of the false and the others derive from the fact that one $(n-1)$-simplex is the symmetric of the other by the central symmetry of contradiction. Let $(V_1,\dots V_n)$ be the vertices of the contrariety simplex. Let $(V'_1,\dots V'_{n-1})$ be their symmetrical vertices in the subcontrariety simplex. As $(V_1,\dots V_n)$ form a n-partition of the true, V_n and the disjunction of $(V_1,\dots V_{n-1})$ are complementary. But as V'_n is the complementary set of V_n, we have that V'_n is equal to the disjunction of $(V_1,\dots V_{n-1})$. Furthermore, if we take the negation of this equation, we find that V_n is the conjunction of $(V'_1,\dots V'_{n-1})$. Therefore each vertex of one $(n-1)$-simplex is barycentrically defined from the vertices of the other $(n-1)$-simplex and consequently each vertex of one $(n-1)$-simplex is in the hyperplane defined by the other $(n-1)$-simplex. So we have proved that one $(n-1)$-simplex is included in the hyperplane generated by the other one and this proves the second part of the theorem. QED.

We can view the non hyperplanar weak n-opposition as a figure constructed from the two $(n-1)$-simplexes living in parallel hyperplanes and whose centers of gravity are symmetrical by respect with the center of the contradiction symmetry. And we can understand these centers of symmetry in the following terms. By definition, a weak n-opposition is formed by a n-partition of a neither false nor true proposition, which we will denote by P, and by a n-copartition of the classical negation of P, denoted *nonP*. Therefore, we can say that the gravity center of the contrariety $(n-1)$-simplex is P and the gravity center of the subcontrariety $(n-1)$-simplex is *nonP*. The symmetry of the gravity centers is evidently the contradiction.

From that interpretation, we can first better understand the non hyperplanar characteristic of the weak n-opposition. But we can also understand the quintessence of the n-opposition. Actually, we can say of weak n-opposition that the n vertices of the contrariety simplex are logically equidistant from P and those of the subcontrariety simplex are logically equidistant from *nonP*. And in applying this remark to the strong case, we can say that the n vertices of the contrariety simplex are logically equidistant from the true and those of the subcontrariety simplex are logically equidistant from the false. But in the strong case, the true and the false, viewed as gravity centers of the two dual simplexes, are identified with the symmetry center of the contradiction symmetry which is also the gravity center of the n-bisimplex.

3. Weak 2-Opposition

The above definitions of strong and weak n-opposition and the theorem of their geometrical characterization are given in this article for $n \geq 1$ while they were given in [Pel] only for $n \geq 3$. In fact, it is easy to visualize these definitions andthe theorem for $n \geq 3$ because they deal with triangles and higher-dimensional simplexes. But for $0 < n < 3$, these definitions and the theorem are more difficult to understand. In this part, we will unknot these tricky cases.

To begin with, let us remark that a 0-opposition is undefinable, because it would involve the notion of (-1)-simplex, which simply does not exist. But 1-opposition and 2-opposition may be defined because the 0-simplex and 1-simplex involved exist and are respectively a point in a 2-dimensional affine space and a segment in a 3-dimensional affine space.

We shall now give as corollary of the above theorem the geometrical realizations of the weak and strong 2-opposition.

Corollary 1. A weak 2-opposition is a logical square whose top segment is the contrariety segment and the bottom segment is the subcontrariety one, whose top vertices imply the bottom one and whose diagonals between symmetrical vertices are contradiction relations. A strong 2-opposition is a segment of contradiction where the two vertices are in contradiction one with another.

Proof: By construction, weak and strong 2-oppositions have a 1-bisimplex as support. By definition, a 1-bisimplex is formed by two symmetrical 1-simplexes, i.e. by two symmetrical segments, and the vertices of the segment of contrariety are relied by implication to the vertices of the segment of subcontrariety. The symmetrical segments are evidently in contradiction.

In the case of weak 2-opposition, the two segments or 1-simplexes are parallel by application of the above theorem, so that the geometrical figure depicted above is simply a logical square with its four types of relations.

In the case of strong 2-opposition, an interesting phenomenon occurs: the 1-bisimplex degenerates into a 1-simplex. In fact, the strong 2-opposition is formed by a 2-partition of the true and its symmetrical 2-copartition of the false. Let A and B denote the two propositions forming the 2-partition of the true. We have then that *nonA* and *nonB* form a 2-copartition of the false. But what does it mean for A and B to form a 2-partition of the true? It means that $A \wedge B = \perp$ and $A \vee B = op$, but that is to say that A and B are contradictory, so that we have the following relations: $A = nonB$ and $C = nonA$. Therefore the 2-copartition of the false is nothing other than the 2-partition of the true, which in turn is nothing other than the contradiction relation. And as the contradiction relation is geometrically a segment, i.e. a 1-simplex, so is the strong 2-opposition and its 1-bisimplex. QED.

And what about the case $n = 1$? This case of formal interest is included in the following corollary.

Corollary 2. A weak 1-opposition is a segment of contradiction.

A strong 1-opposition is the symmetry center of the contradiction relation.

Proof: A 0-simplex is a mere point. By application of the above theorem, a 0-bisimplex is formed of two symmetrical points with no relation of subalternation between them because of their symmetry.

In the weak case, these two points remain "parallel" with only their symmetrical relation of contradiction, so that the weak 1-opposition is the contradiction, as is the strong 2-opposition.

For the strong case, we need to contemplate the 1-partition of the true and the 1-copartition of the false. But a 1-partition of the true is the singleton formed by the true and its dual the 1-copartition of the false is the singleton formed by the false. But these two points, true and false, are identified into the center of symmetry of the contradiction relation, so that the strong 1-opposition is this particular point, from which all the n-opposition theory is constructed. QED.

With these two corollaries, we have finally a complete definition of weak and strong n-opposition. But we have also the interesting case of the weak 2-opposition, which permits us to return, after a long journey, to the logical square. But this time, from the point of view of the simplicial construction, we have a new meaning of this weak 2-oppositional logical square: it is constructed from two parallel and symmetric segments, one of contrariety and the other of subcontrariety. The subalternation liaison connecting face-to-face vertices of the parallel segments comes in a second time and we will not study this part of the logical square. Actually, the two segments of contrariety and subcontrariety represent, respectively, intuitionistic and paraconsistent opposition. Let us denote A and B the vertices of the contrariety segment and A' and B' their symmetric in the subcontrariety segment. In the case of the contrariety segment, we have $A \wedge B = \bot$ and $A \vee B \neq \top$ because we are in a weak 2-opposition where A and B is a 2-partition of a neither true nor false proposition. Therefore this contrariety is intuitionistic as the axiom of excluded middle is not verified. Dually, for the subcontrariety segment, the 2-copartition of the classic negation of the neither true nor false proposition has the following properties: $A \vee B = \top$ and $A \wedge B \neq \bot$. This time, we are in a paraconsistent opposition because the non-contradiction axiom is no longer verified and the induced logic does not collapse.

In order to illustrate that important property of weak 2-opposition, we need the notion of topological space and of Heyting and co-Heyting algebras. Therefore we shall recall in the next section these notions with which not all readers are necessarily familiar.

4. Intuitionistic Heyting Algebra of Open Sets and Paraconsistent Co-Heyting Algebra of Closed Sets

4.1 Topology

In this section, we shall present the intuitionistic logic of the open subsets of a topological space and the dual paraconsistent logic of the closed subsets. For this, we will begin by presenting the notion of a topology.

Definition 4. A *topological space* is a doublet $(X,\mathcal{O})$ where X is a set and $\mathcal{O}$ a subset of $P(X)$, the set of subsets of X, verifying the following axioms:
- $\varnothing$ and X are included in $\mathcal{O}$,
- all unions of elements of $\mathcal{O}$ are also in $\mathcal{O}$,
- all finite intersections of elements of $\mathcal{O}$ are also in $\mathcal{O}$.

Elements of $\mathcal{O}$ are called *open subsets* of X (or "open") and their complementary sets are called *closed subsets* of X (or "closed").

More simply, we can describe a topological space as a space recovered by its open subsets with the three properties that the space itself and the empty set are opens, all unions of opens are opens and all finite intersections of opens are opens. A given space can support different topologies whose refinement spreads from the trivial topology, where the only open subsets are the space itself and the empty set, to the discrete topology, where all subsets are opens.

Let us talk also of the closed subsets. They are the complementary sets of the open subsets and their axioms are the dual axioms of the open subsets, that is:

- $\varnothing$ and X are closed,
- all intersections of closed subsets are also closed subsets,
- all finite unions of closed subsets are also closed subsets.

In the trivial topology, only the space itself and the empty set are closed and, in the discrete topology, all subsets are closed.

Topologically, two operations permit one to transform any subset of a topological space respectively into an open and a closed subset.

Definition 5. Let X be a space with a topology and Y a subset of X.

The *interior* of Y, denoted *int*(Y), is the greatest open subset included in Y. It is the union of all open subsets included in Y.

The *closure* of Y, denoted *clos*(Y), is the smallest closed subset including Y. It is the intersection of all closed subsets including Y.

The *boundary* of Y, denoted *Fr*(Y), is the border between the interior of Y and the closure of Y. In set-theoretical language, we have the following definition:

$Fr(Y) = clos(Y) \setminus int(Y)$. From these three definitions, it follows that the boundary of Y is a closed subset whose interior is empty.

These definitions being given, we can clearly see that the interior of an open subset is the open subset itself and the closure of a closed subset is the closed subset itself. Furthermore, no open subset includes its boundary, except if it is also closed, and all closed subsets include their boundary.

To visualize these operations, we shall take the example of *IR* with its usual topology, whose open subsets are unions of intervals of the following types: $]a,b[$, $]-\infty,a[$ and $]b,+\infty[$, where a and b are real numbers. The closed subsets are finite unions of intervals of the following types: $[a,b]$,

]$-\infty$,*a*] and [*b*,$+\infty$[, where *a* and *b* are real numbers. Let us now take the interval [0,1[, which is neither a closed nor an open subset. Its interior is the open subset]0,1[, its closure the closed subset [0,1] and its boundary is the pair {0,1} (and it is easily verified that the boundary is a closed subset whose interior is empty).

With these operations, each open subset can be closed into a closed subset and each closed subset has an open subset as interior. The border between an open subset and its closure, as well as the border between a closed subset and its interior, is actually the boundary. Furthermore, in combining these operations with the set complementarity, we obtain the two following internal negations or internal complementarities.

Definition 6. Let *X* be a space with a topology, *O* an open subset of *X* and *C* a closed subset of *X*. Let the apostrophe denote the set complementarity operation.

The *open negation* of *O*, denoted *notO*, is the interior of the set complement of *O*, that is *int*(*O′*).

The *closed negation* of *C*, denoted *notC*, is the closure of the set complement of *C*, that is *clos*(*C′*).

Why these complex definitions of open and closed negations? Because by definition, the complementary set of an open subset is closed and the complementary set of a closed subset is open. Except in topologies where all open subsets are also closed, as it is the case with the trivial and discrete topologies, the set complementarity operation does not remain in the same field, i.e. it sends open subsets to closed subsets and *vice versa*. But we can remedy this topological disability of the set complementarity operation by means of the interior and closure operations which assure us that the result of the operations remains respectively open or closed. We shall see examples of these internal negations in the last part of this article.

From these definitions, we can deduce the following phenomena.

- For the open negation, *O* and *notO* are disjuncted but their union is not *X* because *Fr*(*O*) is missing. In fact, *O*, *notO* and *Fr*(*O*) form a 3-partition of *X*.
- For the closed negation, the union of *C* and *notC* is *X* but *C* and *notC* are not disjuncted because *Fr*(*C*) is included in both *C* and *notC*.

To understand the logical properties of these two negations, we shall recall in the following subsection the notions of Heyting and co-Heyting algebras.

4.2 Heyting Algebra and its Dual

A Heyting algebra is an easy tool to understand intuitionism because of the simplicity of that algebra and because of the classical result: all intuitionistic propositional calculi are equivalent to Heyting algebras. Here is the definition of a Heyting algebra.

Definition 7. A *Heyting algebra* is an ordered set $(H,\leq)$ equipped with:

- a minimal element denoted 0 such that, for all element a, we have $0 \leq a$,
- a maximal element denoted 1 such that, for all element a, we have $a \leq 1$,
- for each pair of elements (a,b), an element denoted $inf(a,b)$ with the following properties: $inf(a,b) \leq a$, $inf(a,b) \leq b$ and, for all element c such that $c \leq a$ and $c \leq b$, we have $c \leq inf(a,b)$,
- for each pair of elements (a,b), an element denoted $sup(a,b)$ with the following properties: $a \leq sup(a,b)$, $b \leq sup(a,b)$ and, for all element c such that $a \leq c$ and $b \leq c$, we have $sup(a,b) \leq c$,
- for each pair of elements (a,b), an element denoted $a \Rightarrow b$ with the following property: for all element c, $inf(c,a) \leq b$ if and only if $c \leq (a \Rightarrow b)$.

The translation between a Heyting algebra and an intuitionistic propositional calculus is the following:

- formulas are the elements of the algebra,
- the inference is the order relation,
- $\bot = 0$ and $\top = 1$,
- $A \vee B = sup(a,b)$ and $A \wedge B = inf(a,b)$,
- the implication gives the elements $a \Rightarrow b$,
- the intuitionistic negation of a is defined as $a \Rightarrow 0$.

A fundamental result of category theory is that Heyting algebras are in biunivocal correspondence with categories of open subsets. The correspondence is as follows:

- the open subsets of a topological space X are the elements of the algebra,
- the inclusion is the order relation,
- $\varnothing = 0$ and $X = 1$,

- $A \cup B = sup(a,b)$ and $A \cap B = inf(a,b)$,
- $int(A' \cup B) = A \Rightarrow B$.

From these identifications, we observe that the open subsets of a topological space form an intuitionistic propositional calculus. And the intuitionistic negation of an open subset U, defined by $U \Rightarrow 0$, is by the preceding correspondence $int(U' \cup \varnothing) = int(U')$ and that is just what we have called above the open negation of U. That proves that the open negation is an intuitionistic negation. And we have seen above that U and *nonU* are disjuncted, logically their conjunction is false, but their union is not X in general, i.e. their logical disjunction is not true, so that the open subsets verify the axiom of non contradiction but not the axiom of the excluded middle, because the "excluded middle" here is actually the boundary of U which is not an open subset and so does not appear in the logic of open subsets.

To understand the closed subsets logic, we must recall first that the closed subsets are dual to the open subsets. And this fact has a categorical meaning, so that we need the categorical dual of Heyting algebra, which is called co-Heyting algebra (or Brouwer algebra) whose definition is the following:

Definition 8. A *co-Heyting algebra* is an ordered set $(H,\leq)$ equipped with:
- a minimal element denoted 0 such that, for all element a, we have $0 \leq a$,
- a maximal element denoted 1 such that, for all element a, we have $a \leq 1$,
- for each pair of elements (a,b), an element denoted $inf(a,b)$ with the following properties: $inf(a,b) \leq a$, $inf(a,b) \leq b$ and, for all element c such that $c \leq a$ and $c \leq b$, we have $c \leq inf(a,b)$,
- for each pair of elements (a,b), an element denoted $sup(a,b)$ with the following properties: $a \leq sup(a,b)$, $b \leq sup(a,b)$ and, for all element c such that $a \leq c$ and $b \leq c$, we have $sup(a,b) \leq c$,
- for each pair of elements (a,b), an element denoted $a - b$ with the following property: for all element c, $a \leq sup(b,c)$ if and only if $a - b \leq c$.

The translation between a co-Heyting algebra and a paraconsistent propositional calculus dual of an intuitionistic propositional calculus (see details of this calculus in [Que]) is the following:
- formulas are the elements of the algebra,
- the inference relation is the order relation,
- $\bot = 0$ and $\top = 1$,

- $A \vee B = sup(a,b)$ and $A \wedge B = inf(a,b)$,
- the logical difference gives the elements $a - b$,
- the paraconsistent negation of a is defined as $1 - a$.

By applying categorical duality to the biunivocal correspondence between Heyting algebras and categories of open subsets, we obtain a biunivocal correspondence between co-Heyting algebras and categories of closed subsets. The correspondence is as follows:
- the closed subsets of a topological space X are the elements of the algebra,
- the inclusion is the order relation,
- $\varnothing = 0$ and $X = 1$,
- $A \cup B = sup(a,b)$ and $A \cap B = inf(a,b)$,
- $clos(A \cap B') = A - B$.

Therefore the closed subsets of a topological space X form a paraconsistent propositional calculus. And the paraconsistent negation of a closed subset C, defined by $1 - C$, is by the preceding correspondence $clos(X \cap C') = clos(C')$ and that is actually what we have called previously the closed negation of C. That proves that the closed negation is a paraconsistent negation. As we have seen above, the union of C and *nonC* is X, i.e. logically their disjunction is true, but they are non disjuncted, i.e. their conjunction is not false, because of their mutual frontier. So the closed subsets verify the axiom of the excluded middle but not the axiom of non contradiction. Furthermore, the logic does not collapse although contradictions do exist and that is because contradictions are precisely localized in the boundary.

With these notions, we can exhibit the topological hexagon, the main discovery of this article.

5. The Topological Hexagon

We come now to the principal result of this article: the topological hexagon. This particular hexagon has appeared during the reflection about the weak 2-opposition. It is a good illustration of the combination of intuitionistic and paraconsistent logic inside the weak 2-opposition and more globally inside the logical hexagon. Here is the topological hexagon.

Theorem 2. Let X be a space endowed with a topology. Let U be an open subset of X which is neither closed nor the empty set nor X and let us denote F the complementary set of U, i.e. $F = U'$.

The following hexagon forms a strong 3-opposition:

Figure 8

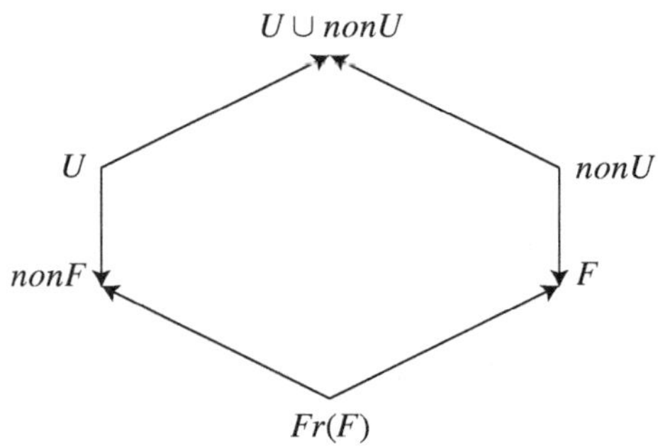

where $nonU$ is the intuitionistic open negation of U, i.e. $nonU = int(F)$, $nonF$ is the paraconsistent closed negation of F, i.e. $nonF = clos(U)$ and the arrows are the inclusion relations. The symmetrical vertices are complementary sets. The triple $(U, nonU, Fr(F))$ forms the triangle of contrariety and the triple $(F, nonF, U \cup nonU)$ forms the dual triangle of subcontrariety.

Proof: The proof is simple because it suffices to prove that $(U, nonU, Fr(F))$ forms a 3-partition of the true, i.e. it is a tripartition of X. By hypothesis, U and F form a bipartition of X because we have denoted F the complementary set of U which is neither the empty set nor X. Furthermore, by definition, $Fr(F)$ is the set difference between $clos(F)$ and $int(F)$ and therefore $(Fr(F), int(F))$ forms a bipartition of $clos(F)$. But, as F is closed, $clos(F)$ is F. And, by definition of the open negation, $nonU$ is $int(U')$, i.e. $int(F)$. We obtain then that $(Fr(F), nonU)$ forms a bipartition of F. In conjunction with the fact that (U,F) is a bipartition of X, this bipartition of F gives birth to a tripartition of X formed by U, $Fr(F)$ and $nonU$. QED.

Let us discuss what this topological hexagon means. First of all, we can observe that the upper part of the hexagon is formed by open subsets while the lower part is formed of closed subsets. This fact explains visually that the open propositional logic is intuitionistic and the closed propositional logic is paraconsistent. The open negation is supported by the contrariety relation, for this reason it verifies the non contradiction axiom. But the excluded middle axiom is not verified by the open negation because *U* and *nonU* are embedded in a tripartition of *X* and therefore the excluded middle is *Fr*(*F*) (which is also *Fr*(*U*)), the tail of the hexagon and the third vertex of the triangle of contrariety. But *Fr*(*F*) is a closed subset. Therefore *Fr*(*F*) does not appear in the open propositional calculus. Dually, the closed negation is supported by the subcontrariety relation which ensures its verifying the excluded middle axiom, but its embedding in the subcontrariety triangle forbids its verifying the non contradiction axiom. The subcontrariety triangle of a strong 3-opposition means that its three vertices form a co-tripartition of the false, therefore only the global intersection of the three vertices is the empty set and not the pair wise intersections. The missing element $U \cup nonU$ does not appear in the closed propositional calculus because $U \cup nonU$ is open. Furthermore, the intersection of *F* and its closed negation *nonF* is *Fr*(*F*) and thus we verify one more time that the contradiction is localized in the boundary.

The second remark is the interpretation of this topological hexagon in terms of 2-opposition. The strong 2-opposition being the contradiction, we find it in the complementary set operation, here visualized as central symmetry. For the weak 2-opposition, we have the following topological square extracted from the topological hexagon:

Figure 9

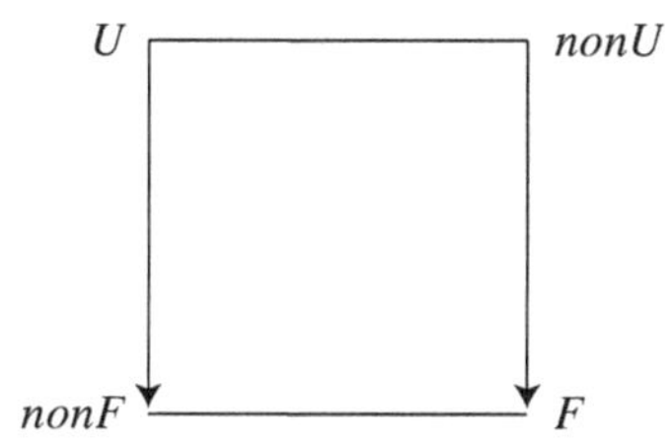

whose top contrariety segment is the open intuitionistic negation, the bottom segment the closed paraconsistent negation, the diagonals are classical negations represented by the set complementarity operation and the subalternation lateral arrows are set inclusions.

The third remark concerns what the topological hexagon becomes when U does not verify the hypothesis of the theorem. To begin with, let us study the case of an open subset U which is also closed, that is the case by example in the discrete topology. As by definition $nonF = clos(U)$ and U is also closed, we have $clos(U) = U$ and then $nonF = U$. By complementarity, we have $nonU = F$. As F is the complementary set of U, $Fr(F)$, which becomes $F \cap U$, is the empty set and $U \cup nonU$, which becomes $U \cup F$, is X. We have then the following degenerated figure:

Figure 10

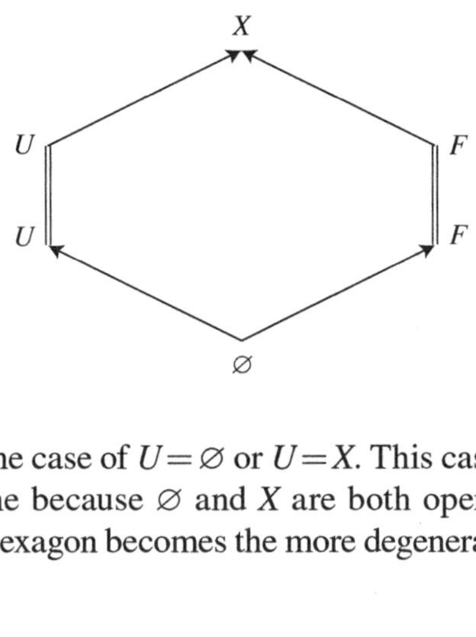

Let us follow with the case of $U = \varnothing$ or $U = X$. This case is a particular case of the preceding one because $\varnothing$ and X are both open and closed. So the above degenerated hexagon becomes the more degenerated following figure:

Figure 11

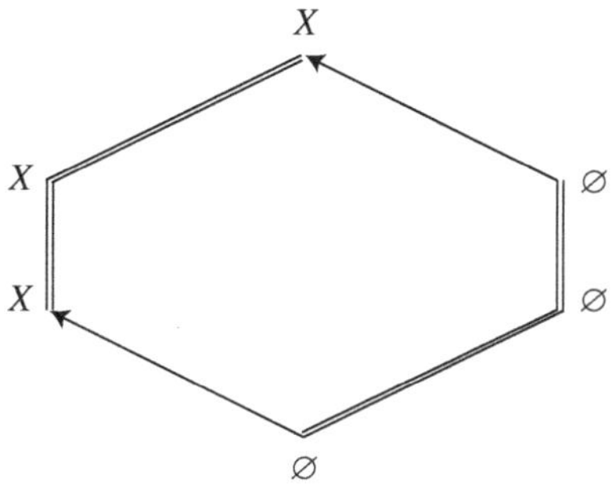

which is in fact the inclusion $\varnothing \subset X$.

6. Illustrations of the Topological Hexagon and Perspectives

For the reader who might be not familiar with topological notions, we need to illustrate the topological hexagon. The following illustration is, we think, a simple one to understand for all readers. We return to the example studied above of *IR* with its usual topology. Let *U* be]0,1[. We have the following instance of the topological hexagon:

Figure 12

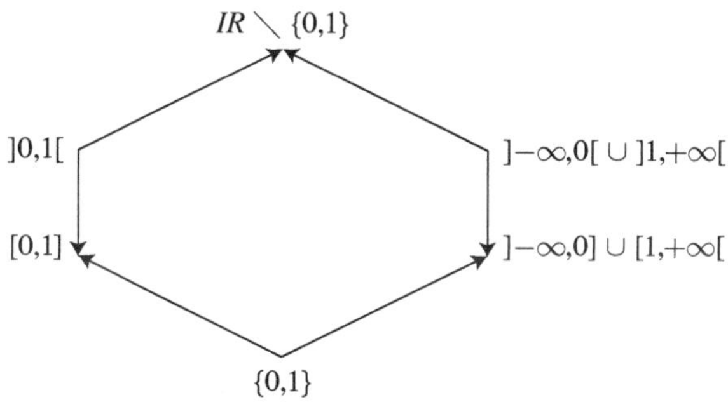

We easily see that the open subset]0,1[together with its open negation]−∞,0[∪]1,+∞[does not recover *IR* because 0 and 1 are missing and also that the intersection of the closed subset [0,1] with its closed negation]−∞,0[∪]1,+∞[is not empty because this intersection is equal to the boundary constituted by 0 and 1.

A still more impressive but very simple example is as follows. This time again, we are in *IR* with its usual topology. Let *U* be [0,+∞], the set of strictly positive real numbers. We have the following instance of the topological hexagon:

Figure 13

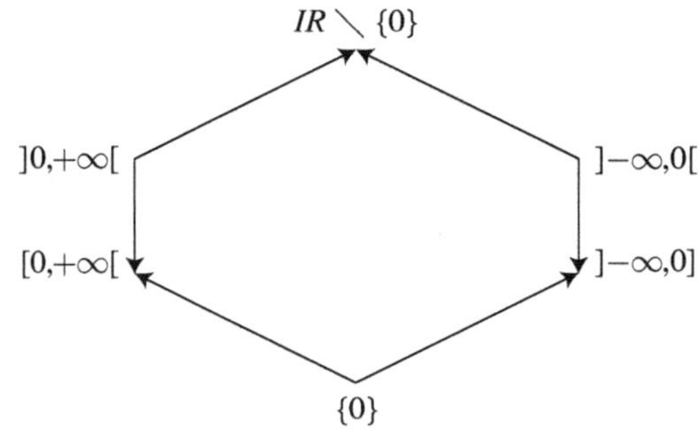

We see visually that the open negation of the set of strictly positive numbers is the set of strictly negative numbers, the "strict" being induced by the nature of open subsets, and that the closed negation of the set of positive numbers is the set of negative numbers. But the union of the sets of strictly positive and strictly negative numbers does not give the set of all real numbers because 0 is missing, while the intersection of the sets of positive and negative numbers is actually 0 which is both positive and negative (in the French school of mathematics) and support the contradiction of the paraconsistent closed propositional calculus.

In another article [Pelj], we illustrate the topological hexagon and the different kinds of 2-opposition in terms of cosmology and the main result of that article is the so-called "Taoist hexagon" constructed as an instantiation of the topological hexagon which illuminates the Taoist cosmology and explains its specifically assumed difficulty of philosophical exposure.

After having explored in this article the lower cases of n-opposition, our future work will be the exploration of the higher-order n-oppositions and their consequences on modal logics in terms of elementary components, classification and comparison between their sets of generators and relations.

References

[Bez] Béziau, J.-Y. (2003). New light on the square of oppositions and its nameless corner, *Logical Investigations, 10*, 218–232.

[Bla] Blanché, R. (1966). *Structures intellectuelles: essai sur l'organisation systématique des concepts.* Paris: Vrin.

[LSD] Luzeaux D., Sallantin J. and Dartnell C. (2008). Logical extensions of Aristotle's square, in: *Logica Universalis, 2* (1), 167–187. Birkhäuser.

[Mor] Moretti, A. (2004). Geometry for Modalities? Yes: Through n-Opposition Theory. In J.-Y.Béziau, A.Costa-Leite and A.Facchini (Eds.), *Aspects of Universal Logic* (= Travaux de logique N. 17). Neuchâtel.

[Mor2] Moretti, A. (2009). *The Geometry of Logical Opposition*, PhD Thesis. University of Neuchâtel.

[Pel] Pellissier, R. (2008). Setting n-opposition. *Logica Universalis, 2* (2), 235–263. Birkhäuser.

[Pelj] Pellissier, R. (2007). Carré cosmo-logique. Introduction cosmologique aux 2-oppositions du carré logique, in: *ELECTROBOLOCHOC 2007.* (To be published).

[Que] da Silva de Queiroz, G. (1998). *Sobre a dualidade entre intuicionismo e paraconsisténcia*, PhD, Universidade Estadual de Campinas (Brasil).

[Ses] Sesmat, A. (1951). *Logique II. Les Raisonnements, la logistique.* Paris: Hermann.

[Sme] Smessaert, H. (2007). On the 3D visualisation of logical relations. (To be published).

Veridications and their Square of Opposition

CORINA STRÖßNER and NIKO STROBACH

Abstract

Veridications are one-place sentence operations. We get a semiconsistent logic $RM1_3$ by adding them to the paraconsistent many-valued logic RM_3. The veridication of $RM1_3$ is truth-functional but has a square of opposition like modal operators.

Keywords: Many-valued logic, Paraconsistent logic, Affirmation, Square of Opposition, Logical Hexagon, Negation

1. What is Veridication?

A *veridication* is a truth-functional one-place sentence operation which gives truth for truth as input and falsehood for falsehood as input.[1] Veridication is trivial for classical logic but not in many-valued logic. Its formal features are recognizable since it shares a lot of properties of modal operations but is purely truth-functional.

1.1 Veridication and Philosophy of Language

Veridication should be understood not only as some formal operation but as a formal way to approach a philosophical problem: Can we talk consistently even if there are true contradictions? The answer is "Yes, by internalization".

1 According to the definition, the trivial one-place operator with identical input and output, is a veridication, too. But we need not pay any attention to it here. The term "veridication" was suggested to us by Bertram Kienzle.

Normal language gives us this possibility: If there is someone claiming that p is true and false and you ask him whether p is true or simply: "Is it a fact that p?" he should answer simply "yes" since you ask for the positive and not for the negative status.[2] This process is an internalization in the following sense: Assertiveness, a feature of a sentence which is usually part of the meta-language, is included in the logical language itself. The result is that ontological contradiction does not necessarily rule out consistent use of language and that there is no need to postulate consistency. A corresponding argument goes for paracompleteness. Take a proposition p which is neither true nor false (since it is non-sense or not determined) and ask if it is a fact that p: The answer is simply "no". Again you get completeness by internalization. The difference is just that it results in a true proposition for paraconsistent values but in a false one for paracompleteness. This means that the simple "It is the case that"-veridication has a different logical strength depending on the context to which it is applied. Other one-place operations like "It is possible that" and "It is necessary that" are rather constant in natural and in formal languages. The following table summarizes the results of natural language veridications combined with classical negation and motivates the formal semantics which are given later in application to many-valued logics:

p is … / not-p is … / Is it…	true/false / true/false	undetermined / undetermined	just true / just false	just false / just true	Negation
the case that p?	YES	NO	YES	NO	
the case that not p?	YES	NO	NO	YES	
not the case that p?	NO	YES	NO	YES	
not the case that not p?	NO	YES	YES	NO	

2 This thought is very old. One could understand Aristotle *Metaphysica* Γ4, 1007a, 8–29 exactly in this way: "[…] even if the same thing is thousand times man and not-man, we must not add, in answering the question whether this is a man, that it is also at the same time not a man […]". Note that this argument does not work for his aim to prove the ontological law of non-contradiction but is rather a rule of conversation.

1.2 Veridication in the History of Logic

Veridication is redundant for classical logic. But things are different in many-valued logics. Since the early days of many-valued logics such one-place truth-functional operations have rarely been studied. Jan Łukasiewicz had one-place truth-functional operators in his many-valued logic but he saw them not as veridications but as operators for modality. In the 1920s he added them to his logic $Ł_3$.[3] This system deals with the third value ½ which means that a proposition is undetermined. Łukasiewicz's operator for possibility, defined as "If not α then α", gives 1 for a proposition with 1 or ½.[4] The operator for necessity gives 1 as output for 1 as input and 0 otherwise. Łukasiewicz's notion of possibility is very problematic: If α is possible and β is possible it is possible that α and β which is implausible since α and β may rule out each other although both are possible apart from each other.[5] His alleged operator for necessity $N\alpha$ should simply be read as a "It is the case that"-veridication: "It is the case that" is true for a true proposition, but not for an undetermined one.

The Russian logician Bochvar suggested another 3-valued logic in 1935. The third value is given for sentences which seem to be meaningless and paradoxical sentences were included among the meaningless. The operators are conservative with respect to classical logic but any meaningless part infects the whole formula. Bochvar uses an assertion operator which is quite the same as a veridication. The assertion $A\alpha$ gives true for true and false otherwise. Bochvar's aim was to avoid paradoxes but it was shown that new paradoxes are possible in his system.[6]

Finally, D'Ottaviano and Da Costa, in 1970, suggested a paraconsistent version of $Ł_3$ which they called J_3.[7] J_3 differs from $Ł_3$ only by having 1 and {1,0} as designated values, where $Ł_3$ has 1, but not ½. However, none of the handful of implications they define is the $RM1_3$ one we will choose. The intuitive application for J_3 they propose are non-trivial scientific theories

3 See Łukasiewicz 1931.

4 Note that, in this case, "then" is not the RM_3 conditional. Like in J_3, the operator we interpret as the weak veridication, is introduced as a primitive sign in $RM1_3$. A useful survey of similar systems is Karpenko 1999.

5 See Urquhardt 2001 and Priest 2001.

6 For a short description of Bochvars logic and the critics see Urquhardt 2001.

7 D'Ottaviano/Da Costa 1970. As to axiomatizations of J_3, see Epstein 1990, ch.9, and Ciuciura 1999.

in the early stage of development with seeming contradictions (that are eliminated later on). They don't pay particular attention to the signs which we call veridications. Although such one-placed operators were used from time to time, veridication is far from being a well-known and well-understood subject of modern logic. So it seems that the study of veridications deserves to be continued.

2. $RM1_3$

$RM1_3$ is a *semiconsistent* logical system: This means *ex falso quodlibet* (EFQ), the implication from contradiction to anything, is not valid but there are sets of formulae which perfectly satisfy classical logic, including EFQ.[8] The main purpose of that system is to show how consistency is possible even with the presence of contradictions and how classical logic can be extracted from paraconsistent systems.[9] We defined $RM1_3$ by adding a veridication to the semantics of RM_3 and distinguishing different terms of designation and validity.

2.1 RM_3

RM_3 is a well-known paraconsistent logic.[10] It has three values, which are given as sets: {1}, {0}, {1,0} and is characterized by the following semantics where disjunction could be defined and others are primitive.

f~	
{1}	{0}
{1,0}	{1,0}
{0}	{1}

f∧	{1}	{1,0}	{0}
{1}	{1}	{1,0}	{0}
{1,0}	{1,0}	{1,0}	{0}
{0}	{0}	{0}	{0}

f→	{1}	{1,0}	{0}
{1}	{1}	{0}	{0}
{1,0}	{1}	{1,0}	{0}
{0}	{1}	{1}	{1}

f∨	{1}	{1,0}	{0}
{1}	{1}	{1}	{1}
{1,0}	{1}	{1,0}	{1,0}
{0}	{1}	{1,0}	{0}

8 The logic $RM1_3$ was first suggested by Strößner/Strobach in an unpublished paper. For description and investigations see Schang 2008.

9 Note that it is not the purpose to replace paraconsistent logic or to avoid antinomies or higher-order paradoxes of self-reference.

10 For further characterizations of RM_3 see Priest 2001, ch.7.

Validity is defined as preservation of designated truth values. The designated values in RM3 are {1} and {1,0}: If all of the premises are either {1} or {1,0}, then the conclusion is either {1} or {1,0}. On this level RM3 and likewise other three-valued paraconsistentent systems which take the third value to be designated cannot differentiate between truth and contradiction.

2.2 *Truth and Validity*

The words "true" and "false" need some further explanation in every many-valued logic. The standard way is to use designated values for truth while undesignated values are taken to be not true. From a meta-logical perspective this is a two-valued thought style or even worse, as Wansing and Shramko put it, "logical two-valuedness is, in fact, reduced to logical *mono-valuedness* if there is just one entailment relation defined as truth preservation from the premises to the conclusion".[11]

In RM_3 there are only (positively) designated values which are those sets containing 1. In this sense RM_3 has only true and untrue propositions. Intuitively it is not far-fetched to use negatively designated values as well which are those sets containing 0.[12] So a proposition is said to be *true*, if it has a *positively designated value* and is said to be *false*, if it has a *negatively designated value*. An inconsistent proposition is positively and negatively designated and in that sense true and false and not just true as it is in RM_3. We have two kinds of designation for three values.

A formula α is ...

- *positively logically valid* iff α is true in every model.
- *negatively logically valid* iff α is not false in any model.
- *strongly logically valid* iff α is positively and negatively logically valid

11 Wansing/Shramko 2008, p. 414.
The subject of different designations and as result different terms of validity is currently often discussed especially in connection with Suszko's thesis. This is an important subject of many-valued logic but since this article is mainly about veridication and the square it is not the place for a detailed discussion about truth-values. For research on validity see for example Wansing/Shramko 2008. Malinowski did interesting investigations about many-valuedness and inferences in Malinowski 1994. Both give also more extensive bibliographies on this subject.

12 "Negative designated" means the same as "antidesignated" in Schramko/Wansing.

But what about validity? We can distinguish between different forms of logical validity. The first one is for formulae which are always true and the second one for formulae which are never false. A strongly logically valid formula has both qualities.

The distinction for the validity of inference is similar:

An inference $\Delta \vDash \alpha$ is ...
- *positively valid* iff α is true if every formula in Δ is true.
- *negatively valid* iff α is not false if no formula in Δ is false.
- *strongly valid* iff the inference is positively and negatively valid.

2.3 Adding Veridication

As we discussed in the first section the veridication "it is the case that" does not care whether the proposition is true and not false or true and also false. The semantics that results from our previous motivation are as follows.

α	$V\alpha$
{1}	{1}
{1,0}	{1}
{0}	{0}

The extension of RM_3 with rules of veridication is the semiconsistent logic $RM1_3$. Adding veridication has the remarkable advantage that the RM_3-conditional is no longer a primitive operator. It could be defined by conjunction, negation and veridication which are primitive.[13]

13 $\alpha \rightarrow \beta$ is the abbreviation for $\sim(\alpha \wedge \sim\beta) \wedge \sim(V\alpha \wedge \sim V\beta) \wedge \sim(\sim V\sim\alpha \wedge V\sim\alpha)$.

2.4 Classical Islands

The metaphor of *classical islands* means that there are formulae of $RM1_3$ which perfectly satisfy the laws of classical propositional logic. We can call such formulae V-formulae. A formula of $RM1_3$ is a *V-formula* if and only if it has a construction tree in which there is a veridication on every branch. A formula with veridication as main connective, let us call this a *veridicated formula*, meets this condition since the main operator is shared by all branches. This is the recursive definition:

V-formulae

1. Every veridicated formula $V\alpha$ is a V-formulae.
2. If α is a V-formula, then so is $\sim\alpha$. If α and β are V-formulae then so is $\alpha\wedge\beta$.
3. No other formula is a V-formula.

V-formulae have only the values {1} and {0} as possible output since veridication gives this output and $RM1_3$ has no operation which could possibly give {1,0} for this input. So if there is a veridication in each branch of the construction tree there is no chance to get {1,0} for such formulae in any valuation. The other values are classical since {1} is equal to classical 1 and {0} to classical 0. $RM1_3$ is conservative for classical input which means that the semantics of connectives in $RM1_3$ for {1} and {0} as input are the same as for 1 and 0 in classical propositional logic pc. Classical logic holds for V-formulae. So any set which contains only V-formulae can be regarded as a classical island. The connection with classical logic can be expressed in a law of classical islands.

Law of classical islands

If $\Gamma\vDash_{pc}\alpha$ then $\Gamma^*\vDash_{[strong]RM13}\alpha^*$ where α^* and every formula in Γ^* is the result of substituting every atomic part of formulae in Γ and α by a V-formula of $RM1_3$ and using the corresponding connectives of $RM1_3$ (e.g. conjunction of $RM1_3$ for conjunction of pc…).

3. New Operators and Old Oppositions

As we did in the first section for natural language we can now combine veridication and negation on a formal level in $RM1_3$.

	α	{1,0}	1	0
	${\sim}\alpha$	{1,0}	0	1
$V\alpha$		1	1	0
$V{\sim}\alpha$		1	0	1
${\sim}V\alpha$		0	0	1
${\sim}V{\sim}\alpha$		0	1	0

Negation

The veridication in the first line is the new primitive operator in $RM1_3$. We need to veridicate the negation of α just as it is done in the second line for saying "*Not* α is true." or "It' s the case that not α." This is a form of negation with paraconsistent characteristics: α and $V{\sim}\alpha$ are allowed to be true in the same model and the conjunction of α and $V{\sim}\alpha$ doesn't imply every proposition. This version of *ex falso quodlibet* fails, it is neither positively nor negatively valid, which is a characteristic feature of paraconsistency. The stronger negation is formed in the following line: This says that the veridication of α is false. It is equivalent to an operation which is sometimes discussed as Boolean negation and unpopular with defenders of paraconsistent logic like Graham Priest.[14] In fact this negation is not paraconsistent: α and ${\sim}V\alpha$ implies any proposition. This inference is strongly valid because the conjunction of α and ${\sim}V\alpha$ is always false and never true. $V\alpha$ has a dual which is a stronger veridication: We get it by negating the veridication of the negation of a sentence. This is seen in the fourth line. An abbreviated term may be useful, so $W\alpha$ is used as an abbreviation for ${\sim}V{\sim}\alpha$.

14 See Priest 1990.

3.1 The Square of Opposition

The relations between the different veridications and negations which have been described so far are well-known for modal operators and quantifiers. They are typical for the square of opposition. Subalternation is given with the strongly valid inferences from Wα to Vα and from ~Vα to ~Wα and we find the three kinds of oppositions: contradiction, contrariety and subcontrariety.

contradiction	one is true and one is false	Vα ,~Vα / Wα, ~Wα
contrariety	at least one is false	Wα, ~Vα
subcontrariety	at least one is true	Vα, ~Wα

These results may be put into the square with the primitive veridication in the I-corner, the paraconsistent negation in the O-corner, the strong veridication in the A-corner and Boolean negation in the E-corner.

Figure 1

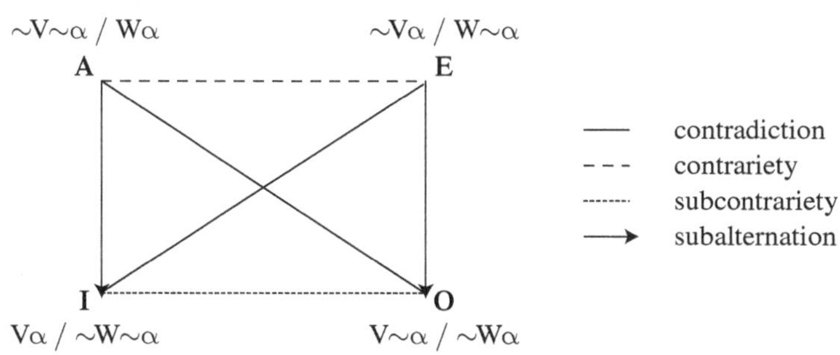

3.2 Extensions of the Square I-Hexagon

The square of opposition was expanded to a hexagon by Robert Blanché.[15] He suggested two new corners. The Y-corner is for terms like "only some" or "possibly but not necessarily" and the U-corner means "all or none" or "necessarily or impossibly". The same could be done in $RM1_3$. Let us define

15 See Blanché 1957.

a new operator which means "it is contradictory that" or "it is inconsistent that". This inconsistency-operator gives {1} for {1, 0} and {0} otherwise. It is expressible as "It is the case that and it is the case that not": Iα abbreviates Vα∧V~α. Iα implies strongly Vα and V~α and ~Iα is implied strongly by Wα and W~α. We find also a further contradiction, Iα to ~Iα, new contraries, Iα to Wα and W~α, and new subcontraries, ~Iα to Vα and V~α.

Figure 2

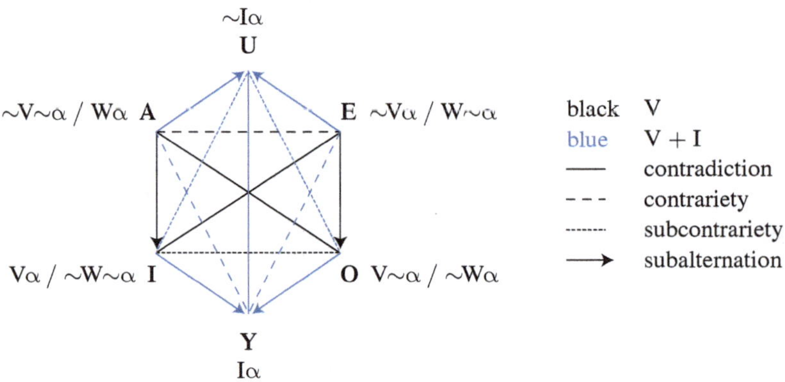

3.3 Extensions of the Square II-Octagon

This hexagon includes no formula without veridication. As a last expansion we try to involve α and ~α. A formula α is implied by Wα and implies Vα. V~α is a negation of α of paraconsistent character, which means that (unlike contradictories) both could be true in one model but (like contradictories) never both be false. This is a subcontrary relation. On the other hand α and ~Vα, a form of Boolean Negation, are contraries for they could not both be true (like contradictories) but both be false (unlike contradictories). The relation between α and ~α needs our special attention since it is not really a contradictory opposition as we would normally expect for negation: Here one is true, the other one is false or both have a third value. It is hard to decide just how contradictory such a relation is. For the purpose of having a name which respects the partly contradictory feature of this opposition we call it *paracontradiction*.

Figure 3

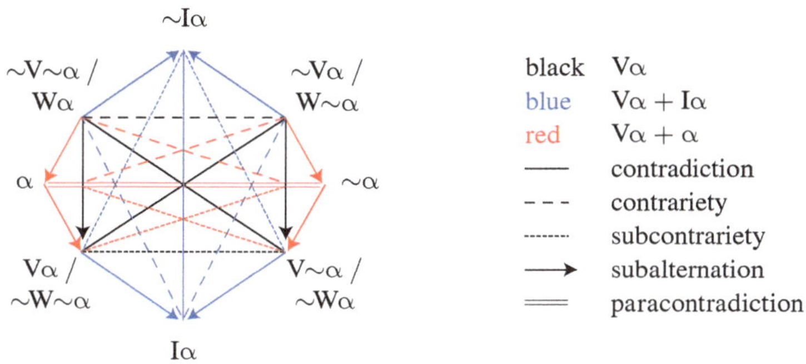

The different oppositions in the octagon are:

contradiction $V\alpha$,$\sim V\alpha$ / $W\alpha$, $\sim W\alpha$ / $I\alpha$, $\sim I\alpha$
paraconsistent contradiction α, $\sim\alpha$
contrariety $W\alpha$, $W\sim\alpha$ / $I\alpha$, $W\alpha$ / $I\alpha$, $W\sim\alpha$ / $W\alpha$, $\sim\alpha$ / $W\sim\alpha$, α
subcontrariety $V\alpha$, $\sim W\alpha$ / $\sim I\alpha$, $V\alpha$ / $\sim I\alpha$, $V\sim\alpha$ / $V\alpha$, $\sim\alpha$ / $V\sim\alpha$, α

4. Oppositions for Veridication and Modality

$RM1_3$ has noticeable similarities to the modal logic S5. The relations of the different operations to one another are almost the same as those of modal operators. The hexagon of veridication can be transferred to modal operators with only one exception: The relation between α and $\sim\alpha$ is a contradictory opposition and not a paracontradiction.

Looking at this one might guess that veridication is just another kind of modal operation but we should keep in mind the essential differences. Veridications are purely truth-functional and the underlying models (just a valuation) are, compared to modal logic (valuation, possible worlds, accesibility), very simple.

Figure 4

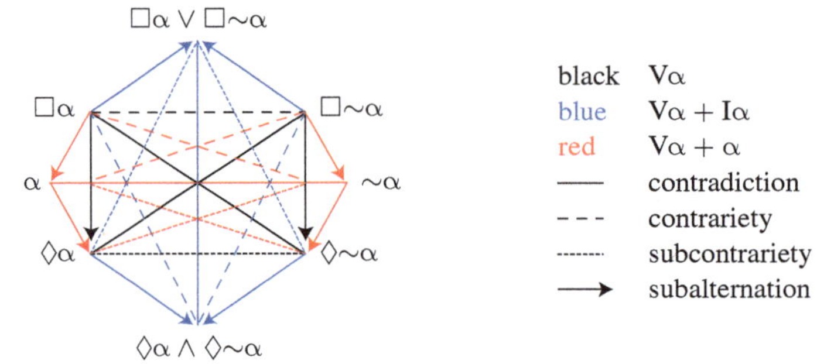

Łukasiewicz's modality gives an octagon as well. The connection between α and ${\sim}\alpha$ is another form of paracontradiction. If α is true then ${\sim}\alpha$ is false and vice versa, or both have a third value, but now this value is ½. In this case paracontradiction is not paraconsistent but paracomplete.

Figure 5

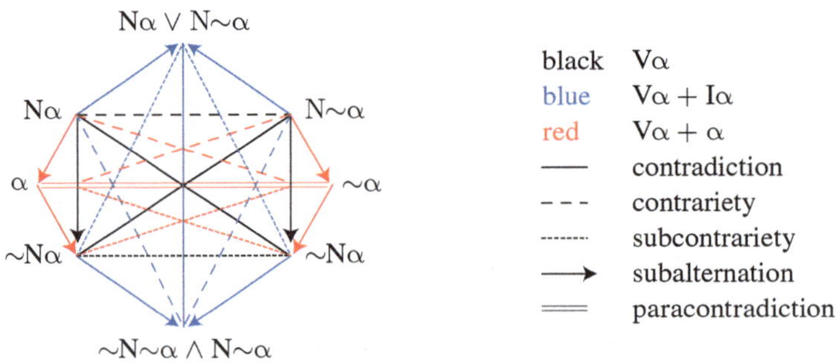

Łukasiewicz's modality, like veridication, is a one-place truth-functional operation for many- valued logic. His necessity and possibility, giving such strange results as $P\alpha \wedge P{\sim}\alpha \vDash P(\alpha \wedge {\sim}\alpha)$ make far better sense as veridications rather than as modalities.

Veridications give rise to many logical inferences and oppositions but are simple in comparison to other operations. A non-trivial kind of veridication can be defined in every many-valued logic and more veridications are possible for each additional value (and so infinite many in fuzzy logic). Since they show typical properties of modal operations they could be very useful for studies in universal logic, opposition theory and the square of opposition.

References

Aristotle (1984). Metaphysics. In Barnes, J. (Ed.) *The complete works of Aristotle*. Princeton: Princeton University Press.

Béziau, J.-Y. (2003). New light on the square of opposition and its nameless corner, *Logical Investigation, 10*, 218–232.

Béziau, J.-Y. (2005). Paraconsistent logic from a modal viewpoint, *Journal of Applied Logic*, 7–14.

Blanché, R. (1957). Sur la structuration du tableau des connectifs interpropositionnels binaires, *Journal of Symbolic Logic, 22*, 17–18.

Ciuciura, J. (1999). History and the Development of the Discursive Logic. *Logica Trianguli, 3*, 3–31. www.filozof.uni.lodz.pl/fileadmin/user_upload/logicatr/Ciuciura3.pdf

Epstein, R. (1990). *The Semantic Foundations of Logic: Propositional Logic*. Dordrecht: Kluwer.

Karpenko, A. (1999). Jaskowski's Criterion and Three-Valued Paraconsistent Logics. *Logic and Logical Philosophy, 7*, 81–86. www.logika.umk.pl/llp/07/ak.pdf

Łukasiewicz, J. (1931). Philosophische Bemerkungen zu mehrwertigen Systemen des Aussagenkalküls. In Berka, Karel / Kreiser, Lothar (eds.) (reprint 1986), *Logik-Texte*. 135–150. Berlin.

Malinowski, G. (1994). Inferential Many-Valuedness. In Jan Woleński (Ed.), *Philosophical Logic in Poland*. Dordrecht (pp. 75–84).

D'Ottaviano, I., & Da Costa, N. (1970). Sur un problème de Jaskowski, *Comptes Rendus de l'Academie des Sciences 270, série A*, 1349–1353.

Priest, G. (1990). Boolean negation and all that. *Journal of Philosophical Logic, 19*, 201–215.

Priest, G. (2001). *An Introduction to Non-Classical Logic*. Cambridge: Cambridge University Press.

Schang, F. (2008). Negation and Dichotomy. Scientific knowledge and common knowledge: the big divide?. International Congress in Bydgoszcz, Poland.
poincare.univ-nancy2.fr/digitalAssets/28002_Schang_Negetion_Dichtomy.pdf

Urquhart, A. (2001). Basic many-valued logic. In Gabbay & Guenther (Eds.), *Handbook of philosophical logic, 2nd edition, Volume 2* (pp. 249–295).

Wansing, H., & S., Yaroslaw (2008). Suszko's Thesis, Inferential Many-Valuedness, and the Notion of a Logical System. *Studia Logica, 88*, 405–429.

Part 5: Linguistics and Semiotics

Histoire d'*O:
Lexical Pragmatics and the Geometry of Opposition

LAURENCE R. HORN

Abstract

The last chapter of Horn (1972) outlines an asymmetry in lexicalization potential for values situated on the square of opposition –in its simplest form, the contrast between *no(ne)* as the contradictory of *some* vs. **nall* as the contradictory of *all* –and offers a Gricean program to account for this asymmetry. In this paper, I return to the scene of the crime, including a (not entirely impartial) review of proto-Gricean approaches to the three-sided square dating back to De Morgan (1858) and a defense of the pragmatic line against critiques dismissing it as inadequate or flawed by Hoeksema 2003, Jaspers 2005, Seuren 2006, inter al.) on the three-sided square.

I argue that the arguments for a Gricean approach have not been fully appreciated, including those based on (i) the relation between the square and quantity scales (Horn 1989) as exemplified across quantifiers, modals, and binary connectives, (ii) the status of the "intermediate" values on the square (e.g. *most/many/few* for the determiners; *likely/unlikely* for the epistemic modalities; *should/ought to/shouldn't* for the modal auxiliaries, *usually/often/rarely* for the quantificational adverbs, etc.) and their implications for a predictive model of lexicalization, and (iii) the widespread cross-linguistic tendency to minimize the expression of subcontrary opposition and maximize that of contrary opposition in natural language as attested by pragmatic strengthening implicatures (as in "neg-raising") and O>E drift, as in the contrary (=[NECESSARY [NOT]]) reading expressed by Fr. "Il ne faut pas que tu meures."

The back cover blurb for Dany Jaspers' (2005) masterful treatise on the linguistic asymmetry of the Square of Opposition begins as follows:

> *Operators in the Lexicon* begins with an old chestnut: why are there no natural single word lexicalizations for negations of the propositional operator *and* and the predicate calculus operator *all:* neither **nand* nor **nall?*

I assume that the "old chestnut" in question concerns not senses 1–5 of the *American Heritage Dictionary* entry (http://dictionary.reference.com/browse/Chestnut) for *chestnut* – relating to trees of the genus *Castanea,* their wood or nuts, the reddish-brown color of those trees or the horse of the same color, or a callus on a horse's foreleg –but rather sense 6: 'An old, frequently repeated joke, story, or song.' Given the limitations of the prose medium, I will assume the intended reference can be further narrowed to the second of the three disjuncts within sense 6, and in particular to the frequently recited story I shall describe (with apologies to the soi-disant Réage 1954) as *l'Histoire d'*O:*

> In a wide variety of languages, values mapping onto the southeast, **O** vertex of the Square of Opposition are systematically restricted in their potential for lexicalization or direct expression; this asymmetry is attributable to the scalar implicature relation obtaining between the two subcontraries.
> (Horn 1972: Chap. 4; Horn 1989: §4.5; Horn 2005: §5; cf. Levinson 2000: 69–71.)

The path between the **I** and **O** subcontraries *some* and *some not* (or *not all*) has proved to be a long and winding road. On the minority view, *some* is two-sided, with upper as well as lower bound, and is thus incompatible with *all.* Some have wishfully read this position into Aristotle –

> On the Aristotelian theory...wherever the affirmative "some are" applies, the negative proposition "some are not" holds also.
> (Dewey 1938: 182)

> Aristotle seems to think that the main function of a particular statement is to describe a situation where the corresponding universal statement is false. His reasoning seems to be: If the universal is true, why assert the particular?
> (Rose 1968: 41)

But such a reading is not tenable, given Aristotle's observation on the nature of the relation between the two subcontraries themselves ("Verbally four kinds of opposition are possible...but really there are only three: for the particular affirmative is only verbally opposed to the particular negative" – *Prior Analytics I,* 63b21 ff.) and his explicit endorsement of the one-way subaltern entailment from **A** to **I** and from **E** to **O**:

> For having shown that it belongs to all, we will have shown also that it belongs to some; similarly, if we should show that it belongs to none, we will have shown also that it does not belong to all.
> (Aristotle, *Topics* 109a3)

Most commentators concur; in the words of the great Arab philosopher Avicenna (ibn-Sīnā):

> If you say "some men are so-and-so", it is not necessary that some others are not so-and-so. If the proposition is about all, it is also about some.
> (Zabeeh 1971: 24)

But even if Aristotle did not support a two-sided *some,* others did. Priority may belong to the early 6th century Buddhist logician Dignāga and his colleagues who, in their *hetu-cakra* or Wheel of Reasons,

> do not admit four kinds of proposition like Aristotle and the Scholastics, but only three, since they interpret 'Some S is P' not as 'at least some' but as 'at least some and not all'...This would give a logical triangle in place of the western logical square.
> (Bochenski 1961: §53E; cf. Tucci 1928)

Such triangles of opposition did not surface explicitly in the West until the mid-19th century when Sir William Hamilton of Edinburgh inaugurated a debate over the proper treatment of the subcontraries. Distinguishing two senses of *some,* the INDEFINITE *(at least some)* and the SEMI-DEFINITE *(some but not all)*, Hamilton (1860: 254) regarded the latter as basic: 'Some, if not otherwise qualified, means *some only* –this by presumption.' On this reading of the particular, the two statements *Some men are learned* and *Some men are not learned* are not only (as for Aristotle) compatible, given that their conjunction is consistent, but logically indistinct. The purported opposition between the two subcontraries, charged Hamilton (1860: 261), was 'only laid down from a love of symmetry, in order to make out the opposition of all the corners in the square of opposition...In reality and in thought, every quantity is necessarily either all, or none, or some. Of these the third...is formally exclusive of the other two.'

As was his wont, Augustus de Morgan was unimpressed with his rival's stance, although he did concede the existence of what we would now view as pragmatic upper-bounding. Some sample passages (here and below, emphasis is added):

> **In common conversation** the affirmation of a part is meant to imply the denial of the remainder. Thus, by 'some of the apples are ripe', it is always intended to signify that some are not ripe.
> (De Morgan 1847: 4)

> *Some,* in logic, means ***one or more, it may be all.*** He who says that *some are,* is not held to mean the rest are not. 'Some men breathe'...would be held false in **common language** [which] usually adopts the complex particular proposition and implies that some are not in saying that some are.
> (De Morgan 1847: 56)

> With logicians the word *some* has in all time been no more than a synonym of *not–none:* it has stood for *one or more, possibly all.* **With the world at large** it is sometimes *possibly all,* sometimes *certainly not all,* **according to the matter spoken of.** But with the logician "some are" is merely and no more than the contradictory of "none are"... *[S]ome* equally contains some–certainly–not–all and some–possibly–all.
> (De Morgan 1861: 51)

As recognized by both De Morgan and Mill (another captain of the anti-Hamiltonian troops), the possibility of upper bounding is subject not only to the vagaries of context and speech level but more specifically to the relative epistemic insecurity of the speaker:

> There are three ways in which one extent may be related to another...: complete inclusion, partial inclusion with partial exclusion, and complete exclusion. This trichotomy would have ruled the forms of logic, **if human knowledge had been more definite...** As it is, we know well the grounds on which predication is not a trichotomy, but two separate dichotomies.
> (De Morgan 1858: 121)

> No shadow of justification is shown...for adopting into logic **a mere sous-entendu of common conversation** in its most unprecise form. If I say to any one, "I saw some of your children today", he might be justified in inferring that I did not see them all, **not because the words mean it,** but because, if I had seen them all, it is most likely that I should have said so: **though even this cannot be presumed unless it is presupposed that I must have known whether the children I saw were all or not.**
> (Mill 1867: 501)

In such remarks, De Morgan and Mill prefigure the two-stage quantity implicature procedure of (neo-)Gricean pragmatics: The use of a weaker value (e.g. *some, most*) suggests that **for all the speaker knows** any stronger value –and especially the strongest value –on the same scale *(all)* could not have been substituted *salva veritate.* Speaker *a*'s utterance of ...W... implicates not $K_a\neg(S)$, i.e. that *a* knows that the stronger counterpart ...S... is false, but only (ceteris paribus) that $\neg K_a(S)$. The strengthening of $\neg K_a(S)$

to $K_a\neg(S)$ is possible only given what Geurts (2009) calls the "competence assumption" (cf. also Horn 2009: §2).

Mill's proto-Gricean allusion to a tacit principle mandating the speaker to use the stronger *all* in place of the weaker *some* when possible, while inviting the hearer to draw the corresponding inference when the stronger term is not used, is echoed even by one of Hamilton's sometime supporters:

> Whenever we think of the class as a whole, we should employ the term All; and therefore when we employ the term Some, **it is implied** that we are not thinking of the whole, but of a part as distinguished from the whole –that is, of a part only.
> (Monck 1881: 156)

The idea that *some* should be assigned a two-sided meaning rather than, or along with, the traditional one-sided interpretation did not die with Hamilton. Ginzberg (1913, 1914) carried the quarrel across the Channel, jettisoning the square of opposition for a triangle of contraries with vertices representing *all, none,* and *exactly some* –'quelques et rien que quelques'. But Couturat (1913, 1914), only too eager to play De Morgan to Ginzberg's Hamilton, sought to dissuade his countryman from following 'le plus mauvais des logiciens' in collapsing the two distinct subcontraries into one basic proposition which is in fact a logical conjunction; he argues that the classical system cannot be perfected by adopting 'précisions' that are inconsistent with its very spirit.

John Neville Keynes echoed Mill in observing (1906: 202–3) that a speaker whose knowledge is incomplete cannot use *some S's are P* with the meaning 'some only'. Unfortunately, many logicians "have not recognized the pitfalls surrounding the use of *some*. Many passages might be quoted in which they distinctly adopt the meaning *–some, but not all*."

To which the great Danish linguist Jespersen (1924: 324) retorted "in the name of common sense", why do logicians "dig such pitfalls for their fellow-logicians to fall in"? Jespersen consequently proselytizes for the "tripartition" of operators in (1) (cf. De Morgan's functional "trichotomy" above) and implicitly proposes his own Triangle of Opposition (1917: Chapter 8) as indicated in (2)

(1) A:	all	everybody	always	everywhere
B:	some/a	somebody	sometimes	somewhere
C:	none/no	nobody	never	nowhere

A: necessity	must/need	command
B: possibility	can/may	permission
C: impossibility	cannot	prohibition

The effect, as with Ginzberg (or Dignāga), is to triangulate the Square:

(2)

Figure 1

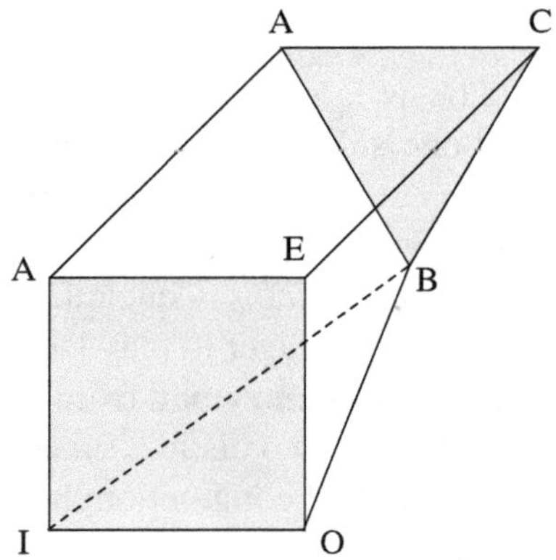

But while Jespersen's B category, the nadir of his Triangle, represents a semantic conjunction (or neutralization) of the **I** and **O** vertices of the traditional Square, it has the lexical membership of the **I** vertex *(some, possible)*. On logical, epistemological, and discourse grounds the identification of **I** and **O** is ultimately untenable, precisely for the traditional reason that the former provides the contradictory of **E**, the latter of **A**. (See Horn 1973 for more on these debates in the light of neo-Gricean pragmatics.)

With characteristic insight, Sapir (1930: 21) opted for a solution midway between the classical Square and the Jespersenian Triangle. His particular subcontraries are neither semantically bilateral nor strictly unilateral. (Again, emphasis is added.)

> 'Not everybody came' **does not mean** 'some came', **which is implied**, but 'some did not come'. **Logically**, the negated totalizer *[not every]* should include the totalized negative, i.e. opposite or contrary *[none]*, as a possibility, but **ordinarily** this interpretation is excluded.

Note especially Sapir's use of *is implied* (vs. *means*) and his qualifier *ordinarily,* emphasizing the essential role of the context in licensing the implication in question.

An expanded development of the triangle was launched in the early 1950's, appropriately enough by three philosophers working independently but exploiting essentially the same geometry. For Jacoby (1950), Sesmat (1951), and especially Blanché (1952, 1953, 1969), the square and triangle in (4) can be combined to form a Hexagon of Opposition on which the diametrically opposed terms are contradictories:[1]

1 It will be noticed that Sesmat's hexagon has the **Y** above and the **U** below, as does the somewhat sketchier model of Hegenberg (1957). I opt here for Blanché's vowel system for its mnemonic value, the **I O U** serving to acknowledge the incurring of an informational debt. Von Wright (1951) presupposes a logical pentagon, with a nadir (= our **Y**) for the conjunction of **I** and **O** but no apex (= **U**) for its contradictory, as in (8) below.

(3)

Figure 2

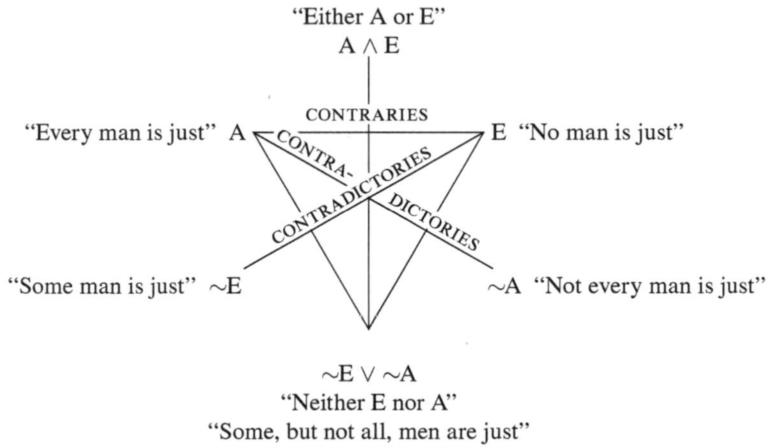

(Jacoby 1950: 44)

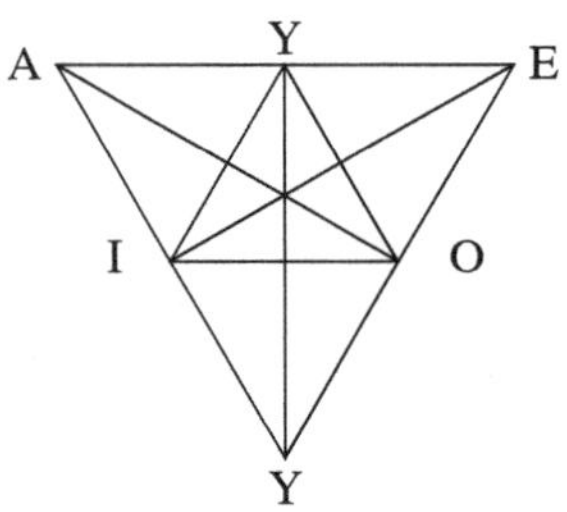

(Sesmat 1951: 450)

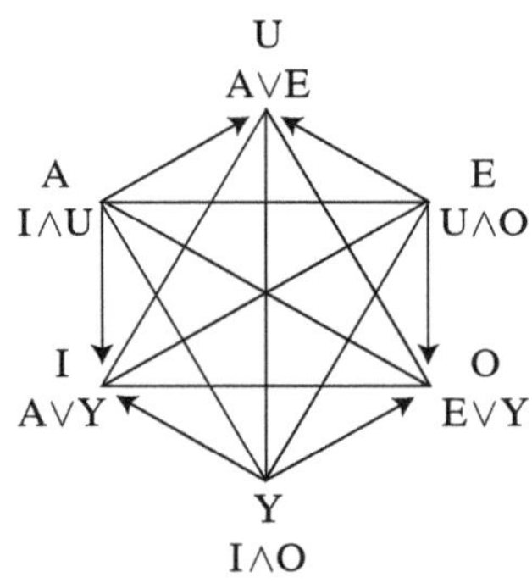

(Blanché 1969: 56)

In earlier work (Horn 1990), I suggested reconfiguring the hexagons in (3) as a Logical Star of David, in which the triple of pairwise contraries **<A E Y>** (exactly one of which must be true in any context) is superimposed on the triple of pairwise subcontraries **<I O U>** (exactly one of which must always be false).

(4) Logical hexagon ⇒ (5) Logical Magen David

Figure 3

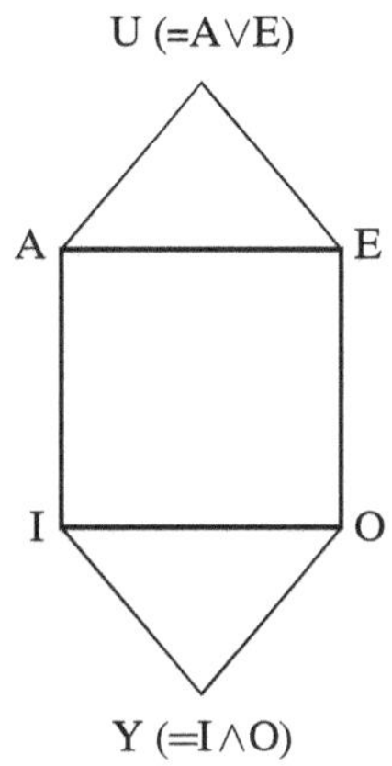

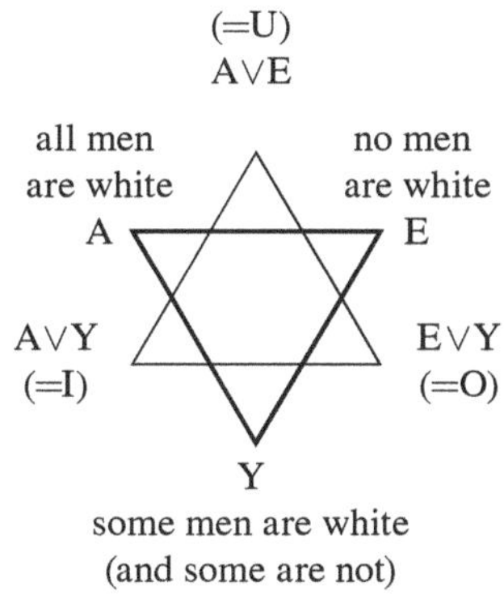

The (neo-)Gricean approach derives the relationship between the subcontraries by means of a pragmatic principle variously defined in terms of strength or quantity:

> One should not make the (logically) lesser, when one could truthfully (and with greater or equal clarity) make the greater claim.
> (Strawson 1952: 178–9, with acknowledgments to 'Mr H. P. Grice')

> One should not make a weaker statement rather than a stronger one unless there is a good reason for so doing.
> (Grice's own 'first shot', 1961: 132)

> Make your contribution as informative as is required (for the current purposes of the talk-exchange).
> (Grice's first MAXIM OF QUANTITY [1967] 1989: 26)

> Make the strongest possible claim that you can legitimately defend!
> (Fogelin's RULE OF STRENGTH, 1967: 20)

This was clearly an idea whose time had come. The Fogelin (1967: 20–22) "rule of strength" is devised contemporaneously with Grice's William James lectures and generates a set of rules of use for the subcontraries:

(6) (i) Do not employ an **I** or an **O** proposition in a context where you can legitimately employ an **A** or an **E** proposition…The *use* of one subcontrary typically suggests the appropriateness of using the other.

(ii) Do not affirm one subcontrary if you are willing to deny the other.
(iii) Subcontraries tend to collapse together.

Based on the rules, Fogelin tries his hand at beating Squares into Triangles:

Figure 4

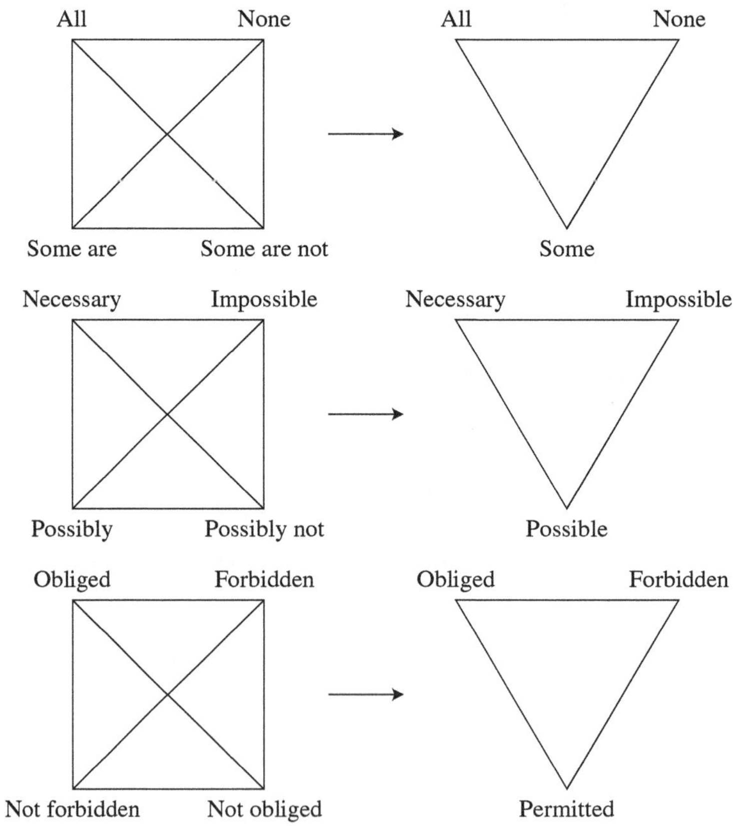

But Fogelin's triangles –unlike those of the Dignāga-Jespersen-Jacoby-Sesmat-Blanché tradition(s) –are pragmatically derived and not semantically driven. The move from **I** and **O** to their conjunction **Y**, as in the pentagonalized Square in (8), is context dependent; it is not that the subcontraries are equipollent but that they will tend (ceteris paribus) to result in a speaker's communicating the same state of affairs.

(8) Extended Square of Opposition

Figure 5

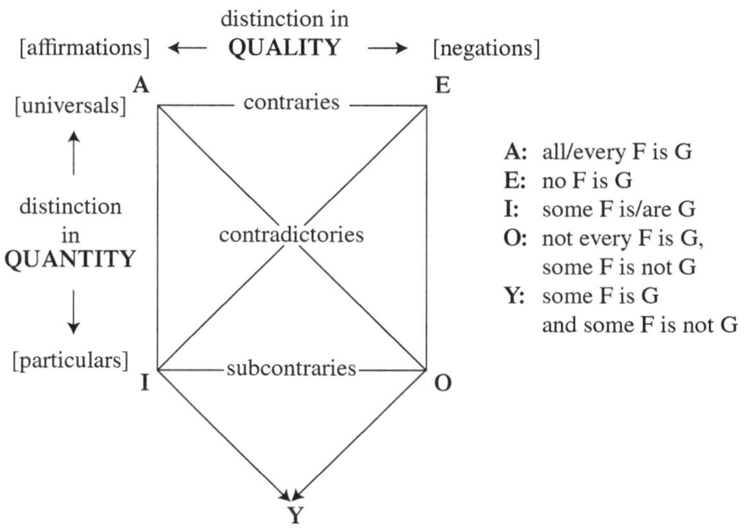

In fact, it can be argued that what we need is not so much a triangle or a pentagon as a defective three-cornered square[2], given that in a wide variety of languages those values mapping onto the southeast corner of the Square are systematically restricted in their potential for lexicalization. This crucial asymmetry was perhaps first recognized by St. Thomas Aquinas, who observed that whereas in the case of the universal negative (**A**) "the word 'no' *[nullus]* has been devised [sic!] to signify that the predicate is removed

2 Cf. Horn 1989: 252–53 for elaboration. I pass over another issue arising in the mapping of quantificational values onto the Square: the role of existential import. Which (if any) of the four statement forms entail or presuppose that the set over which the quantifier ranges is non-null and how does this affect the subaltern and other relations? In particular, if (as in the modern tradition) *all* is import-free while *some* is not, does this vitiate the Square, as Fogelin (1978) maintains? The fact that other operators (binary connectives, adverbs, modals, deontics), for which existential import is irrelevant, can be mapped onto the Square makes Fogelin's move as unappealing as it is unnecessary. This leaves a number of options open for dealing with questions of import and quantification and their relation to the Square; see Horn (1997) for elaboration.

from the universal subject according to the whole of what is contained under it", when it comes to the particular negative (**O**), we find that

> there is no designated word, but 'not all' *[non omnis]* can be used. Just as 'no' removes universally, for it signifies the same thing as if we were to say 'not any' [i.e. 'not some'], so also 'not all' removes particularly inasmuch as it excludes universal affirmation. (Aquinas, *in Arist. de Int.,* Lesson X, Oesterle 1962: 82–3)

Thus alongside the quantificational determiners *all, some, no,* we never find an **O** determiner **nall;* corresponding to the quantificational adverbs *always, sometimes, never,* we have no **nalways* (= 'not always', 'sometimes not'). We may have univerbations for *both (of them), one (of them), neither (of them),* but never for **noth (of them)* (= 'not both', 'at least one…not'); we can find connectives corresponding to *and, or,* and sometimes *nor* (= 'and not'), but never to **nand* (= 'or not', 'not…and'). Schematically:

(9)

	DETERMINERS/ QUANTIFIERS	QUANT. ADVERBS	BINARY QUANTIFIERS	CORRELATIVE CONJUNCTIONS	BINARY CONNECTIVES
A:	all α, everyone	always	both (of them)	both…and	and
I:	some α, someone	sometimes	one (of them)	either…or	or
E:	no α, no one (=all¬/¬some)	never (=always¬)	neither (of them) (=both¬/¬either)	neither…nor (=[both… and]¬)	nor (=and¬)
O:	*nall α, *neveryone (=some¬/¬all)	*nalways (=¬always)	*noth (of them) (=either¬/¬both)	*noth…nand (=[either… or]¬)	*nand (=and¬/¬or)

As argued more fully elsewhere (Horn 1972; Horn 1989: §4.5), the motivation for this asymmetry is pragmatic: the relation of mutual non-logical inference between the positive and negative subcontraries results in the superfluity of one of these subcontraries for lexical realization; given the functional markedness of negation (see Horn 1989 for a comprehensive review), the superfluous, unlexicalized subcontrary will always be **O**.

We shall see below that the missing **O** phenomenon extends to the modals and deontics, and is also reflected in the general tendency toward **O**>**E** drift, wherein lexical items or collocations associated by their compositional form or etymology with the **O** corner of the square move inevitably northward toward **E**. But we will first elaborate the nature of the asymmetry among the quantifiers and related adverbs and connectives in (8), where its operation is particularly robust.

This effect is especially striking when it overcomes morphology and etymology. Thus consider the Old English lexical item *nalles, nealles*, which appears to challenge the constraint blocking **nall*-type determiners, but is in fact attested only with the value 'no, not, not at all', never 'not all'. Other OE quantificational expressions include *næfre* 'never', *næðor* 'neither, nor', *náht* 'nothing', *nán* 'no one, none', and *náhwær* 'nowhere', all occupying the northeast **E** rather than the southeast **O** slot. Along the same lines, Jaspers (2005: 150) points to the Dutch adverb *nimmer* –literally 'nalways', but actually denoting 'never.'

We move now to the binary connectives, which have been situated on the Square since the medievals –cf. Ashworth (1974: 148, 167). Eloquent testimony to the persistent cross-linguistic exclusion of **nand* is provided by the Latin *neque,* which exhibits the *nealles* effect: its **O** morphology ('not' + 'and') belies its **E** ('and not') semantics, i.e. 'neither, nor' (Jaspers 2005: 150). While the *nand gate,* widespread in electronic circuitry, does indeed cover precisely the taboo **O** vertex, this merely underlines the absence of any such lexical item from natural languages. (But see Seuren 2006: fn. 8 for a potential counterexample.)

The modern English connective *nor* is essentially an **E** value, but it is usually restricted to negative polarity contexts –although an untriggered *nor* is not quite as restricted or marginalized as sometimes claimed, as in this passage:

> The two values standardly lexicalized occupy the **A** and **I** slots, represented in English by *and* and *or* respectively. If a third position is represented, typically through negative incorporation, it maps onto the **E** vertex. Jespersen (1917: 108) cites Old English (and Old Norse) *ne:* as a negative phrasal conjunction 'looking before and after', so that *suð ne norð* in *Beowulf* stands for 'neither south nor north'. German offers a (rarely invoked) modern equivalent: *in Wasser noch in Luft* 'neither in the water nor in the air'. (Horn 1989: 256)

But in fact similar occurrences of a Janus-faced *nor* are attested in both literary and colloquial English:[3]

3 The first two examples are cited in the OED, while the last three were provided by Arnold Zwicky and Beverly Flanigan in a thread on this topic on the ADS-L listserv. Other examples can be found by googling <"you nor anyone else" – neither>.

(10) **1872** TENNYSON Great brother, thou nor I have made the world. [OED]
1954 W. FAULKNER A world such as caesar nor sultan nor khan ever saw [OED]
"Kent Smith, nor anyone from that office was present"
"I, nor my management, have ever had any kind of problem with creating a gay character."
"I, nor my host, nor my file server, nor my ISP are responsible for what you do with the patches and ROM images found on this site."

Crucially, while joint denial (p ↓ q, or in bit notation, <0001>) may be lexicalized, the Sheffer stroke (p | q, <0111>) cannot be. But what of exclusive disjunction, <0110>, occupying the **Y** position in the hexagon in (10)?

(11)

Figure 6

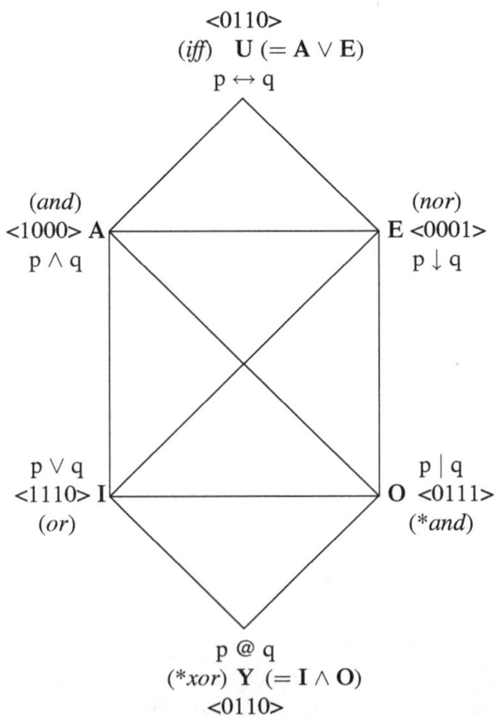

In fact, true exclusive disjunction does exist –just not in natural language. As defined in set-theoretic terms, $x \in A$ XOR B iff $x \in A \cup B \wedge x \notin A \cap B$. But no bona fide representatives of the exclusive disjunction operator have surfaced in **natural** language (Horn 1972, 1989; Gazdar & Pullum 1976; Jennings 1994; Katzir & Singh to appear).

There are, to be sure, pretenders to the throne. While Collinson (1937), Quine (1940), Geach (1972), and even Blanché (1969: 145) assert or presuppose that Lat. *aut* plays exclusive (**Y**) to *vel*'s inclusive (**I**)[4], and while similar claims are made about Finnish or Welsh, closer inspection shows it just ain't so (Horn 1989: 224–26). For example, Latin *aut* –the motivation for my @ ('a(u)t') notation for exclusive disjunction –was employed to oppose two mutually exclusive conditions (= 'p or q, it matters which'), while *vel* – the source of the standard inclusive v connective –was used in free choice contexts ('p or q, it doesn't matter which'). *Aut aut,* damned "ambiguity"!

The pragmatic approach to so-called exclusive disjunction (Horn 1972, 1989) was prefigured in Archbishop Whately's comments on the two disjuncts in e.g. *Virtue tends to procure us* ***either*** *the esteem of mankind* ***or*** *the favor of God* and in the more general (and still apt) considerations of Mill:

> [F]rom one being affirmed we are not authorized to deny the other. Of course, we are left to **conjecture** in each case, **from the context**, whether it is to be implied that the members are or are not exclusive.
> (Whately 1848: 106, emphasis added)

> When we say A is either B or C we imply that it cannot be both...If we assert that a man who has acted in a particular way must be either a knave or a fool, we by no means assert, or intend to assert, that he cannot be both.
> (Mill 1867: 512)

On the neo-Gricean account, **nand* (and **xor*) will be excluded for the same reason as **nall:* given that *p or q* tends to implicate '[for all S knows] not both p and q', the closed set of connectives need admit just the one **I** vertex disjunctive connective.

The asymmetry of (9) extends from the quantificational operators and connectives to other values that can be assigned a logical geometry. Consider, for example, the arithmetical (in)equalities, adjectival comparatives,

4 Here, for example, is Quine (1940: 5) on the "ambiguity": "We must decide whether 'or' is to be construed in an *exclusive* sense, corresponding to the Latin 'aut', or in an *inclusive* sense, corresponding to the Latin 'vel'."

and equatives in (12) (cf. Sesmat 1951: 319 ff. on the relation of these to other values on the hexagon of opposition):

(12)

Figure 7

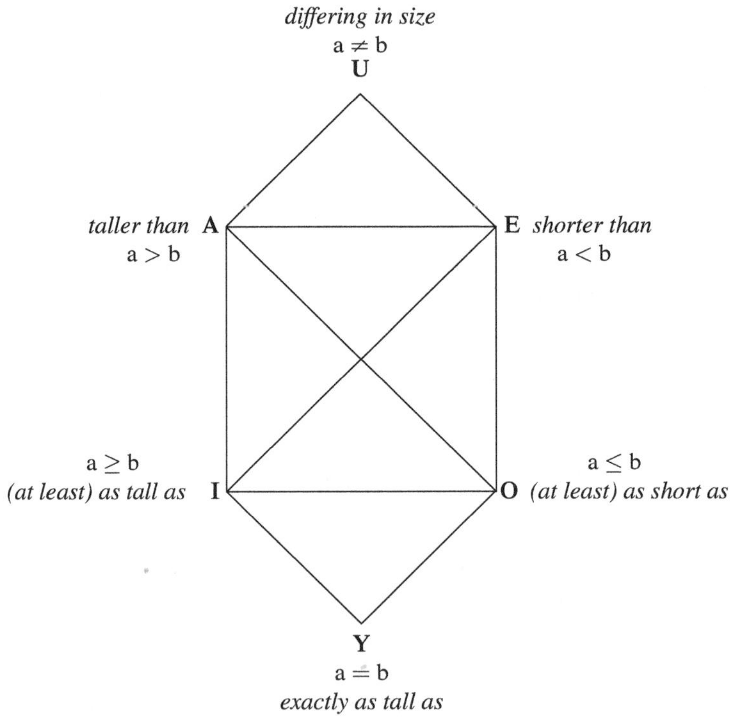

While the **A**, **E**, and **I** values have an unrestricted distribution (Chris can be *taller than, shorter than,* or *as tall as* Robin regardless of their respective heights), the use of the **O** value (e.g *Chris is as short as Robin* presupposes that Chris and Robin are (relatively) short. Earl may be as tall as Muggsy even if they're both unusually short for their comparison set, but Shaq can't be as short as Yao if they're both 7-footers, given the marked nature of the "negative" adjective.

The lexicalization asymmetry of the Square extends to the modals and deontic operators (cf. van der Auwera 1996, Béziau 2003), as illustrated by the fact that the inflected negative in (13b) only allows wide scope

(**E** vertex) negation, i.e. the Roman Catholic reading, while its unlexicalized counterpart in (13a) allows both wide-scope (Catholic) and narrow-scope (Episcopalian) readings of the negation, the **O** version brought out by interpolation as in (14).[5]

(13) a. A priest can not marry. [$\neg\Diamond$ (**E**) or $\Diamond\neg$ (**O**)]
b. A priest {can't/cannot marry}. [only $\neg\Diamond$ (**E**), not $\Diamond\neg$ (**O**)]

(14) A priest can {always/if he wishes/of course} not marry.

The same asymmetry holds for *could not,* although it only shows up in the inflected form, since there is no **couldnot* orthographic lexicalization alongside *cannot:*

(15) a. You could not work hard and still pass. [$\neg\Diamond$ (**E**) or $\Diamond\neg$ (**O**)]
b. You couldn't work hard and still pass. [only $\neg\Diamond$ (**E**), not $\Diamond\neg$ (**O**)]

As with the quantifiers and binary connectives, it is the relation between the modal subcontraries (that of mutual implicature, even if we cannot accept Aristotle's apparent equivalence between them: "whatever may walk...may not walk" –De Int. 21b12) that results in the resistance of the negative (**O**) subcontrary to lexicalization.

What of **A**-vertex modals? *Must* and *should,* whether understood logically, epistemically, or deontically, can only incorporate an inner negation, resulting in an **E** (or, in the case of *should,* **E**-like) meaning:

(16) a. You mustn't go. = 'you must [not go]' [$\Box\neg$(**E**), not $\neg\Box$(**O**)]
b. You shouldn't go. = 'you should [not go]' [basically $\Box\neg$(**E**)]

To be sure, the **A** modal *need* yields a lexicalized **O** value, as in (17a), but given the negative polarity status of *need* as a modal auxiliary (cf. (17b,c)), negation must scope over $\Box$ *(need),* ruling out the **E** reading. (Analogous distributional constraints hold for Dutch *hoeven* and Ger. *brauchen.*)

5 Note that despite the standard dictionary entry for *cannot* in which it is equated to 'can not', the former –as a lexicalization –can only get the **E** reading, never the **O,** regardless of whether the modal is interpreted as an alethic, root, or deontic operator.

(17) a. She needn't go. = 'not [she need go]' [only ¬□(O)]
b. She {needs to/*need} go.
c. %Need you go? [% indicates dialectal variation in acceptability]

Thus, for example, (17'a) with modal *need* has unambiguously wide scope negation, while its non-modal counterpart in (17'b) takes narrow scope negation.

(17') a. He need not leave. [¬□]
b. He needs to not leave. [□¬]

It should also be noted that *needn't* is restricted by semantics (it tends to be deontic) and register (it tends to be constrained to high or written style).

In some languages, including Turkish, ASL (American Sign Language), and LSF *(Langue signée française)*, we find an opaque **E**-valued modal negation that is synchronically unrelated to possibility or necessity (cf. Shaffer 2002); the corresponding **O** form *(≈ need not)* is semantically transparent and indeed non-lexicalized.

As we see from (17) vs. (13)-(16), while the asymmetry in lexicalizing complexes associated with the **A**, **I**, and sometimes **E** vertices as against **O** is equally exhibited across lexical domains, some domains are more equal than others (cf. van der Auwera 2001 for related discussion). The degree of asymmetry varies according to how closed the category is: strongest for connectives *(*nand)* and determiners/quantifiers *(*nall, *nevery, *noth, *nalways),* somewhat weaker for modal auxiliaries (where *needn't* would violate the strong form of the constraint, albeit in a context in which no **E** reading would be possible), and weaker still (though still present) for ordinary adjectives (cf. *impossible* vs. *unnecessary,* where the latter but not the former is restricted to deontic, non-logical contexts).[6] The motivation for this difference in the strength of the lexicalization constraints between closed and open category items is an important one, as an anonymous referee

6 Note also the asymmetry in nominalization potential *(*unnecessity* alongside *impossibility)* and in cross-linguistic parallels (Lat. **innecessarius,* Fr. **innécessaire* vs. *impossibile, impossible).* Similar asymmetries are found in the cross-linguistic expression of modality. Tamil *veeNam* may offer an instance of a lexicalized **O** verb (de Haan 1997: 80), but the facts are more complicated than de Haan suggests (E. Annamalai, p.c.). I hope to address this issue in later work.

points out, but it is one I cannot address here due to both space limitations and the speculative nature of any suggestions I could advance.

What we find in the open-class categories are implicational (rather than absolute) universals. First, the existence of a lexicalized **O** form implies the existence of a lexicalized **E** counterpart but not vice versa. Second, the lexicalized **E** form tends to be more opaque and less constrained (semantically and distributionally) than the lexicalized **O** form (if any). There are also clear differences in number. For example, among English causative verbs incorporating negation, the many verbs in (18a) contrast with the two in (18b):

(18) a. [CAUSE [E]]: 'cause to become not {possible/legal/moral}'

ban	enjoin	interdict	proscribe
bar	exclude	preclude	refuse
deter	forbid	prevent	veto
disallow	inhibit	prohibit	withhold

b. [CAUSE [O]]: 'cause to become not {necessary/obligatory}', '{possible/legal/moral} not...'
excuse exempt

Among the deontics, we have simplex nominal and adjectival realizations for the **A** *(obligation; obligatory/required),* **I** *(permission; permitted/allowed),* and **E** values *(prohibition; forbidden),* but the **O** values can only be complex *(non-obligation, released from obligation,...).* Among the adjectives, we do have **O** candidates like *optional,* Fr. *facultatif,* or Ger. *erübrigen,* but it's often unclear whether a particular modal (whether root, epistemic, or deontic) candidate occupies the **O** or the **Y** slot.[7]

The asymmetry of the Square also extends to "intermediate" values, south of **A/E** but north of **I/O**, where 'not many' can be lexicalized (= *few*) but 'many not' cannot, 'not often/usually not' can be lexicalized (= *seldom, rarely*) but 'often not/not usually' cannot, etc. What is the appropriate generalization? To represent the intermediate values we exploit the fact that the quantifiers, quantificational adverbs, and modal values form scales as defined by unilateral entailment (Horn 1972, following Grice 1989). (In

7 This is a more general problem. Thus, Löbner (1990: 89) nominates a number of potential instances for "Typ 4" (**O**) lexicalizations in German (cf. also Löbner 1987), but on closer inspection several of these nominees *(unnötig, fraglich, bezweifeln)* are either limited to deontic uses or allow non-**O** understandings.

these diagrams, *p* represents an arbitrary proposition containing the scalar value in question and the subscripts indicate variation in strength as determined by unilateral entailment in non-monotone decreasing contexts.)

(19)

Figure 8

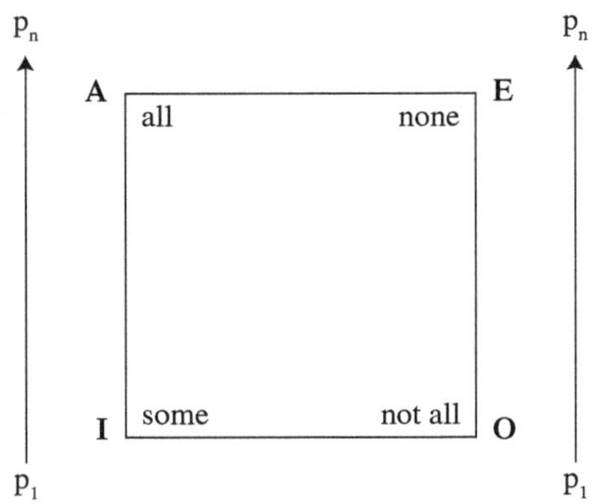

Two terms in any quantitative opposition occupy different positions on a single scale, while two terms in a qualitative opposition will occupy parallel positions (weak, intermediate, or strong) on the corresponding scale. Plotting each scalar value according to its lower bound in the usual way and assigning positive and negative arithmetic values to those positions, we obtain this Arithmeticized Square (Horn 1989: 236 ff.)

(20)

Figure 9

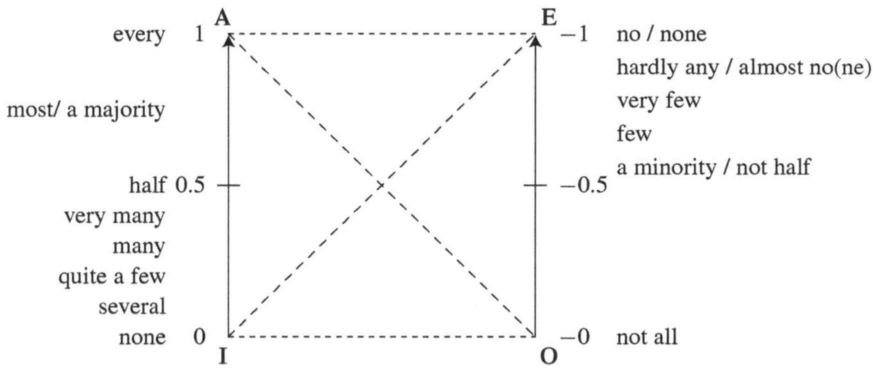

Note that the cardinal values (**A E / I O**) are the strongest/weakest values on their respective scales and the sum of the values of any horizontal pair of quantificational determiners <Q, Q¬>, whether contraries or subcontraries, is always 0.[8]

The crucial factor for determining whether an intermediate value can incorporate an inner or outer negation is where that value is situated with respect to the midpoint of its scale. This is codified by the notion of (IN)TOLERANCE (Löbner 1987: 64): "A quantifier Q is *tolerant* iff Q(P) and Q(~P) is possible at the same time. A quantifier is *intolerant* iff Q(P) excludes Q(~P)". Given quantity scales like those in (21),

(21)	<all, most, many, some>	<no/none, few/not many, not all>
	<always, usually, often, sometimes>	<never, rarely/seldom, not always>
	<(both...) and, (either...) or>	<neither...nor, not both>
	<certain, likely, possible>	<impossible, unlikely, not certain>
	<must, should/ought to, can/may>	<can't/mustn't, shouldn't, needn't>

8 For elaboration, see Horn (1989: 237) and the addenda on pp. xxxiii-xxxiv of Horn (2001).

we can assign the values to the respective positive and negative scalar positions, and obtain the generalizations in (22), as illustrated in (23):

(22) <Q, Q¬> are contraries if Q > .5 and subcontraries if Q ≤ .5.
If Q ≤ .5, the conjunction *Q... and Q¬...* is **consistent**, and Q is **tolerant**.
If Q > .5, the conjunction *Q... and Q¬...* is **inconsistent**, and Q is **intolerant**.

(23) a. **Some** of my friends are linguists and **some** of them are**n't**.
Many of my friends are linguists and **many** of them are**n't**.
He **often** goes to church on Sunday and he often doesn**'t**.
It's **possible** that she'll win, and **possible** that she wo**n't**.
It's **50-50** that it'll land heads, and **50-50** that it wo**n't**.

b. #**All** of my friends are linguists and **all** of them are**n't**.
#**Most** of my friends are linguists and **most** of them are**n't**.
#He **usually** goes to church on Sunday and he **usually** doesn**'t**.
#It's **likely** that it'll land heads, and **likely** that it wo**n't**.
#It's **certain** that she'll win, and **certain** that she wo**n't**.

The key principles of lexicalization (Horn 1972) can now be given as follows:

- When Q is intolerant, it may lexically incorporate its (contrary) inner negation ($Q_I\neg$) but it tends not to lexicalize its outer negation ($\neg Q_I$).
- A tolerant value may incorporate its outer negation ($\neg Q_T$) but bars lexicalization of its inner negation ($Q_T\neg$).

Thus, *seldom* or *rarely* lexicalize the inner negation of the intolerant *usually* (or the outer negation of the tolerant *often*), but there can be no simple lexical realization of *not usually* (or *often not*); similarly for *few (= not many), a minority (= a majority...not),* or *unlikely (= likely...not).*

The effect of these universals is to facilitate the direct expression of contrariety in natural language while blocking the direct expression of subcontrariety. Another reflection of this contrast is the strengthening of sentential (contradictory) negation to express a contrary meaning. In the words of Bosanquet (1911: 281), "The essence of formal negation is to invest the contrary with the character of the contradictory". The result is a variety of cases of contrary negatives in contradictory clothing (Horn 1989: Chapter 5). Some examples from English appear in (24):

(24) a. contrary readings for affixal negation (conventionalized strengthening)

He is unhappy	(unilaterally entails ¬[He is happy])
She was unfriendly	(unilaterally entails ¬[She was friendly])
I disliked the movie	(unilaterally entails ¬[I liked the movie])

b. litotes/understatement in simple denials (online or non-conventionalized strengthening)

He's not happy with it	(approaches 'He's unhappy with it')
I don't like ouzo	(approaches 'I dislike ouzo')
I'm not optimistic that ϕ	(approaches 'I'm pessimistic that ϕ')

c. "neg-raising" effects (strengthened understanding as a convention of usage)

I don't believe it'll snow	(≈ I believe it won't snow)
I don't want you to go	(≈ I want you not to go)
It's not likely they'll win	(≈ It's likely they won't win)

In each case, the negation of an unmarked "mid-scalar" positive scalar predication (at least) implicates a stronger, contrary negation.

Consider in particular the phenomenon of "neg-raising" in (24c), a tendency first recognized by St. Anselm (1033–1109), who pointed out that "we say *non debet peccare* [lit., 's/he NEG should sin'] when we mean *debet non peccare*" (Henry 1967: 193–94; cf. Hopkins 1972: 233–34). But which predicates license this lower-clause understanding of higher-clause negation? In general, neg-raising licensers are weak intolerant positive operators, situated just above the midpoint of the relevant scale (Horn 1989: §5.2). Their intolerance is demonstrated by plugging the NR predicates into the frame of (23):

(23') a. #You **should** marry and you **should not**-marry.

b. #I **believe** it will snow and I **believe** it **won't** snow.

Weak intolerant positive predicates allowing NR readings contrast with tolerant, strong, and/or morphologically negative predicates that block them. Thus NR is licensed by:

(25) ***believe, suppose*** but not ***know, doubt, disbelieve***
want, suggest but not ***insist, forbid, prohibit***
advisable, desirable but not ***obligatory, forbidden***
should, ought to, better but not ***have to, must, can***
likely, probable but not ***certain, impossible***
most but not ***all, many, some, few***
usually but not ***always, often, sometimes, rarely***

As has often been noted, neg-raising functions euphemistically to tone down the expression of negative judgment, a motivation for litotes as well. In addition, as seen in the examples in (25), the weak-intolerance generalization crosscuts the traditional distinction between "lexical" and "functional" operators, as exemplified by the appearance of main verbs, modal auxiliaries, adjectives, determiners, and quantificational adverbs among the neg-raisers, provided they fit the scalar requirement.

A systematic exception to the weak intolerance is the class of strongly intolerant (**A**-position) deontic values, whose members often exhibit neg-raising effects. One classic example is the Fr. *Il ne faut pas que tu meures* (lit., 'it is not necessary that you die', but usually = 'you mustn't die'). This example is described in the eponymous paper by Tobler (1882), who notes that the notoriously "unlogisch" placement of negation was attested by the 14th century alongside the "logical" **O** reading that it eventually comes to evict or block. Another instance of **O>E** drift is provided by the strengthening of negated causatives. Thus while It. *fare* + infinitive on its own conveys a strong causative ('make', not 'let'), its negation is often understood with an **E**-style strengthened meaning:

(26) Il caffè non mi fa dormire. 'Coffee doesn't {make/let} me sleep'

Not only in Italian but in such languages as Japanese, Turkish, Amharic, Czech, Biblical Hebrew, and Jacaltec, the negation of a strong causative (lit., 'not make') may or must strengthen to yield contrary ('make not' = 'not let' = **E**) force. The reverse drift, in which a 'not let' (**E**) causative is understood as 'let not' or 'not make' (**O**), on the other hand, appears to be unattested.

Further instances of the preferential treatment of **E** over **O** values are listed in (27), the first one having already been touched on.

(27) a. **E** readings for **O** forms *(nealles,* Du. *nimmer,* Lat. *neque,* Russ. *nel'zja)*[9]

b. the opacity of **E** forms *(no, nary a;* Ger. *nie;* Fr. *personne, rien, jamais)*

c. the frequent difficulty of negating **A** modals without subsequent drift, e.g.
You are to leave the room. (**A**)
You are not to leave the room. (**E**)

d. an invariant **E** readings for complex adjectives with negation + possibility, e.g. [un-[V [-able]]] = 'incapable of being Ved' (**E**), ≠ 'capable of not being Ved' (**O**)

f. complex adverbs incorporating semantics of **I** *(enough)* or **E** *(too)* but not **O**

g. the existence of "able" polarity and "unable" polarity vs. the non-existence of "needn't" polarity

Thus, the apparent negated universal sequences in (27a) have an **E** meaning instead of the expected **O**. Other lexical instances of **O>E** drift include English *not at all* (cf. the undrifted *not altogether*) and French *pas du tout.* The basic, unanalyzable status characteristic of **E** forms is exemplified in (27b); *nary* –etymologically a reduction of *ne'er a* –is now entirely opaque, while the results of Jespersen's Cycle in French exhibit little trace of their history as positive indefinites that have incorporated negative force (Latin *persona, rem, jam magis* respectively). No **O** values display this degree of semantic opacity. The superficial contradictory negation of **A** values often yield **E** results rather than the expected **O**, another example of induced contrariety (**O>E** drift), as seen in the quasi-modal pair in (27c). When negation combines with a possibility modal within a single lexical item, the result is virtually always an **E** rather than **O** meaning, with negation taking wide scope; an unreadable book is one that cannot be read rather than one that can be not read. Complement-taking adverbs may incorporate **I**-style ('possible') semantics (*tall enough* = 'so tall that ◊') or **E**-style ('impossible')

9 Van der Auwera (2001: 41) mentions additional candidates for this status, including Bengali *nei.*

semantics (*too short* = 'so short that ¬◇') but not **O**-style semantics (**shmoo short* = 'so short that ◇¬'). And as discussed in detail in Horn 1972: Chapter 3, alongside the instances of "able" or "possible" polarity (see (28a)) and "unable" or "impossible" polarity (see (28b)), items whose distribution is restricted to contexts of c-commanding **I** and **E** modality respectively, there is no attested "needn't" polarity with the distribution of *blarf* in (28c).

(28) a. I can (*must, *needn't) finally {afford a new car/tell them apart}.
It's possible for me to {afford a new car/tell them apart}.
b. I couldn't {fathom/make heads or tails of} the proposal.
I'm incapable of {fathoming/making heads or tails of} the proposal.
c. You needn't (*can't, *must) *blarf* that avocado.
It's unnecessary (*possible/*impossible) for you to *blarf* that avocado.

Finally, we touch on one last asymmetry. When the neg-raising effect yields a strengthened **E**-type understanding alongside the literal **O**-type reading we often find that lexical incorporation of the negation renders the quasi-**O** readings inaccessible. Thus:

(29) a. It's {not probable/not likely} that a fair coin will land heads.
(ambiguous; true on outer [contradictory] reading of negation)
b.It's {improbable/unlikely} that a fair coin will land heads.
(unambiguously inner [contrary] negation, hence false)

(30) a. It's not likely that Federer will win and not likely that he'll lose.
(allows tolerant, contradictory reading with outer negation)
b. #It's unlikely that Federer will win and unlikely that he'll lose.
(allows only intolerant [neg-raised], hence anomalous reading)

(31) a. It's not {advisable/desirable} that you go there alone.
(ambiguous)
b. It's {inadvisable/undesirable} that you go there alone.
(only a warning not to)

In each case, as with *can't* or *couldn't,* it is the lexicalization in the (b) cases that forces the contrary **E** readings and blocks the contradictory **O** reading, while both are available in the corresponding non-lexicalized sequences in the (a) counterparts. We could also point to such minimal oppositions as *disprove* and *disbelieve* on the one hand, with $\mathbf{A}\neg = \mathbf{E}$ analyses, as against *disallow* and *disable,* with $\neg\mathbf{I} = \mathbf{E}$ analyses.[10]

The first neo-Gricean account of the asymmetries discussed in this study appeared in Horn (1972), although additional data have been accumulated since. (See in particular the cross-linguistic data surveyed in Horn 1989: §4.5 and the arguments in Jaspers that partially support and partially challenge those analyses.) No natural language instances of **nall* or **nand* have yet surfaced. In many, probably most, languages, neither of the negative positions (**E** or **O**) is lexicalized. But the generalization remains sound: **A**, **I**, and often **E** values may lexicalize, **O** values may not. On the proposed account, the three-cornered square is a consequence of the maxim of Quantity: in any situation in which I possess (and am known to possess) complete knowledge, and in which that knowledge is (and is known to be) relevant to you, I can convey that information to you via a proposition containing one of the three values *all, some* (implicating *some not, not all*), or *none*. The fourth value *(*nall = not all,* implicating *some)* is functionally (although not logically) expendable.

Over the succeeding decades, others have offered their own translations of the story of ***O**. Huybregts' (1979) blocking principle stipulates that *not Q* can lexicalize just when *Q not* can't, but this correlation fails to fully explain just when the latter state of affairs obtains. The monotonicity correspondence universal of Barwise & Cooper (1981: 186–87) predicts inter alia that strong determiners like *every* or *most* can incorporate inner but not outer negation; this correctly allows *neither* and *no* while ruling out **nevery* and **noth.* But without a grounding in the pragmatics of scalar operators, such approaches to the constraints on quantifier lexicalization are ad hoc. Worse still, any treatment based on the semantics of the determiners and quantifiers like that of Barwise & Cooper (1981) fails to generalize to the binary connectives, modals, and other non-quantificational values, much less to the intermediate values. A later proposal along similar lines is that of Hoeksema (1999), who, after reviewing earlier accounts, argues that the non-occurring **O** forms are not blocked by corresponding **I** forms,

10 Further support for the preference for contrary readings in natural language is surveyed in Horn (2008), which includes a closer look at "neg-raising" and related cases of negative strengthening and explores the phenomenon of "virtual contrariety" in cases of double negations that don't quite cancel out.

pace Horn (1972). Rather, he notes, neither of the likely sources that would yield lexicalized **O** quantifiers –the merger of an existential value + negation or the reinterpretation of NPIs –is consistent with what is known of the plausible historical development. Once again, however, the considerations he raises, contributing factors though they may be, fail to extend to other scalar values that manifest an asymmetry in lexicalization.

More recently, Seuren (2003: 13) has advanced a new formalization of the Aristotelian Predicate Calculus on which **A**, **I**, and (sometimes) **A*** (= **E**) will have lexicalized representations but **I*** (the **O** of the standard square) does not. He argues that, contra the Gricean moral to the story I have drawn, "The question [of why there is no **nall*] is superfluous...an artifact of the defective way [the Aristotelian Predicate Calculus] was formalized by Boethius" and the other geometers of the Square. But altering the representation does not explain the asymmetry in the lexical incorporation of negation.[11]

One more approach to the lexicalization asymmetries is worth reviewing briefly here. Moeschler (2007, 2009) disputes the neo-Gricean line of Horn 1972, 1989 on these asymmetries. Like those of Huybregts and Hoeksema, Moeschler's critique is flawed by his exclusive concentration on the quantificational values, ignoring the related asymmetries among the binary connectives, modals, and other operators that can be mapped on the Square, along with the extension of the asymmetries to the intermediate values. In addition, Moeschler (2007: 4) refutes a straw argument from "complexity" that I have never made; indeed, on complexity grounds alone, there is no way to choose between **E** and **O**, since each is equally complex in combining negation with quantification or modality. Yet in each of the relevant cases, the **O** values are more likely to resist lexicalization or direct expression than their **E** counterparts, as the neo-Gricean generalization predicts.

While Moeschler (2007: 5) asserts that "according to Horn, utterances whose truth-conditions are different (I and O) have identical Q-implicatures", the claim is actually that they have complementary Q-implicatures in that the assertion of either will (ceteris paribus) implicate that (for all the speaker knows) the other holds; asserting *some F are G* implicates that (for all the speaker knows) not all F are G, and vice versa. But it is not

11 I must defer any response to the nuanced critiques of Jaspers 2005 and Seuren 2006 for another occasion.

the case that "I and O both implicate the conjunction of I and O." In fact, the **Y**-vertex proposition that some but not all F are G is not claimed to be directly implicated by either an **I** or **O** proposition, but may be **communicated** by what is said together with what is implicated. Nor can Q-implicatures be "just truth-conditional logical implications" (cf. Moeschler 2007: 11), given their cancelability.[12]

Finally, I am not sure what to make of the argument that the implicature-based account should be jettisoned for one invoking the relevance-theoretic notion of explicature on the grounds that "the theory of scalar implicatures is outdated." As will be confirmed by even a quick review of the contents of leading journals, lists of conference and workshop announcements, and recent work like that of Geurts (2009, to appear), this death-knell is untenable as either a theoretical conclusion or a fashion statement.

In sum, we draw this moral from *L'Histoire d'*O:* The relation of mutual quantity implicature holding between positive and negative subcontraries **I** and **O** results in the superfluity of one of the two for lexical realization, while the functional markedness of negation allows us to predict that the unlexicalized subcontrary will always be **O** rather than **I**. The pragmatic account of the asymmetries built into the "three-cornered square" is more general and more explanatory than rival theories that either dismiss the asymmetry as uninteresting or restrict it to the determiners and quantificational operators while ignoring its relevance to lexicalized negation with other operator types (binary connectives, modals, adverbs, predicates) and the intermediate values that can be mapped onto the Square of Opposition.

12 On the claim that "negative particulars do not trigger Q-implicatures or scalar implicatures" (Moeschler 2007: 11), a position also maintained by Chierchia (2004), see Horn (2009: §3) for a detailed response.

Acknowledgments

Earlier versions of some of this material were presented in Berkeley (1970), Austin (1973), Milan (2003), Lansing (2003), and New York (2004). I am indebted to those who attended those presentations, to the participants in the Montreux congress, and to Barbara Abbott, Elayaperumal Annamalai, Jack Hoeksema, Polly Jacobson, Dany Jaspers, Jacques Moeschler, Hans Smessaert, Rashad Ullah, Victor Sánchez Valencia, Raffaella Zanuttini, Arnold Zwicky, and especially the two anonymous referees. Needless to say,...

References

Ashworth, E. J. (1974). *Language and logic in the post-Medieval period.* Dordrecht: Reidel.

van der Auwera, J. (1996). Modality: The three-layered scalar square. *Journal of Semantics, 13,* 181–96.

van der Auwera, J. (2001). On the typology of negative modals. In J. Hoeksema, H. Rullmann, V. Sánchez-Valencia & T. van der Wouden (Eds.), *Perspectives on Negation and Polarity Items,* 23–48. Amsterdam: John Benjamins.

Barwise, J., & Cooper, R. (1981). Generalized quantifiers and natural language. *Linguistics and Philosophy, 4,* 159–219.

Béziau, J.-Y. (2003). New light on the square of opposition and its nameless corner. *Logical Investigations 10,* 218–32.

Blanché, R. (1952). Quantity, modality, and other kindred systems of categories. *Mind, 61,* 369–75.

Blanché, R. (1953). Sur l'opposition des concepts. *Theoria, 19,* 89–130.

Blanché, R. (1969). *Structures intellectuelles,* 2e éd. Paris: J. Vrin.

Bochenski, I.M. (1961). *A history of formal logic,* I. Thomas, trans. Notre Dame: University of Notre Dame Press.

Bosanquet, B. (1911). *Logic, Vol. 1* (2nd ed.). Oxford: Clarendon.

Chierchia, G. (2004). Scalar implicatures, polarity phenomena, and the syntax/ pragmatics interface. In A. Belletti (Ed.), *Structures and beyond* (pp. 39–103). Oxford: Oxford University Press.

Collinson, W.E. (1937). *Indication. Language* monograph No. 17.
Couturat, L. (1913). Des propositions particulières et de leur portée existentielle. *Revue de Métaphysique et de Morale, March 1913,* 256–59.
Couturat, L. (1914). Réponse [to Ginzberg 1914]. *Revue de Métaphysique et de Morale, March 1914,* 259–60.
De Morgan, A. (1847). *Formal logic.* London: Taylor & Walton.
De Morgan, A. (1858). On the syllogism: III, and on logic in general. *On the syllogism and other logical writings,* 74–146. London: Routledge & Kegan Paul.
De Morgan, A. (1861). Hamiltonian logic. *Athenæum* 13 July 1861, p. 51.
Dewey, J. (1938). *Logic: The theory of inquiry.* New York: Henry Holt.
Fogelin, R. (1967). *Evidence and meaning.* New York: Humanities Press.
Fogelin, R. (1978). *Understanding arguments: An introduction to informal logic.* New York: Harcourt Brace Jovanovich.
Gazdar, G., & G.K. Pullum. (1976). Truth-functional connectives in natural language. *CLS 12,* 220–34.
Geach, P.T. (1972). *Logic Matters.* Berkeley: U. of California Press.
Geurts, B. (2009). Scalar implicature and local pragmatics. *Mind and Language, 24,* 51–79.
Geurts, B. To appear. *Quantity Implicatures.* Cambridge: Cambridge University Press.
Ginzberg, S. (1913). Note sur le sens équivoque des propositions particulières. *Revue de Métaphysique et de Morale, January 1913,* 101–06.
Ginzberg, S. (1914). À propos des propositions particulières. *Revue de Métaphysique et de Morale, March 1914,* 257–59.
Grice, H.P. (1961). The causal theory of perception. *Proceedings of the Aristotelian Society, Supplementary Volume 35,* 121–52.
Grice, H.P. (1989). *Studies in the Way of Words.* Cambridge, MA: Harvard U. Press.
de Haan, F. (1997). *The Interaction of Modality and Negation: A Typological Study.* New York: Garland.
Hamilton, Sir William. (1860). *Lectures on logic, Volume I.* Edinburgh: Blackwood.
Hegenberg, Leonidas. (1957). A negação. *Revista Brasieira de Filosofia, 7,* 448–57. Summarized and reviewed by Hector-Neri Castañeda, *Journal of Symbolic Logic, 25* (1960), 265.
Henry, D. (1967). *The Logic of Saint Anselm.* Oxford: Oxford U. Press.

Hoeksema, J. (1999). Blocking effects and polarity sensitivity. In *JFAK: Essays dedicated to Johan van Benthem on the occasion of his 50th birthday,* Amsterdam: Vossiuspers. (http://odur.let.rug.nl/~hoeksema/docs/j50.htm)

Hopkins, J. (1972). *A Companion to the Study of St. Anselm.* Minneapolis: U. of Minnesota Press.

Horn, L. (1972). *On the Semantic Properties of Logical Operators in English,* UCLA dissertation.

Horn, L. (1973). Greek Grice: a brief survey of proto-conversational rules in the history of logic. *CLS 9,* 205–14.

Horn, L. (1989). *A Natural History of Negation.* Chicago: University of Chicago Press.

Horn, L. (1990). Hamburgers and truth: Why Gricean inference is Gricean. *BLS 16,* 454–71.

Horn, L. (1997). All John's children are as bald as the King of France: Existential import and the geometry of opposition. *CLS 33,* 155–79. Chicago: Chicago Linguistic Society.

Horn, L. (2001). [Horn 1989], reissued with new introductory material. Stanford: Center for the Study of Language and Information.

Horn, L. (2005). The border wars: A neo-Gricean perspective. In K. von Heusinger & K. Turner (Eds.), *Where Semantics Meets Pragmatics* (pp. 21–48). London: Elsevier.

Horn, L. (2008). On the contrary. Paper presented at LNAT – Logic Then and Now conference, Brussels, November 2008.

Horn, L. (2009). WJ-40: Implicature, truth, and meaning. *International Review of Pragmatics, 1,* 3–34.

Huybregts, R. (1979). De biologische kern van taal. In R. Huybregts & L. des Tombe (Eds.), *Verkenningen in Taal,* 79–189. Utrecht: Instituut A. W. de Groot.

Jacoby, P. (1950). A triangle of opposition in Aristotelian logic. *The New Scholasticism, 24,* 32–56.

Jaspers, D. (2005). *Operators in the lexicon: On the negative logic of natural language.* (Universiteit Leiden dissertation.) Utrecht: LOT.

Jennings, R.E. (1994). *The genealogy of disjunction.* Oxford: Oxford University Press.

Jespersen, O. (1917). *Negation in English and other languages.* Copenhagen: A. F. Høst.

Jespersen, O. (1924). *Philosophy of grammar.* London: Allen & Unwin.

Katzir, R., & R. Singh. To appear. A note on the absence of XOR in natural language. To appear in P. Égré & G. Magri (Eds.), *Proceedings of MIT-France Workshop on Scalar Implicature and Presupposition.*

Keynes, John Neville. 1906. *Studies and exercises in formal logic.* London: Macmillan.

Levinson, S. (2000). *Presumptive meanings: The theory of generalized conversational implicature.* Cambridge, MA: MIT Press.

Löbner, S. (1987). Quantification as a major module of natural language semantics. In J. A. G. Groenendijk, D. de Jongh & M. Stokhof (Eds.), *Studies in Discourse Representation Theory and the Theory of Generalized Quantifiers* (pp. 53–85). Dordrecht: Foris.

Löbner, S. (1990). *Wahr neben Falsch: Duale Operatoren als die Quantoren naturlicher Sprache.* Tübingen: Max Niemeyer.

Mill, J.S. (1867). *An Examination of Sir William Hamilton's Philosophy* (3d edn). London: Longman.

Moeschler, J. (2007). Why are there no negative particulars? Horn's conjecture revisited. *GG@G: Generative Grammar in Geneva, 5,* 1–13.

Moeschler, J. (2009). Pourquoi n'y a-t-il pas de particuliers négatifs? La conjecture de Horn revisitée. Downloadable at http://www.unige.ch/lettres/linguistique/moeschler/publications.php. To appear in *Actes du colloque sur la quantification.* Caen: Presses universitaire de Caen.

Monck, W.H.S. (1881). *Sir William Hamilton.* London: Sampson, Low.

Oesterle, J. (Ed. & Trans.). (1962). *Aristotle: On interpretation. Commentary by St. Thomas and Cajetan.* Milwaukee: Marquette University Press.

Quine, W. V. O. (1940). *Mathematical logic.* New York: W. W. Norton.

Réage, P. (1954). *Histoire d'O.* Sceaux: Jean-Jacques Pauvert.

Rose, L. E. (1968). *Aristotle's syllogistic.* Springfield, IL: Charles C. Thomas.

Sapir, E. (1930). *Totality. Language* monograph No. 6.

Sesmat, A. (1951). *Logique II.* Paris: Hermann.

Seuren, P. (2003). The logic of language: Aristotelian predicate calculus revisited. Unpublished ms., University of Nijmegen.

Seuren, P. (2006). The natural logic of language and cognition. *Pragmatics, 16,* 103–38.

Shaffer, B. (2002). CAN'T: The negation of modal notions in ASL. *Sign Language Studies, 3,* 34–53.

Strawson, P. F. (1952). *Introduction to logical theory.* London: Methuen.

Tobler, A. (1882). Il ne faut pas que tu meures "du darfst nicht sterben". *Vermischte Beiträge zur französischen Grammatik 1,* (3rd ed.), 201–05. Leipzig: S. Hirzel.
Tucci, G. (1928). On the fragments from Dinnāga. *Journal of the Royal Asiatic Society,* 377–90; 905–06.
Von Wright, G.H. (1951). *An essay in modal logic.* Amsterdam: North-Holland.
Whately, R. (1848). *Elements of logic,* 9th edition. London: John W. Parker.
Zabeeh, F. (Ed. & Trans.). (1971). *Avicenna's treatise on logic.* The Hague: Nijhoff.

How the Semiotic Square Came

SÉMIR BADIR

Abstract

In this paper, I intend to give a presentation of the diagram known in semiotics as "the semiotic square". More precisely, I would like to retrace semiotic square's origins and early history until its "official" presentation in a dictionary of semiotics. One would expect that the logical square should have taken a great part in the invention of the semiotic diagram. However, the trajectory of the semiotic square was more fluctuating and heterogeneous than what is expected *a priori*, coming under the influence not only of logic but also of other sources of knowledge, as anthropology and linguistics. Shaped at the very same moment by a diagrammatical representation the semiotic square allows an underdetermination of the relations of its objects. The object becomes properly structural simply when the square displayed that structure. So the determinations of the structure can be renewed by the way of theoretical discussion around the square itself, constructing and, soon, deconstructing, a concrete geometrical shape. And this is the usual way semioticians think.

Keywords: binarity, Greimas A.J., Rastier Fr., semiotic square, semiotics

I intend to give a presentation of the diagram known in semiotics as "the semiotic square". More precisely, I would like to retrace semiotic square's origins and early history until its "official" presentation in a dictionary of semiotics. One would expect that the logical square should have taken a great part in the invention of the semiotic diagram. However, as we will see, the trajectory of the semiotic square was more fluctuating and heterogeneous than what is expected *a priori*, coming under the influence not only of logic but also of other sources of knowledge, as anthropology and linguistics. I presume that this would be a sufficient explanation to excuse the rather chaotic aspect of this presentation, which I will of course try to keep within the limits of the tolerable.

But first, some words about semiotics. Little needs to be known here, but I do not expect that everyone amongst logicians – I have especially in mind the youngest ones – has even the slightest idea of what semiotics is. Indeed, even though its domain of research was determined in the Ancient Times and its program established at the beginning of the XXth century, semiotics is a rather young discipline: no more than forty years old. This paper is about its first ages, from 1967 to 1970, at the time when semioticians looked for theoretical models and yet unstudied fields of investigation. Basically, semiotic theory intended to be a *theory of meaning*. Its ambition was to spread out the descriptive method of structural linguistics on any kind of languages, either so-called "natural" languages or visual, musical, gestural, literary (and so on) languages. The semiotic square was introduced in this theory as the elementary structure of meaning, that is the structure of any kind of meanings and whatever the field of language described.

During this period, semiotic theory came under the influence of logic, as did its main model, that is to say linguistics. Moreover, logic had precedence over semiotics simply by the way that some of the major semiotic theoreticians, such as Ch. S. Peirce or Ch. Morris, were first of all logicians. However, this direct influence has been more pervasive in Italy than France (the two major countries where semiotics spread out first). As the semiotic square is a French affair, I will only speak then of the indirect influence, that is via linguistics, of logic on semiotics.

The semiotic square is mainly due to A. J. Greimas, who took a leadership role in the discipline in France. But we notice that many of his works about the semiotic square are written in collaboration, firstly with Fr. Rastier (1968), next with J. Courtès (1979), lastly with J. Fontanille (1991, but I do not follow the semiotic square all the way to that point). Three studies must arrest attention:

- *Sémantique structurale* (1966)
- "Pour une théorie de l'interprétation mythique" (1966 [now in Greimas 1970])
- "The Interaction of Semiotic Constraints" (1968; with Fr. Rastier; [also in French in Greimas 1970])

plus, as a coda:

– *Sémiotique. Dictionnaire raisonné de la théorie du langage* (1979; w. J. Courtés).

There will be two simultaneous approaches in my study. One is simply governed by the chronology of these writings, while the other, in a more specific way, deals with the semiotic intersections between linguistic propositions, symbolical notations and diagrams.

Opposition and Correlation

Let us first have a look on two symbolical notations:

Figures 1 and 2: Two Narrative Structures (Greimas 1966: 196 and 198)

$$\frac{\bar{a}}{\overline{\text{non } a}} \ \text{vs} \ \frac{a}{\text{non } a} \qquad\qquad \frac{\bar{p}}{\overline{\text{non } p}} \cong \frac{p}{\text{non } p}$$

They appear in *Sémantique structurale* only two pages away from each other. We can consider them as similar, since we do not pay attention to what those symbolical notations are about. The symbols *a* and *p* appear as interchangeable units, as nothing determines the use of *a* or *p*, and their variations with connectors (*non* and “¯”, either taken separately or one combined with the other) are strictly identical. The only difference between the notations is the use of distinct symbols between the two quotients, *i.e.* *vs* in the first notation, $\cong$ in the second one.

Since those symbols are distinct, one can presume that they are used distinctively. But this is not exactly what happens. There is no such thing as a table or index in *Sémantique structurale* where you could find a strict correspondence between symbols and, say, operations, as opposed to what we are quite assured of finding in a work of logic. Neither will you find any presentation of the symbols that are used as metalinguistic expressions in this work. The only way you have to test your presumption that those two symbols are used distinctively is to verify, for each occurrence of those symbols, whether the co-references between the symbolical notation and an introductory or commentarial sentence are equivalent or not. My examination leads me to conclude that, *generally*, *vs* stands for an *opposition*, ≅ for a *correlation*, whatever opposition and correlation mean (again, this is not specified in Greimas' work). There is however a persistent problem as, in some cases, Greimas refers verbally to those symbols as *articulation*, more often as *categorical articulation*[1]. Does it condemn our attempt to make a distinction? or, as a minor disappointment, does it mean that Greimas renounces, in some cases, to make a distinction between them and instead regards the use of the two symbols indifferently? Of course, a better suggestion would be that articulation is a semantic category comprising both opposition and correlation, but this is purely hypothetical – and a risky hypothesis, since it has implications not only for the interpretation of symbols but also for the interpretation of the theory itself. More reasonably, I will stick to facts: (i) in *Sémantique structurale* the use of symbols is not controlled by a verbal metalanguage ; (ii) the co-reference is variable between symbolical notations and verbal expressions, even if – I prefer to assure the reader about this – one can easily interpret, for each particular case, what specific symbolical notations refers to; the fluctuation is only noticeable in consideration of second-order language (the language referring to its own expressions).

To which causes can we ascribe those indiscriminate uses? We can surely put it down to the novelty of the practice, of semiotics in general, and of symbolical notations inside semiotic works in particular. Moreover, the semiotician borrowed the symbolical notations from several sources. Two sources are unquestionable, since Greimas quotes both many times, namely the French anthropologist Lévi-Strauss and Roman Jakobson, the Russian linguist exiled in America.

In many of Lévi-Strauss' works, there are symbolical notations, such as:

Figure 3: Canonical Relations (Lévi-Strauss 1958: 252)

$$F_x(a) : F_y(b) \cong F_x(b) : F_{a-1}(y)$$

1 In French: *articulation catégorique* ; nowadays, French logicians would say instead *articulation catégorielle*.

This notation, excerpted from *Anthropologie structurale* (1958), would itself be borrowed from logico-mathematical language: the units of this formulæ are built on the model of the symbolization of logico-mathematical functions. However, logicians and mathematicians would be surprised to learn that in *Le Cru et le Cuit* (1964), the symbol "≅" has been replaced by the symbol ": :" while no change in the meaning is noticeable between the formulæ. There is no doubt that, from one book to another, Lévi-Strauss had the right to choose different symbols to refer only to one operation, that is simply "relations", without saying anything about it. The change should be noticed by the exegetes and accounted for coherence, or as some author's influence happening in the interval, or any other hypothesis. But it would be, from Greimas' part, an interpretation which demands some commentary as to his use in one and the same book of the one or other symbol as though they are.

In addition, in Jakobson, whose *Selected Writings* were published in 1962, we can find many symbolical notations, some with the symbol ": :," others with the symbol "=". The verbal expressions referring to those notations are, indifferently, "set of ratios" (Jakobson 1962, p. 420), "equivalence" (Jakobson 1962, p. 518) or "mutually proportional relations" (Jakobson 1962, p. 644).

Thus, in the works of the two authors that could most probably have been a source of influence for Greimas for his own symbolical notations, there is a possibility of confusion, that is an unexpected and implicit equivalence, between symbols such as ": :" and "≅", as well between operations such as correlation and opposition. The symbols in these sources are numerous, disparate and ambivalent. Greimas himself does not seem to put the use of these symbols in order: symbols are simply witness to the intersections between his own work and those linguistic and anthropological sources; they are not investigated by any theoretical thought. Logicians could consider that something seems awkward here. I would say that this lack of definition about symbols indicates only that semiotic theorization happens somewhere else. Now we are going to try to find where, since rigour in Greimas' work cannot be doubted.

Locking Binarity

In contrast, what is markedly new in Greimas' work is the elaboration of other graphical representations, such as tables, trees and schemas[2]. One of those schemas is clearly, retrospectively, a prefigurement of the semiotic square, its first draft:

Figure 4: Structure of Bororo's "Natural" Culture (Greimas [1966] 1970, p. 228; my translation)

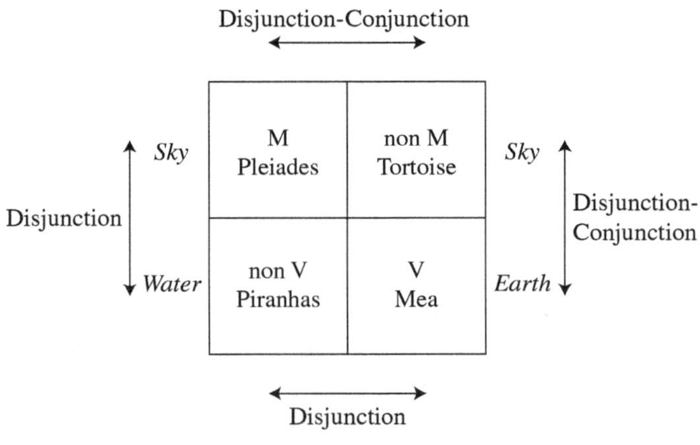

In the study from where this schema is taken, Greimas has no new material to lay forward about Bororo's culture; the only interest he takes in the subject is based on an analysis of some anthropological material which was presented in a study by Lévi-Strauss (the one published in *Le Cru et le Cuit*). So, what Greimas intends to do, is to raise the level of the analysis to a systematisation matching to his "semantic" theory (he will call the same theory a *semiotic* theory two years later). The schema reproduced above comes up in the last pages of the paper, so it could appear as the very result

2 See, for instance, the table in Greimas 1970, p. 214, or a tree in Greimas 1970, p. 196. It is interesting to notice that this tree is an analytical presentation of one of Lévi-Strauss' works, although there is no such thing as a tree diagram in Lévi-Strauss.

of the "semantic" analysis of the Bororo's myths, actually the original contribution of the Greimas' study in comparison with that of Lévi-Strauss.

Of what nature is this contribution? An analytical one, as it is explicitly stated by Greimas. But also, clearly, a renewal of its presentation. More than this, I will lay stress on the conjunction of analytical result and graphical presentation. It must be obvious when you look at the schema, even without knowing anything about semiotics, that the analysis is based on a binary principle, that an analytical result from a semiotic point of view is a binary one. This is what the schema expresses better than any previous forms of language. Each element, verbal or properly graphical (line, geometrical figure or arrow), is coupled to another element. The systematisation of the binary principle gives a *structure* to the analysis. It turns the analysis itself into a structure, and one that is, by the best means, displayable by a schema.

Notice the natural categories outside the square: *sky, water, earth.* Those categories are generally disposed in a tripartite division; or, at least, they should be associated with a fourth term: *fire.* But this is not what happens here. They are included in two bipartite oppositions. It is very doubtful that we can put this down to Bororo's culture, since one term, namely *sky*, is split into two categories. It is the semiotical analysis only, not the analysed material, that allows for this binarism[3].

We can consider that the square diagram brings to binary analysis a sort of locking. Such locking makes that binarism is both the means and the goal of the analysis. From this point of view, the square offers a major advantage in comparison with other graphical diagrams, such as trees or tables, and with symbolical notations. A square can not degenerate in a opened binary series: it is closed on its four sides (or four angles) and can not be developed.

Furthermore, the present square allows us to respond to the problem that we have observed in the choice between correlation and opposition in the symbolical notations: as the conjunction can be equivalent to correlation and disjunction equivalent to opposition, the conjunction – disjunction relation makes them complementary. One could say that this is possible because no symbol is used but instead a closed category of two verbal opposite terms.

3 It is a very well-known procedure, named *catalyse* by structuralists.

Reinventing the Structure

Now here it is the first "official" occurrence of the semiotic square:

Figure 5: "The Elementary Structure of Meaning" (Greimas & Rastier 1968, p. 88)

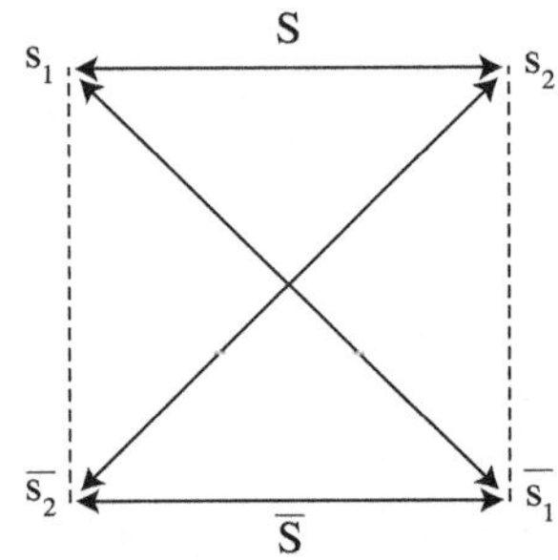

It came up in 1968 in a paper originally published in English in the *French Yale Studies* journal. The article was signed by Greimas and by François Rastier, then a young bachelor close to Greimas circle[4]. Most of the French readers know it by its inclusion in Greimas's work *Du Sens*, published in 1970. The authors claim that this square is "only an adapted formulation of that formerly proposed by one of the present authors" (Greimas & Rastier 1968, p. 88) and they refer then not to the square that we have seen in the previous section but, without quoting any precise page, to *Sémantique structurale,* where there is no mention of any semiotic square, whatsoever. What is more, there is no diagram at all in *Sémantique structurale.* The only "formulation" in this book is the symbolical notation, such as those reported in the first section of this study.

Nevertheless, let us compare this square with the former one, which came out two years earlier. First, this square's object is a high-level of generalisation, as it is nothing less than the "elementary structure of meaning". In regard to this first point, it is true that it is comparable to some symbolical notations in *Sémantique structurale* that share the same general object. Secondly, the binarity is once again strengthened, as every element of the

4 Rastier has since distanced himself from greimassian semiotic. He is well known for a semantic theory, called « Sémantique interprétative », based on analyses of texts. In English, see Rastier 2002.

diagram – lettered symbol, indexed numeral, or line – enters in relation with one other element and constitutes with it a pair of elements. Thirdly, the "diagonal" relations have been added. Fourthly, and this is perhaps the main point, the former diagram was still, in some aspects, a table (because of the four squares and the possible reading of the diagram in lines and columns); the present one could not be taken anymore as a table; it is a graphical representation where each element has the same functional weight. Thus the square can possibly to become a graphical format of its own kind.

This is something we can also say about the logical square: it is its own format. What I mean by "format" is some expressive configuration that you can recognize at first glance. You could change something in a diagram's format, as for instance the choice of the particular lettered symbols, without disturbing its recognizability. The logical square is a diagram that logicians can recognize through out many authors and different interpretations. So will the semiotic square, at least amongst semioticians.

This may appear of no importance at all. One can say that, about the logical square as well as the semiotic square, what it is meaningful is what the graphical representation refers to, and not what it looks like. It is its adequation with its representative function, and not its adaptation to some "format" property. And yet I found some allusions, from Greimas & Rastier themselves, that seem to indicate that they care about the semiotic square status as pure expression.

For example, Greimas & Rastier suggest that "this new presentation makes it isomorphic to the logical hexagon of R. Blanché [...] as well as to the structures called, in mathematics, the Klin group, and, in psychology, the Piaget group" (1968, p. 88). Their suggestion would be completely unfounded if it is interpreted as a comparison of specific objects content in the diagram, since mathematical, psychological, logical and semiotical objects do not have so much in common. It would not make any sense to compare their own faculty of representativeness, for as nothing is said about the objects, it would be a remark far from relevant. What is suggested by the authors is simply to compare the use of different graphical *shapes*.

There is another indication that the semioticans focused on the graphical shape of the semiotic square. This indication is written in a footnote directly connected to the diagram: "If the existence of this kind of relation seems unquestionable, the problem of its orientation ($s_1 \rightarrow \bar{s}_2$ or $\bar{s}_2 \rightarrow s_1$) is not determined yet. We will not discuss it, since the solution is not required in the rest of the demonstration"[5]. The kind of relation, allegedly "unquestionable", is indeed brought up for discussion in *Sémantique structurale*, notably through the functions of correlation and opposition that are evoked in the examination of symbolical notations. But the question of *orientation* is, for that part, entirely new. This question emerges only with graphical representation. And it is indeed remarkable that this question can be formulated in the same time as it is suspended. The graphic is a tool for theoretical discovery: the relation of implication is highlighted simply by the locking of the binarity into the semiotic square.

Thus, by way of presenting its object, *i.e.* the elementary structure of meaning, the semiotic square becomes, in the same time, the "right shape" for this structure: every straight line cutting through two of its angles has to be an elementary relation of meaning, even if the necessity of it had not appeared before.

Epilog

Of course, it is because of this relation of implication that the semiotic square deserves to be compared with the standard logical square ("standard" as it appears in handbooks without any further reference; see, for instance, De Praetere 1998). The semiotic relation of implication would take the place of the relation of sulbaternation.

There are two possible orientations of this relation, considering whether the inference concerns true or false propositions. One says that truth comes down, and falsity comes up (see Figure 6).

How would that appear in the semiotic square? Well, eleven years later, in a dictionary written with J. Courtés, Greimas has solved the question (see Figure 7).

As we can see, the falsity has definitely increased – but someone must be joking here. More seriously, it should be obvious that the relata of a semiotic square are not responding as logical propositions are in the logical square. Relata of meaning have their own structure of relations that does not correspond to the structure of logical relations. How that could be falls outside

5 This footnote is an addition of the French version, in Greimas 1970: 137.

Figure 6: Logical Inferences (De Praetere 1998)

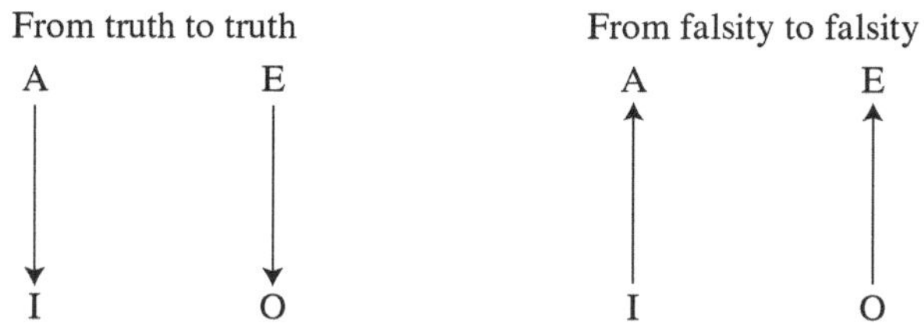

Figure 7: Carré sémiotique (Greimas & Courtés 1979, p. 31)

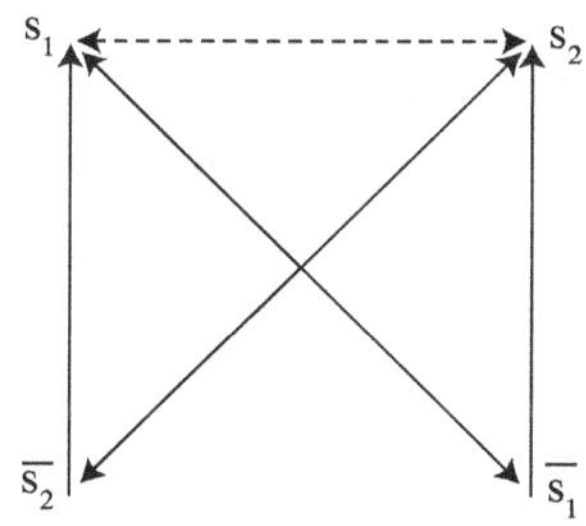

the matter of this paper (and it would not be an easy question for any semiotician). In my historical point of view, I would just underline that a logical relation has been taken in consideration, then its role in the structure and its function completely changed. That change is obviously due to the fact that the elementary structure of meaning has its own constraints. And yet, the semiotic relation of implication would never have appeared without the homologation between graphical representations, logical or semiotic, based on their similar format.

Moreover, we can notice that the binarity is not applied with the same systematicity as before. That is a main aspect of its evolution – but we leave the history of the semiotic square at this point. Semioticians will "deconstruct" the structure with the same enthusiasm they built it. Even the structural principle is not completely safe in contemporary semiotics. But what I would like to stress is that "deconstruction" would not have happened without the tools of graphical representation.

To sum up: We have seen that Greimas looked for some non verbal means, such as symbolical notations. Those means were used to express general, as well as particular, considerations implying the "structure" of the objects. They have been borrowed from authors such as Lévi-Strauss and Roman Jakobson who are not semioticians, but who are the main sources for semiotics at its inception. A square diagrammatical representation appeared at the very same moment. The square has the advantage over symbolical representations to allow an underdetermination of the relations of its objects. The object became properly structural simply when the square displayed that structure, at the same time binarily closed and underdetermined. So the determinations of the structure can be renewed by the way of theoretical discussion around the square itself, constructing and, soon, deconstructing, a concrete geometrical shape. By that means, we see that the graphical representations have participated to the conceptualisation in semiotics at its first ages, as they are a main conceptual instrument in contemporary semiotics. But we also see that graphical representations, simply by the way of their format, can serve as vehicles for exportation of concepts from a theory to another, especially in the case when different disciplinary traditions are concerned; that was the case here, through the square, between logic and semiotics.

References

De Praetere, Th. (1998). *Théorie de l'argumentation et éléments de logique.* In http://www.icampus.ucl.ac.be/DROI1111/document/logique/4bINFI.htm

Greimas, A. J.(1966). *Sémantique structurale, Recherche de méthode.* Paris: Larousse.

Greimas, A. J. (1970). *Du Sens.* Paris: Seuil.

Greimas, A. J., & Courtés, J. (1979). *Sémiotique. Dictionnaire raisonné de la théorie du langage.* Paris: Hachette (1993).

Greimas, A. J., & Fontanille, J. (1991). *Sémiotique des passions*, Paris: Seuil.

Greimas, A. J., & Rastier, Fr. (1968). The Interaction of Semiotic Constraints, *Yale French Studies, 41*, 86–105.

Jakobson, R. (1962). *Selected Writings* (2 vols.). La Haye: Mouton.
Lévi-Strauss, Cl. (1958). *Anthropologie structurale.* Paris: Plon (1974).
Lévi-Strauss, Cl. (1964). *Mythologiques, 1: Le cru et le cuit.* Paris: Plon.
Rastier, Fr. (2002). *Semantics for Descriptions.* Chicago: Chicago University Press [with the collaborations of M. Cavazza, & A. Abeillé].

Part 6: Cognitive and Social Sciences

The Cognitive Dimensions of Contrariety

IVANA BIANCHI[1] and UGO SAVARDI[2]

Abstract

Within the framework of the Cognitive Sciences, the authors propose a shift from the main stream of research on opposition which, since the seventies, has focused on the study of antonyms as linguistic structures. A new approach which deals with contrariety as a perceptual structure has been developed by the authors over the last 10 years. The chapter focuses on two aspects of this broad research project, both concerning the putative dimensional structure underlying and connecting two contrary extremes (and somehow also implied in the model of the square). The first is a description of the internal structure of contrary properties in terms of phenomenological psychophysics and the second discusses the issue of whether contraries, from a cognitive point of view, lie on a common continuum.

The proposal is introduced with a presentation of what we consider to be the elements which make Aristotle's analysis of contrariety potentially interesting for cognitive scientists.

Keywords: perception and cognition of contrariety, perceptual structure of dimensions, unidimensionality of contraries, spatial opposites.

1 Associate professor of General Psychology, Department of Educational Sciences, University of Macerata, Italy. E-mail: ivana.bianchi@unimc.it

2 Full professor of General Psychology, Department of Philosophy, Education and Psychology, University of Verona, Italy. E-mail: ugo.savardi@univr.it

1. Introduction

The analysis of opposition, which has its roots in classical philosophy and has been developed in various logical theories since then (as this book clearly testifies), in the last decades has also become an issue for cognitive scientists.

The main stream of research on opposition has, since the seventies, focused on the study of antonyms from the point of view of linguistics, psycholinguistics and semantics. Although some authors define 'antonymy' as a species of 'opposition' (e.g., Cruse, 1986; Lehrer and Lehrer, 1982; Lyons, 1977), an increasing number of researchers have shifted their attention towards a general phenomenon of opposition, using the term 'antonym' as a cover term for all kinds of opposites (see, for instance, Jones, 2002; Muehleisen 1997; Willners, 2001; Willners and Paradis, 2006). These studies have concentrated, variously, on the lexical, semantic and pragmatic aspects of opposition in natural language and considerable efforts have been devoted to creating exhaustive classifications. In spite of a great deal of work on this subject, researchers admit that they are still puzzled as to why opposites have a special status in human cognitive organization (Croft and Cruse, 2004; Cruse, 1986; 2000; Fellbaum, 1995; Jones, 2002; Muehleisen, 1997; Murphy, 2003; Willners, 2001). Opposites are intuitively and naturally understood and learnt, even in early language acquisition (Kagan, 1984; Miller and Fellbaum, 1991; Muehleisen, 1997). Cruse (2000, p. 167) and Jones (2002) noted that this is the only semantic relationship "to receive direct lexical recognition in everyday language. In this sense it is more widespread and more primal than other relations holding between words" (Jones, 2002, p. 8). It is an extremely robust relationship of the mental lexicon, "a key feature of everyday life" (Jones, 2002, p. 7).

The reason for this special status – which some authors have referred to as the 'unique fascination' of antonyms (see Cruse, 1986, p. 197; Jones, 2002, p. 1) – has not yet been fully understood in traditional linguistic and psycholinguistic studies of opposites and this has stimulated, within the framework of Cognitive Sciences, fresh approaches reaching beyond old classifications towards new questions and new methods of research.

Kelso and Engstrøm's (2006) approach, for instance, represents a recent attempt to put forward a new theory aimed at explaining the concept of complementary pairs inherent in human brains and behavior, based on Coordination Dynamics. Their questions only in part overlap with those of other authors in the domain of psycholinguistics or cognitive semantics, for example, 1) Why do human beings routinely divide the world into contraries? 2) Why are contraries so often interpreted as being mutually exclusive? 3) What is the ultimate nature of contraries and contrariety? Are contraries physical phenomena, mental phenomena, or somehow *both*? (Engstrøm and Kelso, 2008). The search for answers led them to focus on the dynamics of specific mechanisms in the brain, on which contrariety is considered to be based. Without going into the details of their proposal, what we want to emphasize is that this is a novel perspective on contraries, founded on experimental studies and mathematical models with the aim of understanding how and in what circumstances there is a system which produces complementary pairs that follow repeatable and predictable dynamic laws.

In yet another approach to contrariety, the investigation shifted from regarding opposition as a linguistic structure to contrariety as a perceptual structure (for an overview, see Bianchi and Savardi, 2008a, b; Savardi, 2009; Savardi and Bianchi 2000a). As shown by means of systematic explorations of the perceptual rules underlying the recognition of contrariety by naïve participants in various perceptual domains (visual, acoustical, motor gestures), it was found that people have an immediate understanding of the nature of contraries. They can spontaneously *see* where and how two configurations are contrary to each other, as well as being able to *see* where and how two configurations are identical, similar, or different to each other. They can easily "do the opposite" of a gesture, as well as draw a shape which is opposite to one presented to them. In literature in the field of psycholinguistics, it has been emphasized that people have an intuitive understanding of whether two words are opposite or not, and if so, whether they are 'good' or 'bad' antonyms (for a recent overview on the issue see Paradis and Willners, 2011; Paradis, Willners and Jones, 2009). The results from perceptual research on contraries reveals that it goes further than this. The ease with which people recognize contrariety between visual configurations, body postures or acoustic events suggests that it is a structure that is self-evident in the phenomenal aspect of objects, bodies and places, and is somehow linked to how the human perceptual system works. A new cognitive theory of contrariety, grounded in perception, has thus been put forward, based on evidence that a) people seem to follow specific "rules" when asked to produce the contrary of a shape or a gesture, etc. (Bianchi and

Savardi, 2008a, pp. 143–152); b) people have a precise perception of what is contrary as opposed to what is similar or different – traces of this can be found both in phenomenal tasks (for an overview see Bianchi and Savardi, 2008a; see also Bianchi and Savardi, 2008c; Savardi, Bianchi and Bertamini, 2010) and in change detection tasks (see Bracco, Bianchi, Chiorri, Burro and Savardi, 2009); c) the perception of contrariety is embedded in human direct experiences of space, both in terms of experiences of one's own body as part of the spatial environment and in terms of the perceptual shape of the environment as it appears from the point of view of the eye (on this see Savardi and Bianchi, 2000b; Savardi and Bianchi, 2009; Bianchi, Savardi, and Kubovy, 2011).

So far we have mentioned three different approaches to the understanding of opposites developed within the field of Cognitive Sciences which, despite profound differences, all take the hypothesis that contrariety is a central structure for the human cognitive system seriously. We will now focus on the third approach, which grounds contrariety in direct perceptual experiences. We will introduce our analysis by presenting what is, from our point of view, the link connecting our perceptual approach to contrariety with classical (in particular Aristotelian) philosophical theories on contraries. We will then focus on the idea of dimensions as linear structures underlying two opposites. This is a widespread idea in psychological theories and applications (and is interestingly related to ancient models of negation as developed along unidimensional continua). We will discuss the idea of dimensions in the light of some findings which emerged from our experimental research on the issue.

2. From Contrariety in the Works of Aristotle to Contemporary Cognitive Sciences: What's in It for a Psychologist of Perception?

The idea that relationships are direct perceptual data filtered into the field of Experimental Psychology from Philosophy with the theoretical standpoint of Gestalt Psychologists along with the relative experimental evidence. There is an interesting path which leads from Aristotle to Brentano and which thus bridges the gap between Aristotle's way of thinking and the phenomenological Psychology which developed in the late 19th century as a result of Brentano's theories (Antonelli, 2003; Mulligan and Smith, 1988; Runggaldier, 1989; Smith, 1988). We are not going to reconstruct this bridge here. What we are going to briefly make reference to are, specifically, the elements which make Aristotle's analysis of contrariety interesting for cognitive scientists, in particular from the point of view of a perceptual approach to contrariety.

Evidence of the idea that contrariety is grounded in perception is as old as the roots of philosophical and scientific thought (see Bianchi and Savardi, 2008a, pp. 15–23; Migliori, Palpacelli and Bernardini, 2009; Savardi and Bianchi, 1996). The relevance of contrariety in perceptual experiences was already clearly recognized by pre-Socratic philosophers. It is certain that Aristotle, while discussing the basic role of contrariety in logic (see the square of opposition), at the same time (particularly in the *Categories*) put forward a proposal for taxonomy in which contrariety appears as the key relationship organizing any phenomenal type of variation.

Before going on, it's worth noting that the fact that we can analyze ideas originating a hundred or a thousand years ago and regard them as useful for the understanding of human perception and cognition should not arouse suspicion or cause wonder "unless one assumes that the world of direct perception (of colors, sounds, shapes, or motion) has radically changed over the course of the years. The hypothesis that our perception of the outside world might dramatically change over time and be influenced by cultural changes is, of course, tenable. This however would make it very hard to understand why, in such a maelstrom of changes, some fragments of reality remain unaltered. We should be able to explain why there are some descriptions of perception written two hundred or two thousand years ago which fit in with current perceptual experiences" (Bozzi, 1969, pp. 17 f.). The presence in ancient writings of accurate descriptions of perceptual phenomena which have been re-described in contemporary psychology is not something new (see, for instance, Bozzi, 1990; Gregory, 2000; Pittinger, 1989; Ross, 2000; Wade, 1998).

The *Categories*, in this sense, make a very interesting reading for psychologists who want to understand the cognitive rules of natural classification processes (as discussed in Savardi and Bianchi, 1996). In fact, the Aristotelian system of classification is not very different from the *feature-based* methods of classification developed more then two thousand years later in experimental Psychology (e.g. Jolicoeur, Gluck and Kosslyn, 1984; Rosch, 1975; Rosch and Mervis, 1975; Tversky, 1977). What is still fresh in Aristotle's system and particularly relevant for our discourse here is his way of dealing with variations (or what in the terminology of many cognitive models are called *distinctive features*). He proposed that every change is between contraries (*Cat.* 5,4a 30ss; 10, 12b, 19–21) and indeed he brought every type of non-identity present on any level of taxonomy back to contrariety. He explicitly distinguished between different species within the same genus basing this distinction on contrariety (*Metaph.*, Z, 12, 1037b, 20). Even when referring to small variations perceived between similar things and the variations perceived between completely different things he once again referred to the idea of contrariety. This is consistent with the definition of contrariety that he gave. He spoke in terms of 'maximum difference' (*Metaph.*, D, 10, 1018a 27–28; *Cat.* 6, 6a 17–18; *Metaph.* I, 4, 1055 a 27–28; *Post. An.* I 4, 73b 21–22), where *maximum* refers to an extreme degree or intensity of the variation and *difference* means a special kind of variation. In effect, Aristotle considered *difference* to be separate from *diversity* based on the number of characteristics which co-vary and thus distinguish two objects. For *diversity*, and for things which can be regarded as being *diverse* from each other, there is no need for there to be any aspect in common. The concept of *difference*, conversely, implies that two things differ with regard to some aspects, but otherwise necessarily have many aspects in common (see *Metaph.*, X (I), 3, 1054b 23–26). Using a language more familiar to cognitive scientists, we could translate this distinction between *difference* and *diversity* in terms of unidimensional *versus* multidimensional variations. By considering contrariety as the maximum variation within a dimension (i.e the maximum difference) and defining diversity in terms of the sum of differences (i.e. the sum of contrarieties in many dimensions), contrariety forms the basis of both diversity and difference. Moreover, by regarding intermediates in terms of contrariety (*Metaph.* I, 7), even minor variations from a particular quality lead back to the idea of contrariety.

To cut a long story short, at least two basic aspects of Aristotle's analysis of contrariety are interesting for cognitive psychologists in this context. Firstly, in his taxonomic system, the relationship of contrariety, defined as maximum difference within the genre, works as a 'reagent' in the process of perceptual discrimination between non-identical facts and, at the same time, as a cognitive 'binder' in the organization of taxonomy. Second, by specifying that every variation is between contraries, the whole set of possible variations between two objects (not only maximum differences) is related. Aristotle's system suggests that contrariety might be thought of as the basic relationship on which natural recognition of any kind of variation (minimum, maximum, or multidimensional) is grounded.

In a previous work (Bianchi, Bressanelli, Nucci and Savardi, 2003), we proposed that a stimulating experiment might be that of replacing the construct of *distinctive features*, as put forward by Tversky in his contrast model (1977), with that of *contrary features*, as derived from Aristotle.

Here we will focus a) on the idea that contraries are characterized by the fact that they admit intermediates and b) on the idea of unidimensionality as related to the definition of contraries as *maximum differences* along a single dimension (and not as *maximum diversity*). Interestingly, this latter idea is in agreement with the default idea of dimensions which is widely used nowadays in the Cognitive Sciences, as being the area comprised between two opposite endpoints. This obviously is also related to the geometrical definition of the 4 sides of a square, each connecting two vertices. Our task here will be to present some experimental developments on the structure of dimensions, based on evidence from our experimental work on the perception of contraries, which shows that a great deal of experimental research is still needed to clarify the psychological nature of dimensions.

3. Contraries and Dimensions

The idea that two contrary properties lie on the same continuum is taken for granted in many fields of psychological research or practice. Among the methods which have been widely used is the semantic differential method (Osgood, Suci and Tannenbaum, 1957). People are asked to rate the qualities of a stimulus using bipolar scales identified by opposite adjectives placed at the two extremes of each scale. These are divided by a midpoint (zero value) on the left and right of which are two sub-sections expressing the gradations of one or the opposite property. The structure of the Likert scales is also usually presented with verbal labels expressing gradual contrariety: "totally in agreement", "in agreement", "uncertain", "in disagreement", "totally in disagreement". These are conceived as consecutive, cumulative steps in a single additive dimension. In the case of Kelly's Theory of Personal Constructs too and the corresponding Repertory Grid interview technique (Kelly, 1955; Jankowicz, 2003) – which over the last 40 years has been applied in various areas of psychological assessment and research – constructs are by definition bipolar and scalar.

The fact that so many methods assume the unidimensionality of opposites is not surprising if one takes into account that the idea that opposites lie on unidimensional underlying continua has been undisputed in the literature on semantic or linguistic theories of opposites since the early seventies. This was, for example, the basis of the concept of markedness (Clark, 1973). According to this concept, dimensions are generated by the unmarked pole of each pair. Gradations of these unmarked poles are in fact expected to potentially cover all the variations of a given property, thus giving rise to a unidimensional counterpart of the pair. For example, in the pair *high-low*, *high* is the unmarked pole and *height* is the corresponding dimension. In spite of the difficulties encountered by researchers when operationalizing the concept of markedness (for an overview see Haspelmath, 2006), the presupposition of an implicit unidimensional continuum corresponding to the two opposite poles has persisted as a kind of default assumption.

This is curious since – as we stated in the introduction to this paper – linguists and psycholinguists have spent a great deal of time trying to account for the differences between the various different types of opposites: complementary opposites, gradable opposite, reversive opposites, directional opposites, direct opposites, indirect opposites, near opposites, good opposites, etc. (Croft and Cruse, 2004; Cruse, 1986, 2000; Fellbaum, 1995; Jones, 2002; Lyons, 1977; Muehleisen, 1997; Murphy, 2003; Paradis, 1997, 2001; Willners, 2001). But the fact that there are various types of opposites has not caused researchers to wonder whether this might imply or indicate varying structures related to the supposed corresponding dimensional continua. Thus, an acknowledgment of the existence of linguistically and psychologically differing species of opposites (and the efforts made to provide an exhaustive classification) has coexisted with the assumption that "antonyms name opposite sections of a single scale" (Lehrer and Lehrer, 1982, p. 484; see also Cruse, 1986; Kennedy and McNally, 2005).

Another important area in Cognitive Sciences where references to dimensions are central concerns research on conceptual spaces. Gärdenfors (2000, 2004, 2007), for example, maintains that a conceptual space can be defined as a collection of one or more quality dimensions. "The coordinates of a point within a conceptual space represent particular instances of each dimension, for example, a particular temperature, a particular weight, and so forth." (Gärdenfors, 2000, p. 6). Dimensions form the framework used to assign properties to objects and to specify the similarity relationships between them. The interpretation of dimension that Gärdenfors has in mind is certainly psychological, or phenomenal as he also calls it (2000, p. 8; 2004, p. 14), and in this sense their structure "should be determined by psychophysical measurements that determine how our concepts are represented" (2007, p. 169).

One of the questions that Gärdenfors did not address directly is whether the shape of dimensions has to be modeled as bipolar or otherwise. Miller (1996, p. 193) is clearer on this issue. When modeling the psychological shape of dimensions, he tries to account for a unidimensional definition of dimensions (e.g. he talks of the *size* and the *height* of a cup) and, at the same time, for a direct experience of bipolarity in the perception of qualities: "the N attributes of a given referent define an abstract N-dimensional space (a hyperspace); any particular instance can be thought of as a point in that space, its location given by its value on each dimension. (...). The size (...) of the cup, for example, can be *large* or *small* (modifiers), its height (...) can be *tall* or *short* (modifiers), and so on. Each modifier specifies the position of a referent along a particular dimension; the full set of modifiers specifies a particular location in the N-dimensional space; differences between referents are given by their different locations in that space".

Gärdenfors (2000; 2007) seems to admit that dimensions might be bipolar, as well as admitting that some structures of dimensions are ordinal and others topological. Two of the examples of dimensions that he refers to, *pitch* (defined by low and high tones) and *brightness* (defined by two endpoints, black and white), are of this kind. But from his point of view, this remains a local solution which covers some individual quality dimensions but is not a general solution.

This is therefore the first question that remains open and needs to be addressed when studying the perceptual structure of dimensions: i.e. is there perceptual evidence that *long* and *short*, or *large* and *small*, or *full* and *empty*, etc. map inverse orderings of a common underlying continuum?

Another question that needs to be addressed concerns the internal structure of dimensions, as defined based on the nature of the three relative components: the two opposite poles and the intermediates. We will start from this last point and then move back to the first.

4. The Internal Perceptual Structure of Dimensions

In linguistics the question of the internal structure of dimensions has been studied taking into account the possibility of applying certain and not other lexical degree modifiers to the two opposite adjectives (on this, see for instance Kennedy and McNally, 2005; Paradis, 1997, 2000a, 2000b, 2001). For example, graded adjectives do not admit the application of modifiers such as 'completely' (e.g. *completely wide* or *completely narrow* are infelicitous) while they admit scaling degree modifiers such as 'very' or 'fairly' (e.g. *very wide* or *fairly wide*). On the other hand, adjectives expressing a notion of 'either-or' can combine with absolute modifiers (e.g. *totally dead* or *almost dead*).

A radically new approach to the study of this issue was presented by Savardi and Bianchi and defined by Kubovy (in The MIT *Encyclopedia of the Cognitive Sciences*, 2002) as a clear example of what "Phenomenological Psychophysics" is meant to be. Our methods, in fact, fit in with the definition of phenomenological psychophysics formulated by Kubovy and Gepshtein (2003, p. 45) "...our methods are *phenomenological*; they rely on the reports of observers about their phenomenal experiences. They also are *psychophysical:* they involve systematic exploration of stimulus spaces and quantitative representation of perceptual responses to variations in stimulus parameters."

In the study we are going to refer to, an analysis of 37 basic spatial dimensions (e.g. *large-small*, *obtuse-acute*, *vertical-horizontal*, *far-near*, *full-empty*, etc…,) was carried out. This analysis aimed to define the perceptual internal structure of each dimension by exploring the characteristics of the objects that are perceived as being situated at one or the other pole and those of objects that are perceived as being at neither one pole nor the other (intermediate properties). Different structures were thus defined, based on 3 aspects: the metric asymmetry/symmetry of the two poles; the metric extension and anisotropy of intermediates; and the qualitative structure of the poles and intermediates (i.e. whether they consist of one or more qualitatively different states).

In the following pages, we will briefly explain the methods used and the results found.

4.1 Two Tasks

The studies (Savardi and Bianchi 2003; Bianchi and Savardi, 2008a, pp. 59–77; Bianchi, Savardi, and Kubovy, 2011) were conducted on a set of pairs of contrary properties obtained from responses collected in the first phase of the research. We hypothesized that these spatial pairs would correspond to perceptually defined dimensions, which could be described based on the range of perceivable gradations of two opposite properties (e.g. *small* as opposed to *large*) and of their intermediates, if any (e.g. objects or places that are neither *small* nor *large*). Each dimension was thus described in terms of the characteristics of each pole and of the instances that are perceived as intermediates. It is important to note that the present definition of a dimension is in perceptual terms, independently of the existence of correspondent linguistic terms.

Given that the aim of the study was to come up with a general description of the structures of these dimensions, the analysis needed to focus on a broad set of spatial experiences. For this reason, in all phases of the research, participants were not presented with a limited set of stimuli, but were asked to focus on as many different spatial experiences concerning day to day life as they could. In order to avoid variability in the results due to the fact that participants could refer to different 'samples', the research was performed using inter-observational groups of 3 people. Thus, participants first shared within the group their experiences of the objects or environments they were focusing on and then agreed on a common description. The major assumption underlying this methodological choice was the inter-subjectivity and inter-observability of phenomenological experiences (Husserl, 1931/1967; Köhler, 1929/1947). This assumption is accepted as part of the methodology of investigation of perceptual experiences as developed within the field of experimental phenomenology (see Bozzi, 1978; Bozzi and Martinuzzi, 1989; Kubovy, 1999, 2002; Savardi and Bianchi, 2008).

Two tasks were considered, from which a quantitative and qualitative description of the phenomenological structure of the dimensions emerged.

Task 1) Quantitative description. Participants were presented with scales of gradation (see Fig. 1) and told that the space between the two poles represented the total possible range of experiences (e.g. all possible visual experiences of, for instance, *small* or *large* or anything in between). They were first asked to mark on the top bar the boundary that separates *small* from *large*, i.e. the proportion that is perceived as expressing gradations of *small* and the proportion perceived as expressing gradations of *large*. They were then asked to mark on the lower bar the range of things that were *neither small nor large*, and to position this area in relation to the mark on the top bar by deciding how much of the intermediate space to ascribe to one pole and how much to the other.

The participants' responses were analyzed by computing:

1. (on the lower bar) the proportion of the dimension covered by the intermediate region, and the proportion of dimension covered by more polarized properties. This proportion was used as a measure of the degree of *polarization* characterizing each dimension;
2. (on the top bar) the proportion of the dimension covered by pole A and pole B. This proportion was used as a measure of the *symmetry/asymmetry* of the dimension in relation to the two poles.

Task 2) *Qualitative* description. The second task was designed to integrate results from Task 1 (concerning the proportional metric extensions of the three components) with data regarding quality. For instance, while a door

Figure 1: Example of the Table Used in Task 1.

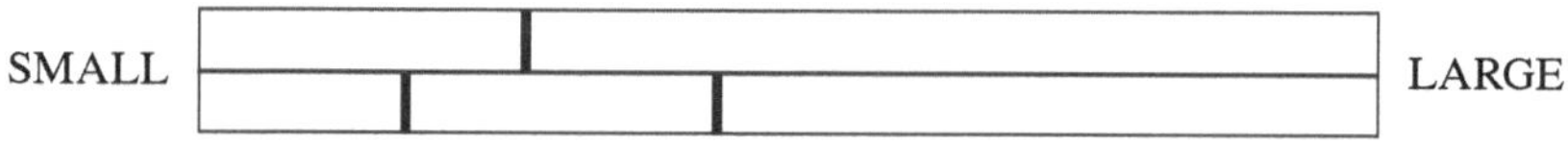

The thicker lines simulate participants' responses.

can be *open* to various different degrees, when it is perceived as *closed*, there is only one possibility.

Participants were asked to make the following distinctions:

- for poles, the distinction between single experiences and ranges of experiences and, in the latter case, between *bounded and unbounded ranges* of experiences. This further distinction is between ranges having a 'final state', showing the property at the maximum possible degree (e.g. a wide open door) and those where this final state is not identifiable (e.g. different heights of a building, where any of these heights may appear to be the maximum possible height);
- for intermediate stages, the distinction between the *existence or non-existence* of properties which are neither one pole nor the other. If these properties were considered to exist, participants were then asked to distinguish between intermediates referring to *ranges of experiences* (even very limited ranges) or to a *single experience.* This further distinction aimed to identify which of the intermediate areas described in Task 1 were to be considered as cases of a first or second 'type' of intermediacy. These two types were proposed (Savardi and Bianchi, 2000a) in order to distinguish between two different conditions of intermediacy. The first type refers to those conditions where the property does not belong to either of the two poles (for instance *tepid* is neither an experience of *cold*, nor an experience of *hot*). The second refers to a condition where the intermediate state belongs to one of the two poles (i.e. something that is *ajar* is experienced as being *open*, but is also perceived as being in a state similar to *closed*).

4.2 Results

By combining the data from research on the three components, various structures of dimensions were derived (see Fig. 2). Whether the six structures found for spatial contraries can be extended to other perceptual domains is to be verified through further empirical work. The description of these structures in terms of fuzzy functions is a possible way of enriching the qualitative description of the three components using data on their quantitative extension. The curves serve, in fact, to define the perceptual structure of the spatial dimensions analyzed, with respect to four criteria: 1) the symmetry/asymmetry of the poles; 2) the degree of polarization; 3) the isotropy/anisotropy of the intermediates; 4) the shape of the three components.

If we compare, in Figure 2, the structure *high-low* with the structure *open-closed*, we notice that the two dimensions share an asymmetrical structure, even though the latter is more asymmetrical. The intermediate region is less extended in the latter and almost entirely overlaps the curve of *open* (that is, a door which is ajar is perceived as being already open). For *high-low* the state of being *neither high nor low* is mainly perceived as being a form of *height*, but is sometimes perceived as *low*. The structure for *full-empty* is very different. It is again asymmetrical, but with the intermediate region covering more than half of the entire extension of the dimension. *Supported-unsupported* offers an example of an opposite structure: very symmetrical, without an intermediate region.

There were three main findings regarding the set of dimensions analyzed (37 in total) that we would like to emphasize:

a) The spatial dimensions showed highly polarized structures. Even though different degrees of polarization were found, in only 5 dimensions was the proportion of the dimension relative to the experiences perceived as gradations of the two poles found to be less extended than that found relative to properties identified as 'neither one pole, nor the other'. It would be interesting if further research revealed that a high degree of polarization is not only typical in spatial dimensions but is a general condition of perceptual experiences in various domains. This would confirm the hypothesis that opposition has a basic role in structuring the direct experience of variations. If perceptual experience is fundamentally based on properties characterized in terms of one pole or another, then the recognition of opposition between objects,

Figure 2: Fuzzy Functions of *High-Low, Open-Closed, Supported-Unsupported, Ascending-Descending, Above-Below,* and *Full-Empty,* Based on the Three Components: Pole A, Pole B, Intermediates.

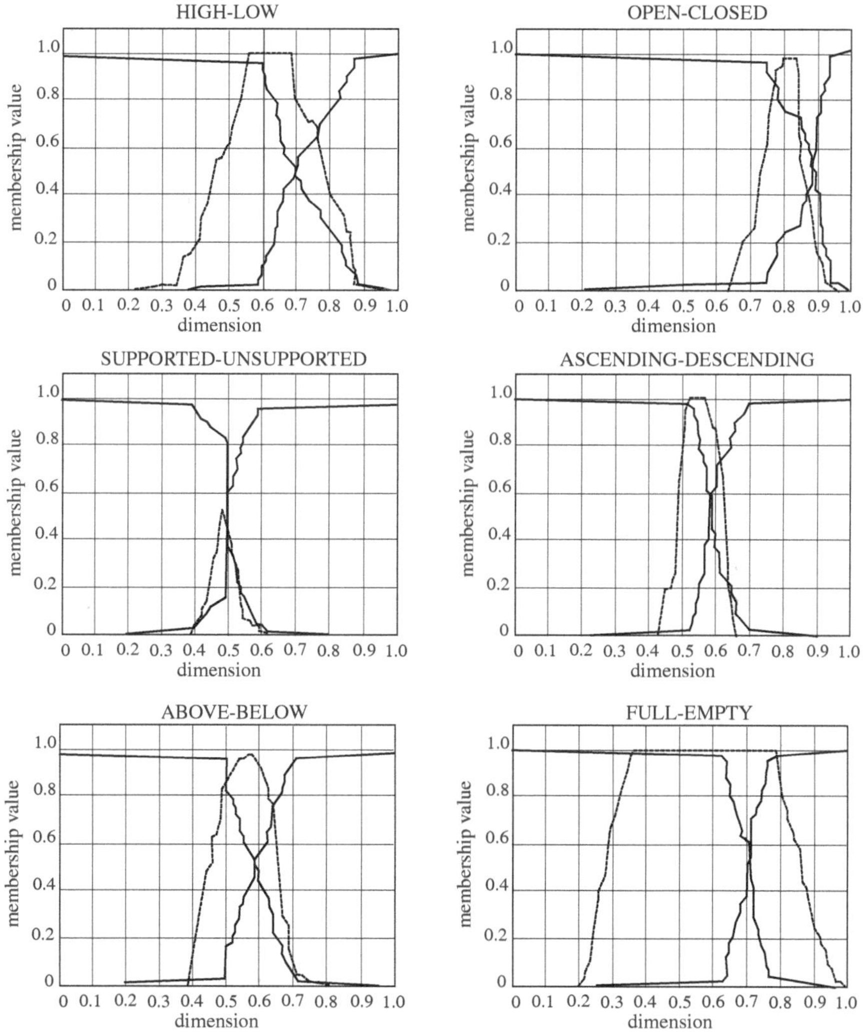

The shape of each function (in the y axis, reaching 1 or close to 1) represents, for the two poles, the characteristic of being single properties, bounded ranges or unbounded ranges; for the intermediate component, the characteristic of being intermediates of the 'first type' or of the 'second type'. The extension of each curve in the x axis expresses the proportion of the dimension covered by each component.

properties or different aspects of the environment becomes a common condition.

b) Intermediates were found to be a well-defined component. The study demonstrated that they can be highly accurately defined, both in terms of their extension and their proximity to the two poles. Results from Task 2 added a qualitative description of their nature as a single property or range of properties. In addition to the finding that intermediates seem to have a generalized anisotropic structure, it is also interesting to note that the isotropy/anisotropy of the intermediates was found to be not simply in correspondence with the asymmetry/symmetry of the dimension. This confirms that these two measurements express two different aspects of the structure within the dimensions. The relationship between the two provides an interesting point for further investigation. It would also be interesting to establish a relationship between the structural features of intermediates (i.e. various different proportions, a 'first' or 'second' type of intermediacy, singular states or ranges of states, isotropic or anisotropic structures) and the existence, in everyday language, of specific terms to refer to them.

c) Results revealed the dimensions analyzed to be variously but in any case generally asymmetrical. This generalized asymmetry is in contrast with the symmetry that statistical methods based on scales of opposites generally assume – e.g. the semantic differential method (Osgood, Suci and Tannenbaum, 1957) or the Likert scales (1932).

5. The Failure of Unidimensionality of Opposites

In another research we are going to present here (Burro, Savardi, and Bianchi, 2008; Savardi, Bianchi, and Burro, 2009; Bianchi, Savardi, and Burro, 2011), we used a metrical measurement method based on the direct perception of the presence of certain properties in various concrete objects in order to examine whether *high-low, wide-narrow, long-short, large-small,* would emerge as unidimensional scales (Luce, Krantz, Suppes and Tversky, 1990).

We decided to use the Extended Logistic Model (ELM, Andrich, 1978a, b; 1988; Rasch, 1960/80) as this method makes it possible to define which objects incorporate a given characteristic (e.g., length), and the extent to which they do so. Participants were asked to rate the extent to which the target properties were present in each of the objects presented. The ELM calculates two indices (expressed in logit) on a continuum: a β value indicating the discriminative ability of the participants, and a δ value which represents, in this specific case, how easy or difficult it is to perceive a specific spatial characteristic in a particular object (Andrich, 1978a, 1978b, 1988). Bond (2001) proposed a procedure that starts from the δ-logit values – e.g., for *long* and *short* – and converts them into their opposites. The degree to which these values approach the y=x identity function can then be estimated. When they significantly approximate the identity function, we can conclude that the two scales measure the same quality and lie at opposite poles of the same dimension; when they significantly deviate from the identity function beyond a specific confidence interval, we can conclude that the two opposite scales measure different qualities.

In two different phases, participants were shown 24 drawings of objects (grey scale, 6 x 5 cm), used as 'evokers' of real objects. The task required participants to attribute a score expressing the extent to which the property analyzed (e.g. *high, low, large, small, wide, narrow…*) was present in each object.

Various metrical scales were obtained from the ratings given by participants, each scale relative to one of the properties. These expressed the participant's ability to capture a given spatial characteristic and the degree to which an object itself was good at expressing that property. For purposes of comparison, Bond's procedure (described above) was applied to the two scales referring to the various pairs of opposite.

As shown in the graphs (Figure 3), no relationship of identity was found for the contraries examined: each of the linear functions calculated and shown in the figure fell outside the critical range (in Figure 3 the area lying between the hatched lines) – i.e., outside the area of an acceptable 95% probability that the two scales compared measure the same quality.

For other dimensions, such as *vertical-horizontal*, the possibility of using the term unidimensionality is even more problematic. In this case there is a quite large subset of stimuli (almost half of the set of stimuli presented) in which participants could not recognize either *vertical* or *horizontal* (these were the stimuli in which a bar was tilted by around 30° to 70°).

Figure 3: The Relationship between *High/Low, Large/Small, Long/Short* and *Wide/Narrow* (in δ_i Logit), According to the Results of the Study.

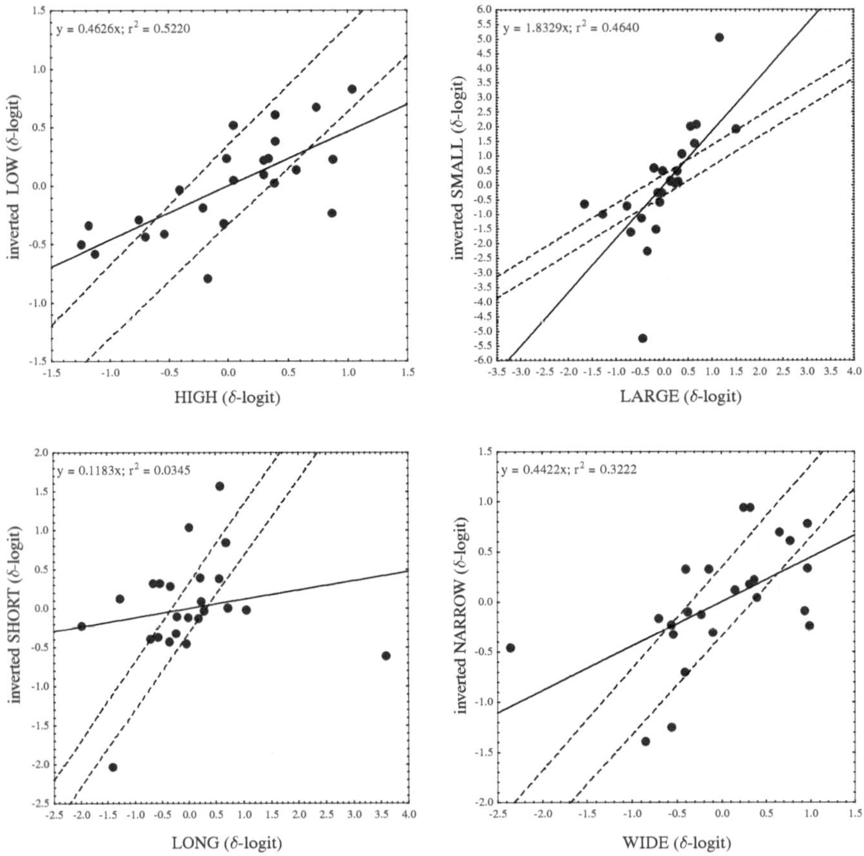

In each graph the area between the broken lines represents the area where the hypothesis that the two scales measure the same characteristic is accepted (95% confidence band).

The domain of the two properties is thus completely disjointed and there is a big gap in between the areas of application of the two opposite properties. This is hardly compatible with the idea of an underlying continuum.

A difficulty arose also for pairs such as *open-closed*, where there is a small area of overlap between the domains of application of these two opposite qualities and in most cases participants perceived either one or the other property – thus again excluding the idea that the two opposites correspond to inverse measurements of the same characteristic.

In conclusion, these results show that there are more or less extended gaps in the supposed common continuum underlying the two opposite properties and that, for the stimuli where an overlap of the two ratings is in fact present (as for those shown in Figure 3), this does not correspond to inverse degrees of the same characteristic.

What might be the cause of this lack of unidimensionality? Participants estimated to what extent a given object appeared to them to be, for example, *high* or *low*, or *large* or *small*. They did not simply quantify the metric distance between the final endpoints of the objects along the gravitational axis (in the case of *high* and *low*). In fact, the order of the 24 objects within each pair of opposite scales reveals that frequently they do not fit in with what was to be expected in physical terms. For instance, it appeared to be more difficult to perceive the property *high* in the slide than in a pair of scissors or a pipe. It is of course evident that a slide is *physically* much higher than the pair of scissors or the pipe. However, our results demonstrated that *perceptually*, it is the most difficult stimulus in terms of showing the property of being high. When considering the inverse ratings of *low*, the slide was ranked between the washbasin and the car (and their ratings indicate that they are not stimuli where it is easy to perceive the property *low*), while the pipe, the binoculars and the scissors are at the other end of the scale with ratings that indicate that they are stimuli where it is easy to perceive the property *low*.

These results suggest that participants referred to the global perceptual aspect of the objects. If, perceptually, properties have a multi-dimensional nature, objects might differ significantly in terms of being good at showing a certain property, not because of, for example, their physical height but because a cluster of properties is present in one object as compared to another and this emphasizes the salience of "being *high*".

We studied the hypothesis regarding an N-dimensional characterization of each property with respect to our list of primitive spatial characteristics by asking adult participants to describe whether each of the 74 properties in our set was *congruent* (similar to), *incongruent* (contrary to) or *independent* as compared to each of the other 73 properties (Savardi and Bianchi, 2000a, pp. 103–119; Bianchi and Savardi, 2008a, pp. 85–93). We are not going to enter into the details of the results of this study here. We

Figure 4: Properties Described as Incongruent to *In Front* by at least 90% of the Participants.

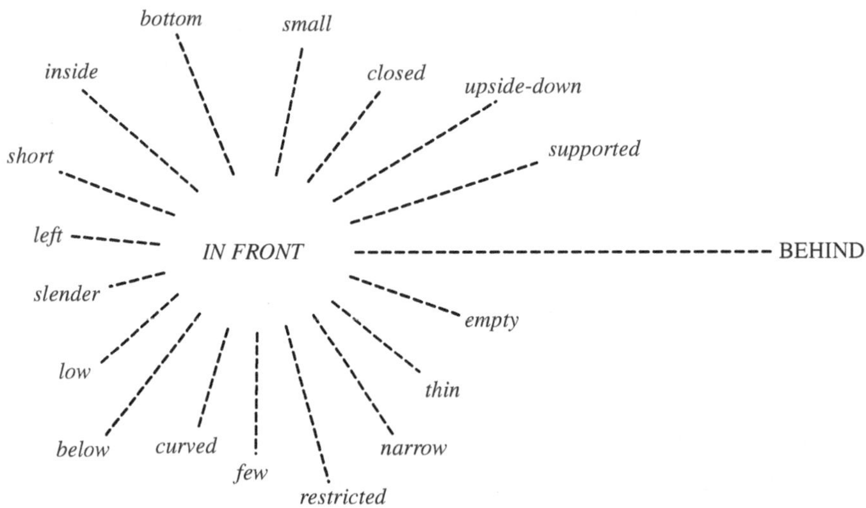

The length of the broken lines is directly proportional to the average degree of incongruity.

would however like to mention two major findings which give an idea of the numbers of instances confirming the hypothesis of an N-dimensional characterization of properties.

First, only 7.96% of the total number of comparisons between properties were described by at least 70% of the participants as independent, i.e. as lacking any relationship. The remaining 92.04% of the comparisons were described either as congruent or incongruent by at least 50% of participants. Second, incongruences (contrarieties) were recognized non only intra-pair but also when properties belonging to other contrary pairs were compared. Figure 4 shows, for example, the network of properties that are incongruent to *in front*. The length of the broken lines in the figure is directly proportional to the average degree of incongruence described. As one can see, *in front* is maximally incongruent to *behind*, but there is a quite rich array of other incongruences as indicated by 90% of the participants.
This is a first step towards the idea of an 'N-dimensional' rather than simply unidimensional space of contraries. Another aspect leading to the same

Figure 5: The N-Dimensional Space of Contrary Properties for the Dimension *In Front-Behind*.

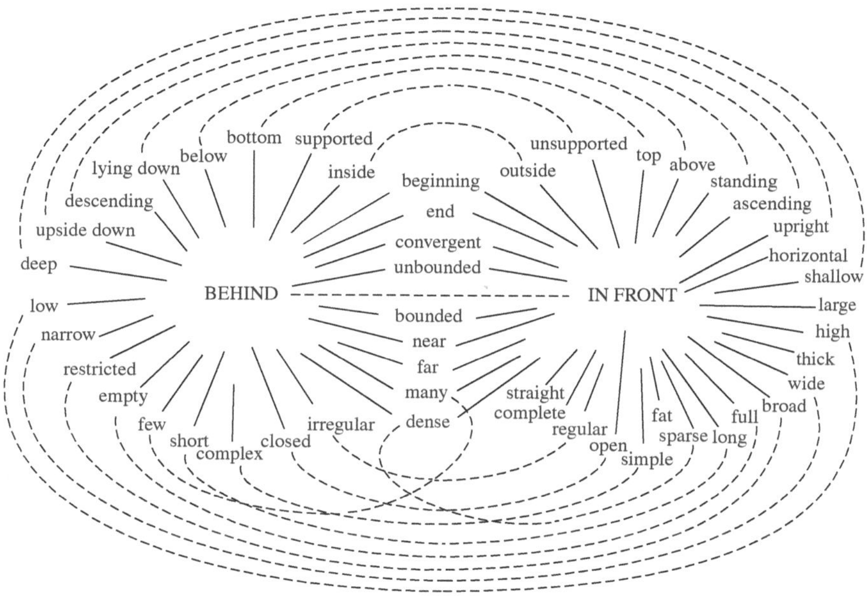

The continuous lines represent the spatial properties described as congruent to *behind* and *in front*, respectively. The broken lines represent the N-dimensional network of contraries radiating out from the basic pair *in front- behind.*

conclusion is suggested by the network of *congruent* properties 'blooming' around the basic contrary pair. Consider, for example the two contraries *in front* and *behind* and the properties described as congruent to each of them (see Fig. 5, the continuous lines). As one can see, there are a certain number of properties described as being congruent to one of the two poles but not the other and a certain number of properties which are congruent to both the poles (in Fig. 5, the latter are vertically aligned in the middle of the figure). This is the map of the field of rays forming the N-dimensional space of congruent properties around the two contrary poles.

However, a second field of rays is immediately created and this is represented by the broken lines. If you follow the broken lines, you notice that *in front* and *behind* are not linked to each other by only one relationship of contrariety (the one represented by the horizontal broken line at the center of the figure). As the many broken lines radiating out from the two poles reveal, there is a very dense network representing further layers of contrariety. Again an N-dimensional space of contraries rather than a simple unidimensional contrariety emerges. We are inclined to think that the articulated phenomenal nature of each property might be one of the principal causes of the lack of direct correspondence between opposite ratings and thus of the lack of unidimensionality discussed so far.

6. Conclusion

The square of opposites is the epitome of human interest for the understanding of the logical structures of opposition. Our aim has been to point out that there is also a corresponding desire in the field of Cognitive Sciences to understand the cognitive structures underlying the recognition of opposition. We have also shown that this second stream of research can be grounded in a series of intuitions expressed by Aristotle, first and foremost in the definition of contrariety as particular species of opposition. Our purpose was to look at potential contributions from the field of experimental psychology in terms of identifying the *perceptual* structures which human direct experience of contrariety is grounded on. From our point of view, limiting an analysis of contrariety to the boundaries of language means basing that analysis on a false assumption, i.e. that contrariety is a *linguistic relationship*, existing between words rather than a *directly perceived relationship* existing between properties, objects or parts of objects. In general, the need to ground language in perception, or – more generally – of grounding cognition in perception and action has been recently stated (for an overview see Pecher and Zwaan, 2005) with respect to other cognitive functions and processes (situating abstract concepts, embodiment in metaphorical imagination, embodied sentence comprehension). In a sense, our experimental research project investigating contrariety in terms of perceptual experience can be considered an independent project carried out within the tradition of experimental phenomenology. It however shifts the plane of analysis in the same direction: from language to the perceptual structure of contrariety.

In this chapter, the implications of this new view of contrariety have been discussed with special reference to the operationalization of the psychological concept of dimension (showing the contribution that this kind of investigation can make), as well as to the definition of the internal structure of contraries and their unidimensionality or non- unidimensionality.

References

Andrich, D. (1978a). A rating formulation for ordered response categories. *Psychometrika, XLIII (4)*, 561–573.

Andrich, D. (1978b). Application of a psychometric rating model to ordered categories which are scored with successive integers. *Applied Psychological Measurement, II*, 581–594.

Andrich, D. (1988). A general form of Rasch's extended logistic model for partial credit scoring. *Applied Measurement in Education, I*, 363–378.

Antonelli, M. (2003). La relazione intenzionale in Franz Brentano [Intentional relations in Franz Brentano]. In U. Savardi & I. Bianchi (Eds.), *Le relazioni empiriche. Per una scienza delle relazioni in Psicologia [Empirical Relations. Towards a Science of Relationships in Psychology]* (pp. 67–86). Milano: FrancoAngeli.

Aristotle (ed. 1984). Categories. In J. Barnes (Ed.), *The Complete Works of Aristotle: The Revised Oxford Translation.* Vol. I. Princeton: Princeton University Press.

Aristotle (ed. 1984). Metaphysics. In: J. Barnes (Ed.), *The Complete Works of Aristotle: The Revised Oxford Translation.* Vol. I. Princeton: Princeton University Press.

Bianchi, I., Bressanelli, D., Nucci, M., & Savardi, U. (2003). Se Aristotele, allora Tversky. Spunti per una fenomenologia sperimentale delle relazioni [If Aristotle, then Tversky. Suggestions in experimental phenomenology of perception]. In U. Savardi & I. Bianchi (Eds.). *Le Relazioni Empiriche. Per una scienza delle Relazioni in Psicologia* (pp. 171–196). Milano: FrancoAngeli.

Bianchi, I., & Savardi, U. (2008a). *The perception of Contraries.* Roma: Aracne.
Bianchi, I., & Savardi, U. (2008b). Contrariety as perceptual relationship. *Gestalt Theory, 2,* 106–120.
Bianchi, I., & Savardi, U. (2008c). The relationship perceived between the real body and the mirror image. *Perception, 5,* 666–687.
Bianchi, I., Savardi, U., & Burro, R. (2011, in press). Perceptual ratings of opposite spatial properties: do they lie on the same dimension? *Acta Psychologica,* DOI: 10.1016/j.actpsy.2011.08.003.
Bianchi, I., Savardi, U., & Kubovy, M. (2011, in press). Dimensions and their poles: A metric and topological theory of psychological dimensions. *Language and Cognitive Processes,* DOI: 10.1080/01690965.2010.520943.
Bond, T. (2001). *Applying the Rasch Model. Fundamental Measurement in the Human Sciences.* London: Lawrence Erlbaum Associates.
Bozzi, P. (1969). *Unità, identità, causalità* [*Unity, Identity, Causality*]. Bologna: Cappelli ed.
Bozzi, P. (1978). L'interosservazione come metodo per la fenomenologia sperimentale [Inter-observational method as a method for experimental phenomenology]. *Giornale Italiano di Psicologia, 5,* 229–239.
Bozzi, P. (1990). *Fisica ingenua. Studi di Psicologia della percezione* [Naïve Physics. Studies on the Psychology of perception]. Milano: Garzanti.
Bozzi, P., & Martinuzzi, L. (1989). Un esperimento di interosservazione [An experiment of inter-observation]. *Rivista di Psicologia, LXXIV,* 11–46.
Bracco, F., Bianchi, I., Chiorri, C., Burro, R., & Savardi, U. (2009). Investigating contraries by means of change detection. In U. Savardi (Ed.). *The perception and cognition of contraries* (pp. 93–112). Milano: McGraw-Hill.
Burro, R., Savardi, U., & Bianchi, I. (2008). Are Opposites Unidimensional? *Gestalt Theory, 2,* 191–195.
Clark, H. H. (1973). Space, time, semantics, and the child. In Timothy Moore E. (Ed.), *Cognitive development and the acquisition of language* (pp. 27–62). London: Academic Press.
Croft, W., & Cruse, D.A. (2004). *Cognitive Linguistics.* Cambridge: Cambridge University Press.
Cruse, D. A. (1986). *Lexical semantics.* Cambridge: Cambridge University Press.
Cruse, D. A. (2000). *Meaning in language: An introduction to semantics and pragmatics.* New York: Oxford University Press.

Engstrøm, D., & Kelso, J. A. S. (2008). Coordination Dynamics of the Complementary nature. *Gestalt Theory*, *30*, 121–134.
Fellbaum, C. (1995). Co-occurrence and antonymy. *International Journal of Lexicography*, *8*, 281–303.
Gärdenfors, P. (2000). *Conceptual spaces*. Cambridge: MIT Press.
Gärdenfors, P. (2004). Conceptual Spaces as Framework for Knowledge Representation. *Mind and Matter*, *2* (2), 9–27.
Gärdenfors, P. (2007). Representing actions and functional properties in conceptual spaces. In Ziemke T., Zlatev J., Frank R.M. (Eds.), *Body, language, and mind: Embodiment* (pp. 167–196). Berlin: Walter de Gruyter.
Gregory, R. L. (2000). A memo to *Meno*. *Perception*, *29*, 631–633.
Haspelmath M. (2006). Against markedness (and what to replace it with). *Journal of Linguistics*, *42* (1), 25–70.
Husserl, E. (1931/1967). *Méditations cartésiennes. Introduction à la Phénoménologie*. Paris: Vrin.
Jankowicz, D. (2003). *The Easy Guide to Repertory Grids*. New York: Wiley.
Jolicoeur, P., Gluck, M. A., & Kosslyn, S. M. (1984). Pictures and names: Making the connection. *Cognitive Psychology, 16* (2), 243–257.
Jones, S. (2002). *Antonymy: a Corpus-based Perspective*. Amsterdam: Routledge.
Kagan, J. (1984). *The nature of the child*. New York: Basic Books.
Kelly, G. A. (1955). *The Psychology of Personal Constructs*. New York: Norton.
Kelso, J. A. S., & Engstrøm, D. (2006). *The complementary nature*. Cambridge, MA: The MIT Press.
Kennedy, C., & McNally, L. (2005). Scale structure and the semantic typology of gradable predicates. *Language, 81* (2), 345–381.
Köhler, W. (1929). *Gestalt Psychology*. New York: Liveridge. (Trad. it. 1947) La Psicologia della Gestalt. Milano: Feltrinelli.
Kubovy, M. (1999). Gestalt psychology. In R. A. Wilson & F. C. Keil (Eds.), *The MIT Encyclopedia of the Cognitive Sciences* (pp. 346–349). Cambridge: MIT Press.
Kubovy, M., & Gepshtein, S. (2003). Perceptual grouping in space and space-time: an exercise in Phenomenological Psychophysics. In R. Kimchi, M. Behrmann, C. R. Olson (Eds.), *Perceptual organization in vision: Behavioral and neural perspectives* (pp. 45–85). Mahwah: Erlbaum.

Kubovy, M. (2002). Phenomenology, psychological. In L. Nadel (Ed.), *Encyclopedia of cognitive science* (pp. 579–586). Houndmills, Basingstoke, Hampshire: Macmillan.

Lehrer, A., & Lehrer, K. (1982). Antonymy. *Linguistics and Philosophy, 5*, 483–501.

Likert, R. (1932). A Technique for the Measurement of Attitudes. *Archives of Psychology, 140*, 1–55.

Luce, R. D., Krantz, D. H., Suppes, P., & Tversky, A. (1990). *Foundations of measurement: Vol. 3. Representation, axiomatization and invariance.* San Diego: Academic Press.

Lyons, J. (1977). *Semantics.* Cambridge: Cambridge University Press.

Migliori, M., Palpacelli, L., & Bernardini, M. (2009). The relation of contrariety in the ancient thought and in the Aristotelian formalization. In U. Savardi (Ed.), *The perception and cognition of contraries* (pp. 3–28). Milano: McGraw-Hill.

Miller, G. (1996). *The Science of Words.* New York: Scientific American Library.

Miller, G. A., & Fellbaum, C. (1991). Semantic networks of English. *Cognition, 41*, 197–229.

Muehleisen, V. (1997). *Antonymy and Semantic Range in English.* Doctoral Dissertation in Philosophy, Field of Linguistics. Evanston, Illinois.

Mulligan, K., & Smith, B. (1988). Mach and Ehrenfels: The Foundations of Gestalt Theory. In B. Smith (Ed.), *Foundations of Gestalt Theory* (pp. 124–157). Munich and Vienna: Philosophia.

Murphy, L. (2003). *Semantic Relations and the Lexicon: Antonyms, Synonyms and other Semantic Paradigms.* Cambridge: Cambridge University Press.

Osgood, C.E., Suci, G.J., & Tannenbaum, P. H. (1957). *The Measurement of Meaning.* Urbana: University of Illinois Press.

Paradis, C., & Willners, C. (2011, in press). Antonymy: from conventionalization to meaningmaking. *Review of Cognitive Linguistics, 9* (2).

Paradis, C., Willners, C., & Jones, S. (2009). Good and bad opposites: using textual and psycholinguistic techniques to measure antonym canonicity. *The Mental Lexicon, 4* (3), 380–429.

Paradis, C. (1997). *Degree Modifiers of Adjectives.* Lund: Lund University Press.

Paradis, C. (2000a). It's well weird. Degree modifiers of adjectives revisited: the nineties. In J. Kirk (Ed.), *Corpora Galore: Analyses and Techniques in Describing English* (pp. 147–160). Amsterdam and Atlanta: Rodopi.
Paradis, C. (2000b). Reinforcing adjectives: a cognitive semantic perspective on grammaticalization. In R. Bermudez-Otero, D. Denison, R. Hogg, & C. McCully (Eds.), *Generative Theory and Corpus Studies* (pp. 233–258). Berlin: Mouton de Gruyter.
Paradis, C. (2001). Adjectives and boundedness. *Cognitive Linguistics, 12* (1), 47–65.
Pecher, D., & Zwaan, R. A. (2005). *Grounding Cognition. The Role of Perception and Action in Language, Memory and Thinking.* Cambridge: Cambridge University Press.
Pittenger, J. B. (1989). Detection of violations of the law of pendulum motion: Observers' sensitivity to the relation between period and length. *Ecological Psychology, 2,* 55–81.
Rasch, G. (1960/1980). *Probabilistic models for some intelligence and attainment tests.* Copenhagen: Danish Institute for Educational Research.
Rosch, E. (1975). Cognitive representation of semantic categories. *Journal of experimental Psychology: General, 104* (3), 192–233.
Rosch, E., & Mervis, C.B. (1975). Family resemblances: Studies in the internal structure of categories. *Cognitive psychology, 7* (4), 573–605.
Ross, H. E. (2000). Cleomedes (c. 1st century AD) on the celestial illusion, atmospheric enlargement, and size-distance. *Perception, 29*, 863–871.
Runggaldier, E. (1989). On the Scholastic or Aristotelian Roots of "Intentionality" in Brentano. *Topoi, 8*, 97–103.
Savardi, U. (Ed.) (2009). *The perception and cognition of contraries.* Milano: McGraw-Hill.
Savardi, U., & Bianchi, I. (1996). Qualità e contrarietà, da una prospettiva fenomenologica alle categorie di Aristotele [Qualities and contrariety: from phenomenology to Aristotle's categories]. *Annali dell'Istituto di Psicologia,* Università di Verona, 161–191.
Savardi, U., & Bianchi, I. (2000a). *L'identità dei contrari* [The Identity of Opposition]. Verona: Cierre.
Savardi, U., & Bianchi, I. (2003). Metodo interosservativo ed eventi sotto osservazione: l'ostensione dell'esperienza di spazio nella fenomenologia sperimentale della percezione [Inter-observational method and events under observation: the extension of spatial experience in the experimental phenomenology of perception]. In U. Savardi & A. Mazzocco (Eds.),

Figura e sfondo. Temi e variazioni per Paolo Bozzi [Figure and background. Themes and variations for Paolo Bozzi] (pp. 107–130). Padova: Cleup.

Savardi, U., & Bianchi, I. (2008). The minimal features of the experimental Phenomenology of Perception. *Teorie & Modelli, 2–3*, 79–94.

Savardi, U., & Bianchi, I. (2009). The spatial path to contrariety. In U. Savardi (Ed.), *The perception and cognition of contraries* (pp. 61–92). Milano: McGraw-Hill.

Savardi, U., Bianchi, I., & Bertamini, M. (2010). Naive prediction of orientation and motion in mirrors. From what we see to what we expect reflections to do. *Acta Psychologica, 134* (1), 1–15.

Savardi, U., Bianchi, I., & Burro, R. (2009). From opposites to dimensions: filling in the gaps. In U. Savardi (Ed.), *The perception and cognition of contraries* (pp. 275–294). Milano: McGraw-Hill.

Smith, B. (1988). The Soul and Its Parts. A Study in Aristotle and Brentano. *Brentano Studien, 1*, 75–88.

Tversky, A. (1977). Features of similarity. *Psychological Review, 84*, 327–352.

Wade, N. (1998). *A natural History of Vision*. Cambridge: MIT press.

Willners, C. (2001). Antonyms in context. A corpus-based semantic analysis of Swedish descriptive adjectives. *Travaux de l'Institut de Linguistique de Lund 40,* Department of Linguistics, Lund University.

Wilners, C., & Paradis, C. (2006). Antonymy and negation: The boundedness hypothesis. *Journal of Pragmatics, 38*, 1051–1080.

Human Deontic Reasoning in the Deontic Square

SIEGHARD BELLER[1]

Abstract

The deontic square of oppositions describes relations between four deontic concepts: ban, permission, obligation, and release from obligation. It is argued that social rules imposing constraints on individual actions constitute the fundamental concept for the system of these four deontic modalities, and that people reason from such rules according to the relations defined by the deontic square. Empirical data support the proposed representational and inferential principles and demonstrate people's high deontic competence.

Keywords: Social rules; representational principles; deontic square of oppositions; deontic inferences; deontic reasoning capabilities; psychological theories

Introduction

Robinson Crusoe, a mariner from York and the only survivor of his shipwrecked sailing vessel, lived 28 years "all alone on an uninhabited island on the coast of America, near the mouth of the great river of Oroonoque", as we are told in the 1661 novel by Daniel Defoe. On his island, nobody dictated to him what to do or not to do, and accordingly, Robinson Crusoe could act at his own discretion in each situation. This is clearly not true for us. Other than Robinson Crusoe, we live in groups with a differentiated

1 Correspondence concerning this article should be addressed to Dr Sieghard Beller, Department of Psychology, Abteilung Allgemeine Psychologie, University of Freiburg, D-79085 Freiburg (Germany), email: beller@psychologie.uni-freiburg.de.

social organization, we maintain relations of different kinds to various people, and therefore our behaviour as individuals is flanked by numerous explicit and implicit social rules that impose normative constraints on our actions. Barely any field is exempt: There are rules for political events (e.g., minors must not vote), rules that regulate our leisure activities (e.g., only with a ticket, may you enter the concert), rules that affect our mobility (e.g., car drivers must stop at red traffic lights), rules for managing hazards (e.g., protection gloves are not needed outside the danger zone, but are required inside), and so on. Being engaged in deontic reasoning means nothing less than thinking about what is *forbidden* or *allowed*, *obligatory* or *not obligatory* according to such rules.

The corresponding four deontic modalities *ban*, *permission*, *obligation*, and *release from obligation* constitute a square of oppositions (e.g., Anderson 1956). The square defines relations between all possible pairs corresponding to assertions typically made in deontic logics (e.g., Chellas 1980; von Wright 1981). Ban and obligation form a pair of contraries: In order to avoid a deontic dilemma, an action must not be forbidden and obligatory simultaneously. If an action is forbidden, then – in a consistent system of norms – it is not obligatory according to the contrary relation, and so its omission must be possible, meaning that the action *need not* be taken. Conversely, if an action is obligatory, then it is not banned, meaning that the action *may* be taken. Therefore, along the subaltern paths of the square, the top implies the bottom. Ban and permission as well as obligation and release from obligation are pairs of contradictories; exactly one of each pair is true in a particular situation: An action is either forbidden or allowed; it is either obligatory or not. Therefore, permission corresponds to the absence of a ban ('*may* X ≡ ¬(*must not* X)'), release from obligation corresponds to the absence of an obligation ('*need not* X ≡ ¬(*must* X)'), and vice versa. Finally, permission and release from obligation are subcontraries; at least one of them holds in each situation: An action *may* be taken if there is no ban, it *need not* be taken if it is not obligatory, or both if there is no deontic rule at all. These relations hold in a consistent system of norms, that is, in a system that is free from deontic dilemmas.

If both the commission of an action (X) and its omission (¬X) are taken into consideration, the two contraries, and analogously the two subcontraries, are *interchangeable*: To be obligated to take an action means that the omission of the action is forbidden (i.e. '*must* X ≡ *must not* (¬X)'), while a forbidden action means that its omission is obligatory (i.e. '*must*

Figure 1

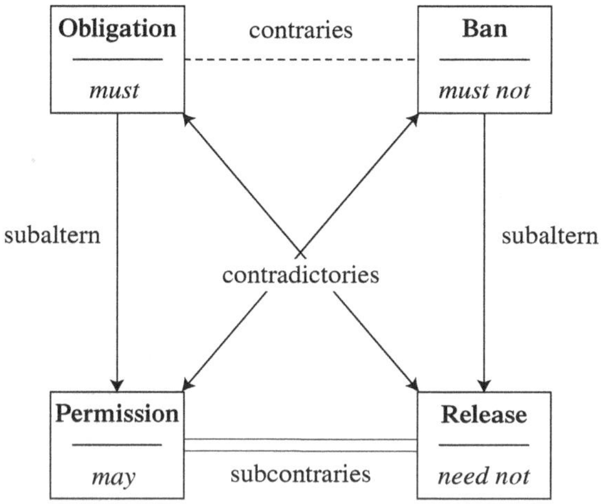

not X ≡ *must* (¬X)'). And, to be permitted to take an action means that the omission of the action is not necessary (i.e. '*may* X ≡ *need not* (¬X)'), while a release from obligation means that its omission is permitted (i.e. '*need not* X ≡ *may* (¬X)').

The question that is addressed in this article is: Do all these relations between the deontic modals help us to understand how people reason about social rules? In the psychological theorizing on human deontic capabilities, the potential of the deontic square of oppositions has not been fully exploited until the present day (see Beller 2008, 2010). For example, evolutionary approaches argue that special deontic reasoning mechanisms evolved in order to cope with several adaptive problems in the history of mankind: Mechanisms for detecting cheaters (Cosmides 1989) and free-riders (Hiraishi and Hasegawa 2001) on social contracts for mutual cooperation, mechanisms for detecting what is forbidden for subordinate individuals in dominance hierarchies (Cummins 1996b), or mechanisms for detecting and handling hazards in dangerous environments (Fiddick 2004). The evolutionary approaches characterized several deontic domains and their possible evolutionary roots, but they did not answer the question of how the system of deontic modalities is semantically structured.

Pragmatic reasoning schema theory (Cheng and Holyoak 1985; Holyoak and Cheng 1995) takes up this question by proposing two reasoning schemas for the deontic domain, one for obligation and one for permission. For example, the rules of the permission schema are:

P1: If the action is to be taken, then the precondition *must* be satisfied.
P2: If the action is not to be taken, then the precondition *need not* be satisfied.
P3: If the precondition is satisfied, then the action *may* be taken.
P4: If the precondition is not satisfied, then the action *must not* be taken.

Obviously, this schema refers to the four deontic modalities. However, the interrelations of the modalities are not considered further. As a consequence, some possible deontic inferences – for instance those that follow the rules of interchangeability – are not covered by the proposed rules of the schema. In other words: The schema approach is incomplete, at least in its current version.

The most elaborated approach, the mental model theory of deontic reasoning (Bucciarelli and Johnson-Laird 2005), includes the relations from the square, but it considers only three deontic modalities: It describes people's understanding of and reasoning about three types of statements: 'A permits B', 'A obligates B', and 'A prohibits B'. The fourth modality, release from obligation, is omitted and is handled as a type of permission.

These considerations show that the square of oppositions is a useful theoretical device to detect gaps in psychological approaches to deontic reasoning. But, one might go one step further and ask: Do people follow the relations of the square when they reason from social rules? In the rest of this article, it is argued that people interpret social rules according to a small set of representational principles, that their deontic inferences from these rules generally follow logical principles, and that they in fact consider the interrelations between the four deontic modalities as defined by the square of oppositions.

Representing Deontic Rules

Deontic rules are intended to regulate a person's actions under certain conditions. It is proposed that people's representations of such rules follow three principles (Beller 2001, 2003, 2008): Rules define social constraints on actions, their application conditions must be considered exhaustively, and there is a strong relation between the condition side of a rule and the action constraint.

Constraint principle. If there is no social or moral context, individual behaviour is not constrained and people can act at their own discretion (like Robinson Crusoe on his island). Starting from this "unregulated background", social rules introduce constraints on actions and thereby define which actions are regarded as socially not acceptable. The four deontic modalities all deal with establishing or lifting such constraints, but only two of them explicitly denote a constraint: ban and obligation. As these two are interchangeable, it is sufficient to use one of them as basic representational concept. In the following, the notion of a ban is taken as basic and is represented as

'forbidden(Action)'.

Exhaustivity principle. In order to determine whether or not a particular action is forbidden in a specific situation, it is necessary to take into consideration *all* conditions that put a ban into force. Accordingly people have to represent each ban together with all its conditions exhaustively. The whole set of conditions can be represented in terms of necessity and sufficiency as a disjunction of conjunctions of single factors. Please note that the exhaustivity principle does not imply that people in fact have complete knowledge about the conditions of a deontic rule; it only implies that they assume – until there is evidence to the contrary – that they have considered all relevant conditions. Otherwise deontic inferences from a social rule would not be possible.

Equivalence principle. If all relevant conditions for a particular ban are considered, an equivalence relation between the conditions and the banned action is justified: If at least one condition is met, then the action is forbidden; otherwise it is allowed.

According to these principles, a deontic rule regarding one particular action can be represented by the following schema:

$$\text{'Conditions} \leftrightarrow \text{forbidden(Action)'}$$

The following three examples illustrate how this schema can be applied.

(1) Having a ticket is often the only condition for being admitted to a concert, so that entering the concert without a ticket is not allowed. This simple rule is represented by the following ban:

$$\text{'}\neg\text{ticket} \leftrightarrow \text{forbidden(entering)'}$$

(2) Sometimes, however, more than one condition has to be considered: The musicians, for instance, are typically allowed to enter the concert hall without a ticket. In order to consider this aspect as well, the rule has to encompass two conjunctive conditions: A person is not allowed to enter if (and only if) he or she has no ticket *and* is also not one of the musicians, which is represented as:

$$\text{'(}\neg\text{ticket} \wedge \neg\text{musician)} \leftrightarrow \text{forbidden(entering)'}$$

(3) Spectators of soccer matches, on the other hand, are often inspected not only for their tickets but also for weapons. In order to consider this aspect instead, the rule has to encompass two disjunctive conditions: Persons are not allowed to enter the soccer stadium if (and only if) they have no ticket *or* have a weapon, which is represented as:

$$\text{'(}\neg\text{ticket} \vee \text{weapon)} \leftrightarrow \text{forbidden(entering)'}$$

Five remarks are necessary at this point: First, while the condition side of a rule can involve various numbers of preconditions until the exhaustivity criterion is met, the deontic side 'forbidden(Action)' refers to one particular action only. Consequently, every deontically regulated action requires only one rule. Second, the schema above uses "ban" as basic deontic concept, which represents an action constraint explicitly. A weak definition of permission is associated with this representation; actions are allowed if they are not explicitly forbidden (i.e. '*may* X $\equiv \neg$(*must not* X)'). Third, from a formal point of view, each of the other three modalities could equally be used to represent a corresponding constraint – at least implicitly – if it is combined with the exhaustivity and equivalence principle. Rule (1), for instance, might equivalently be represented by the permission 'ticket $\leftrightarrow$ permitted(entering)'. Depending on which deontic aspect a person focuses on – ban, permission, obligation, or release from obligation – the representation of the deontic rule might thus look differently, but the basic construction principles remain the same. Fourth, deontic rules do not describe which action a person *actually takes* or which conditions *actually hold* in a specific situation. As long as people have the ability and the freedom to commit or omit an action, they can conform to a rule or violate it. Therefore, the ban 'forbidden(X)' does neither imply 'not-X' nor 'X', and the same holds for the other modalities. Finally, in the real world, a deontic rule might conflict with other rules. In such a case, people are faced with a deontic dilemma, which makes it impossible to draw any definite inference about what is forbidden or allowed. In order to avoid such a situation, the deontic conflict has to be solved, for example, by removing one of the conflicting rules, or by rank-ordering the respective rules and applying only the strongest one. In the following part on deontic inferences, it is assumed that the deontic rules are free from dilemmas.

Inferences from Deontic Rules

From a deontic rule of the general form 'Conditions $\leftrightarrow$ forbidden(Action), three types of inferences can be derived: Forward inferences and two types of backward inferences (Beller 2008).

Forward inferences are inferences from the condition side of the rule to the deontic status of the regulated action: If the conditions are fulfilled then the ban applies, and if the conditions are not fulfilled, then the regulated action is not banned according to the equivalence principle and the contradictory relation. Consider the ticket example (1):

'$\neg$ticket $\leftrightarrow$ forbidden(entering)'

The following statements can be derived from this rule by forward inferences: Given that a person has *no* ticket, this person must not enter (or must stay outside, according to the interchangeability of the modal terms), and, according to the subaltern relations, he or she need not enter (i.e. may stay outside). On the other hand, a person *with* a ticket may enter (i.e. need not stay outside), and, as there is no obligation to enter, this person need not enter (i.e. may stay outside).

Backward inferences are inferences from the action side of the rule to the condition side and may occur in two types. *Deontic-to-factual* inferences are derived from a deontic statement about the regulated action: If an action is banned then the conditions of the ban are fulfilled as otherwise the action would be allowed. If, however, a potentially banned action is not forbidden then the conditions of the ban are not fulfilled. Suppose you know that the ban (1) applies to a person so that entering is forbidden. From this you can conclude that the condition is fulfilled: If a person must not enter, then this person has *no* ticket, as otherwise entering would be allowed. Analogously, suppose you know that ban (1) does not apply to a person so that entering is not forbidden. From this you can conclude that the condition of the ban is not fulfilled. Therefore, if a person may enter, then this person *has* a ticket, as otherwise entering would be forbidden.

The second type of backward inferences, which may be called *factual-to-deontic* inferences, is derived from information on whether or not the deontically constrained action is taken. If a banned action is taken, then – in order to conform to the rule – the conditions of the ban must not be fulfilled. If the banned action is not taken, then the conditions may but need not be fulfilled; at least as long as these conditions do not concern actions that are deontically regulated by another rule. If a person in example (1) enters the concert, then fulfilling the entry conditions is deontically necessary, in other words: the person must not be without a ticket (i.e. must have a ticket). If the person stays outside then the person may, but need not be without a ticket. Note that the modal terms *must not*, *must*, *may*, and *need not* are not used in an epistemic sense here, but still in a deontic sense, that is, with reference to the underlying deontic rule. The fact that a ticket would be deontically necessary for an entering person does *not* imply that the person in fact conforms to the rule. In other words, it is possible that the person has a ticket and thus conforms to the rule or that the person has no ticket and thus violates the rule. The backward factual-to-deontic inferences therefore clearly differ from the first two types of inferences (forward and backward deontic-to-factual), which are validity preserving, and this difference might be reflected in different difficulties that people have with the three types of inferences.

Empirical Evidence for the Representational and Inferential Principles

An early piece of evidence for the proposed representational and inferential principles comes from a study of Klaczynski and Narasimham (1998). They found that people tend to go beyond the explicitly mentioned conditions of a deontic regulation, if they possess knowledge about alternative conditions. In other words, they indeed try to consider all factors, which they regard as deontically relevant, in their representation of the rule in question as suggested by the exhaustivity principle. More direct evidence is provided by a recent study on conditional promises and threats (Beller, Bender and Kuhnmünch 2005). A conditional promise, for example, has the form "If you do P for me, then I will reward you with Q". People interpreted the deontic aspect of such speech acts according to the constraint, exhaustivity and equivalence principle. Their deontic inferences corresponded to a social rule that obligates the speaker to cooperate if and only if the addressee previously cooperated:

$$\text{'cooperation}_{\text{Adressee}} \leftrightarrow \text{forbidden}(\neg\text{cooperation}_{\text{Speaker}})$$

The obligation to cooperate was associated with a corresponding permission ('*must* X' implying '*may* X' according to the subaltern relation), and, in the absence of the obligation, cooperation was regarded as not necessary but as allowed (according to the subcontrary relation). This pattern of results could also be confirmed in a cross-cultural study with participants from Germany, China, and the Kingdom of Tonga (Beller, Bender and Song 2009).

The most direct test of the proposed principles is provided by three experiments on people's understanding and inferential use of conditional permissions and obligations with a total of 195 participants (one experiment is reported in Beller 2003, the two other experiments are reported in Beller

2008). All three experiments used the same three deontic scenarios, each of which referred to the primary condition of "having a ticket" for admission to a concert. In one scenario, this was the only condition mentioned, in the other two scenarios a second condition was introduced: either an alternative ("being a musician") or an additionally necessary one ("having no weapons"). While the three scenarios differ in the number and the type of preconditions for the deontic rule, the interpretation should always follow the same principles: constraint, exhaustivity and equivalence. The participants were required to solve two types of tasks: a reformulation task and various inference tasks. The reformulation task required participants to choose the best reformulation of the complete deontic rule among several given formulations. The inference tasks implemented the three types of inferences (forward, backward deontic-to-factual, and backward factual-to-deontic). Each inference was tested in two answer modes, using either the modal "must" or the modal "may", in order to check whether people use deontically corresponding formulations with "may" and "must" as interchangeable.

The results clearly supported the representational and inferential assumptions. The relation between the condition side of the deontic rule and the regulated action was predominantly reformulated as an equivalence in all three scenarios (84.6% across all three experiments). This provides clear evidence for the equivalence principle and shows at the same time that the participants regarded the conditions mentioned in the respective task as an exhaustive description of the rule. The experiments also checked whether people applied the rules of interchangeability of the deontic modals correctly. Across the three types of inferences, a very high consistency between deontically corresponding pairs of tasks was found (82.1% consistent answers in the two answer modes "must" and "may"). In drawing deontic inferences, participants used the deontic operators *may* and *must* to a large degree as logically interchangeable. With regard to the three types of inferences, the following pattern emerged: Forward inferences from a deontic rule followed the deontic prediction to a very high degree (85.9% correct). The backward deontic-to-factual inferences appeared to be equally easy (87.4% correct). While these validity preserving inferences were thus drawn with high accuracy, the overall accuracy of the backward factual-to-deontic inference tasks was lower (63.4% correct).

Examining participants' written explanations suggested two distinct reasons for this decrease in accuracy. Remember that the backward factual-to-deontic tasks require drawing a deontic inference from information on whether or not the deontically regulated action is taken. Let us consider at first the case that a deontically regulated action *is* taken: For instance, if a person in example (1) enters the concert, then he or she *must* have a ticket in order to conform to the rule. In this case, some people seemed to confuse *deontic* necessity with *epistemic* necessity. The mere fact that a person enters the concert includes the *possibility* that this person has *no* ticket thereby violating the rule. In this sense, having a ticket is not a necessary implication of entering the concert, although for guests it is a deontically necessary precondition for being allowed to enter. The second case, that the deontically regulated action is *not* taken, was different: If a person in example (1) stays outside, then he or she *may*, but *need not* have a ticket. Here, some participants took people's *intentions* into consideration. They argued that people who are outside but *want* to attend a concert need a ticket, while those who stay outside for other reasons do not need one. As long as this motivational background remains unclear, it cannot be decided whether or not a ticket is deontically necessary, resulting in an indefinite answer.

Concluding Remarks

Deontic rules are at the heart of living together in groups and are part of what one generation hands down to the next. The acquisition of social rules and the adequate dealing with them are so important for people's social activities that the underlying deontic capabilities undoubtedly constitute an essential part of human social cognition.

From developmental studies we know that deontic capabilities are not available "in one block" at a certain age as the evolutionary view might suggest, but co-develop with children's linguistic capabilities. Modal expressions with deontic meaning appear at the age of two-and-a-half years in children's speech production (e.g., Bliss 1988; Stephany 1986; Wells 1979). From three years up, children are already quite good at detecting violations of deontic rules (e.g., Cummins 1996a; Harris, Núñez and Brett 2001; O'Brien, Dias, Roazzi and Cantor 1998). However, the full comprehension of the system of the four deontic modalities is the result of a process of several years (e.g., Bascelli and Barbieri 2002; Hirst and Weil 1982) that is not yet finished at the age of twelve (Coates 1988).

Nevertheless, children learn how to interpret and to reason from deontic rules, and as adults they are able to do this with remarkable precision. The presented data show that people's interpretation of social rules follows the constraint, exhaustivity and equivalence principle, that their deontic inferences systematically follow logical principles, and that they consider the interrelations between the four deontic modalities as defined by the square of oppositions. The observed deviations from the deontic predictions were not random, but result from comprehensible reasons – the close relation to epistemic modality and additional information about the persons involved, such as intentions.

Over the years, many rules are internalized to such a large degree that people follow them without external control. It is therefore not unlikely that even Robinson Crusoe, alone on his island, still complied with some of the rules he grew up with.

References

Anderson, A. R. (1956). The formal analysis of normative systems. In N. Rescher (Ed.), *The logic of decision and action* (pp. 147–213). Pittsburgh: University of Pittsburgh Press.

Bascelli, E., & Barbieri, M. S. (2002). Italian children's understanding of the epistemic and deontic modal verbs *dovere* (must) and *potere* (may). *Journal of Child Language*, *29*, 87–107.

Beller, S. (2001). A model theory of deontic reasoning about social norms. In J. D. Moore & K. Stenning (Eds.), *Proceedings of the Twenty-Third Annual Conference of the Cognitive Science Society* (pp. 63–68). Mahwah, NJ: Lawrence Erlbaum.

Beller, S. (2003). The flexible use of deontic mental models. In R. Alterman & D. Kirsh (Eds.), *Proceedings of the Twenty-Fifth Annual Conference of the Cognitive Science Society* (pp. 127–132). Mahwah, NJ: Lawrence Erlbaum.

Beller, S. (2008). Deontic norms, deontic reasoning, and deontic conditionals. *Thinking and Reasoning*, 14, 305–341.

Beller, S. (2010). Deontic reasoning reviewed: Psychological questions, empirical findings, and current theories. *Cognitive Processing, 11*, 123–132.

Beller, S., Bender, A., & Kuhnmünch, G. (2005). Understanding conditional promises and threats. *Thinking and Reasoning, 11*, 209–238.

Beller, S., Bender, A., & Song, J. (2009). Conditional promises and threats in Germany, China, and Tonga: Cognition and Emotion. *Journal of Cognition and Culture, 9*, 115–139.

Bliss, L. S. (1988). Modal usage by pre-school children. *Journal of Applied Developmental Psychology, 9*, 253–261.

Bucciarelli, M., & Johnson-Laird, P. N. (2005). Naïve deontics: A theory of meaning, representation, and reasoning. *Cognitive Psychology, 50*, 159–193.

Chellas, B. F. (1980). *Modal logic: An introduction.* Cambridge, UK: Cambridge University Press.

Cheng, P. W., & Holyoak, K. J. (1985). Pragmatic reasoning schemas. *Cognitive Psychology, 17*, 391–416.

Coates, J. (1988). The acquisition of the meanings of modality in children aged eight and twelve. *Journal of Child Language, 15*, 425–434.

Cosmides, L. (1989). The logic of social exchange: Has natural selection shaped how humans reason? Studies with the Wason selection task. *Cognition, 31*, 187–276.

Cummins, D. D. (1996a). Evidence of deontic reasoning in 3- and 4-year-old children. *Memory & Cognition, 24*, 823–829.

Cummins, D. D. (1996b). Evidence for the innateness of deontic reasoning. *Mind & Language, 11*, 160–190.

Fiddick, L. (2004). Domains of deontic reasoning: Resolving the discrepancy between the cognitive and moral reasoning literatures. *The Quarterly Journal of Experimental Psychology, 57A*, 447–474.

Harris, P. L., Núñez, M., & Brett, C. (2001). Let's swap: Early understanding of social exchange by British and Nepali children. *Memory & Cognition, 29*, 757–764.

Hiraishi, K., & Hasegawa, T. (2001). Sharing-rule and detection of free-riders in cooperative groups: Evolutionarily important deontic reasoning in the Wason selection task. *Thinking and Reasoning, 7*, 255–294.

Hirst, W., & Weil, J. (1982). Acquisition of epistemic and deontic meaning of modals. *Journal of Child Language*, *9*, 659–666.

Holyoak, K. J., & Cheng, P. W. (1995). Pragmatic reasoning with a point of view. *Thinking and Reasoning*, *1*, 289–313.

Klaczynski, P. A., & Narasimham, G. (1998). Representations as mediators of adolescent deductive reasoning. *Developmental Psychology*, *34*, 865–881.

O'Brien, D., Dias, M. G., Roazzi, A., & Cantor, J. B. (1998). Pinocchio's nose knows: Preschool children recognize that a pragmatic rule can be violated, an indicative conditional can be falsified, and that a broken promise is a false promise. In M. D. S. Braine & D. P. O'Brien (Eds.), *Mental logic* (pp. 447–457). Mahwah, NJ: Lawrence Erlbaum.

Stephany, U. (1986). Modality. In P. Fletcher & M. Garman (Eds.), *Language acquisition* (pp. 375–400). Cambridge: Cambridge University Press.

von Wright, G. H. (1981). On the logic of norms and actions. In R. Hilpinen (Ed.), *New studies in deontic logic* (pp. 3–35). Dordrecht: Reidel.

Wells, G. (1979). Learning and using the auxiliary verb in english. In V. Lee, (Ed.), *Language development* (pp. 250–270). London: Croom Helm.

n-Opposition Theory to Structure Debates

ANTOINE SEILLES and JEAN SALLANTIN

Abstract

2007 was the first international congress on the "square of oppositions". A first attempt to structure debate using n-opposition theory was presented along with the results of a first experiment on the web. Our proposal is to define relations between arguments through a structure of opposition (square of oppositions is one structure of opposition). We will try to answer the following questions: How to organize debates on the web 2.0? How to structure them in a logical way? What is the role of n-opposition theory in our context?

Keywords: online debate, opinion, opposition

Introduction

In the era of participatory democracy and web2.0, there is no solution to debate on the web. Existing solutions are very limited (chats, forums...). Many sites exist for Web-users to express their opinion but can they build structured arguments and confront them? We try to answer these questions by offering new software. We ask ourself what kind of logic can describe the arguments of a debate. We are concerned with the role of n-opposition theory.

In this paper we compare the results of three experiments. First we introduce some important notions, then we present our experimental protocol and finally we present detailed discussion system. We also explain how we use the n-opposition theory to structure debate and reason about it. To conclude, we discuss experiments' contexts and results.

1. Toward a Web2.0 Debate

The work we present began in 2007. Today it is related to the Intermed project. The Intermed project aims to provide a set of interoperable tools to debate on the web. This first chapter presents our actual model of debate and define some important notions.

1.1 Modeling

For our first experiment (Betapolitique 2007) we had not created a debate model. We had a set of documents (political articles of betapolitique website www.betapolitique.fr) and users had the opportunity to annotate and express their point of view. Annotation is defined in the next part of this chapter. Expressions of point of view or argument comprise a highlighted part of text, a judgment and a comment.

The set of documents for our second experiment (ECAP 2008 www.lirmm.fr/ECAP08) are the abstracts of the congress speakers. Arguments are more structured. Selected text are now rephrased. By rephrasing, a user is able to express his understanding of the text.

We had the opportunity to work with sociologists of research institutes for our third experiment (the first Intermed experiment). N.Desquinabo thesis is about interlocutivs kinds in TV shows (Desquinabo, 2005). N.Desquinabo tasks' in the Intermed project revolve around interlocutivs kinds in online debates (Desquinabo, 2008). This collaboration lead us to produce a first relational model (Figure 1.) based on the following principles:

- A **debate** is firstly defined by a subject and reflects a problem (to be resolved or restate …).
- Each debate is associated with a **body of documents** called references.
- Documents can be put in **categories**.
- Participants can create **group** to defend a same opinion.
- Participants can link **arguments** to a document.
- **Relations** ties arguments together.

We use the concept of argumentation from Vignaux (1988) as a starting point. An argumentation is a set of reasoning to enhance a thesis. In a dialectic discussion between two players, there is a binary opposition, affirmation and negation, victory and defeat. Argumentation consists in confronting for and against arguments to find an issue.

1.2 Arguments and Annotation

First we present our mecanism to write arguments based on annotation. An argument is a statement supporting a thesis or against a thesis. A thesis is a point-of-view regarding a debate. A thesis can be, for example, an opinion (for or against Nanotechnologies?) or an important conclusion (Nanotechnologies are dangerous). Arguments are judgments expressed regarding a statement. A set of arguments for a thesis is called an argumentation.

Today forums, chats or emails are used to debate via the web. We have decided to build a different system using annotation because it can precisely localize the part of a text that a user wants to comment (or argue upon).

Definition of **Annotation**: An annotation is a data attached to a document or a part of a document.

We had first developed ad-hoc annotations systems for our first and second experiment platform (Betapolitique was built with SPIP www.spip.net, and ECAP website was built with XWiki www.xwiki.com). For our third experiment, we have developed a solution based on Annotea protocol (www.w3.org/2001/Annotea) and Annozilla (Annozilla is a client for the Annotea protocol developped by Mozilla for firefox web browser annozilla.mozdev.org). This solution is currently a prototype.

There are other annotation systems existing (we recommend the lecture of Hughes et al., 2006) but choosing Annotea was done for the following reasons:

- It allows to annotate HTML documents.
- Annotea is a w3c web standard protocol. This guarantees interoperability.
- Annotations can be stored in RDF database.
- This is an asynchronous solution based on client-server architecture.

We add to the current Annozilla implementation, the possibility for an administrator, to design forms for annotation on a specified category of text.

In our system, an argument is an "illocutionary-act". The statement is selected in a document and must be rephrased to be judged and justified by a comment. Judgment allows to express easily how the statement is for or against the thesis. Rephrasing highlights wich part of the statement is supported or not. User can justify his point of view using a comment wich becomes a new statement in the system.

1.3 Experimental Protocol

This section deals with experimental protocol we have identified from our experiments. Judgments and relations between judgments are defined with respect to the community of users, the subject of the debate, and the categories of documents. A body of documents is created. Users read documents and can annotate to express their point of view and argue. They select and rephrase a statement in a document, state their judgment on it and justify it. Then, each argument can be discussed. When the protocol ends, a synthesis of the debate must be produced.

1. Define a subject for the debate.
2. Create forms to argue (forms for annotation)
3. Reading documents
4. Production of arguments
5. Discussion of arguments (there is a loop between the fourth and fifth step)
6. End of the experiment, production of a synthesis

In a following section of this document, we present three experiments. Criterias given to compare these experiments are:

- Subject of the debate
- Categories of document
- Do users know each other
- Links between judgments
- Aim of the debate (Is this debate going to produce a new document?)
- Duration of the experiment

- Content of justifications (Do comments present frequent patterns to justify a given judgment?)

Aim of each experiment:

1. The first experiment was done as a test to see if it is natural to use judgments to annotate.
2. The second experiment was done as a test to see if users rephrase their selection and justify their judgment.
3. The third experiment was done to test the structure based on the n-opposition theory.

2. Relations between Arguments and n-Opposition Theory

The annotation forms we have designed force users to make a judgment. We propose to use structures of the n-opposition theory to define relations between judgments.

This section presents the n-opposition theory and structures we designed for our three experiments.

2.1 The n-Opposition Theory

To define relations between judgments, we design a n-opposition structure based on n-opposition theory Moretti (2005). This theory is a geometrical generalisation of Aristotle's classical opposition theory. The most famous structure of Aristotle's opposition theory is Aristotle's logical square of oppositions.

Figure 1 shows four kinds of relations (arrows and lines) between four judgments, traditionally represented by letters A,E,I,O. A and E represent universal judgments (A is the affirmative universal judgment, E is the negative universal judgment). I and O represent particular judgments (I is the affirmative particular judgment, O is the negative particular judgment). Arrows between universal and particular judgments represent relations of

Figure 1: Traditional Aristotle's Logical Square of Opposition

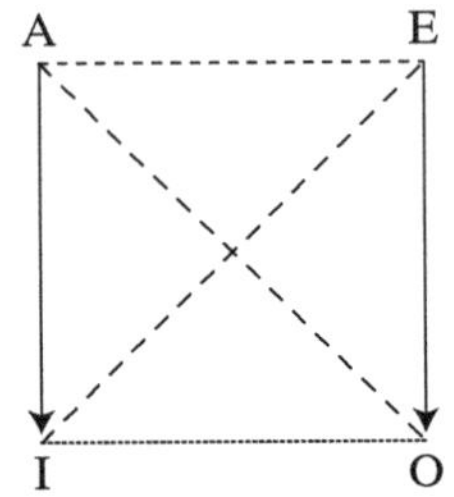

Figure 2: Aristotle's Square of Opposition in Predicate Logic

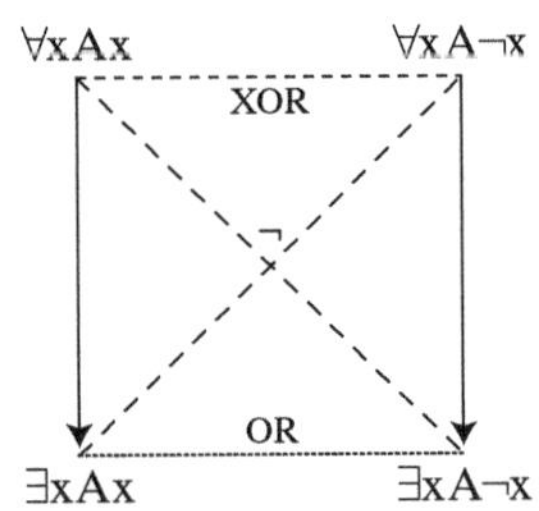

sub-alternation. The upper line between A and E represents a relation of **contrariety**. The bottom line between I and O represents a relation of **sub-contrariety**. Diagonals represent **contradiction**. Figure 2 shows an example of square of oppositions in predicate logic.

Following definitions are quoted from Luzeaux et al. (2007).

- **sub-alternation**: Sub-alternation is defined as the impossibility of having the first term true without having also the second true.
- **contrariety**: Contrariety between two terms is the impossibility that both terms are true, but the possibility that both terms are false.
- **sub-contrariety**: Sub-contrariety is the impossibility that both terms are false, but the possibility that both are true.
- **contradiction**: Contradiction between two terms is defined as the impossibility for them to be both true and both false.

Today, our system doesn't assist users because we test if users naturally respect the structure (the structure isn't explicitly proposed to users). If a structure appears to be adapted for a debate and its users, it will become possible to assist users bysignaling them when they violate the structure.

The following section presents structures we tested in our experiments.

2.2 *Structures of Opposition*

We defined six judgments for our first experiment. Two for a **theoretical** level: **I agree, I don't agree**. Two for a **practical** level: **it works, it doesn't work**. Two for an **emotional** level: xtbfI like, **I don't like**. We suggest a structure in which emotional level is a consequence of theoretical level and practical level is a consequence of emotional level. We considered instinct can be educated by theory and acts are consequences of instinct.

Figure 3: Structure of Opposition for the Betapolitique Experiment

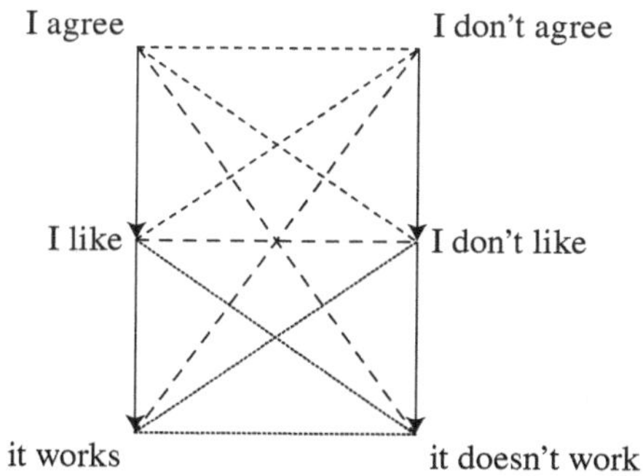

Luzeaux et al. (2007) show how to translate a n-opposition structure into an epistemistic logic. We had found modalitites of a defeasible logic introduced in Antoniou et al. (2000) organized in a n-opposition structure, according to their definitions. This structurelooks like the one we designed for the experiment. Defeasible reasonings deal with conflictual statements. Defeasible logics are introduced by Nute (1994).

A defeasible theory is a triple (F,R,>) where F is a set of facts, R a set of rules and > a superiority relation. There are three kinds of rules, strict rules, defeasible rules and defeaters.

1. Facts are a set of literals and represent indisputable knowledge.
2. Strict rules are denoted by $A \rightarrow B$ and are interpreted in the classical sense.
3. Defeasible rules are rules that can be defeated by contrary evidences.
4. Defeaters are rules that cannot be used to draw any conclusions. Their only use is to prevent conclusions. In other words, they are used to defeat some defeasible rule by producing evidence to the contrary.
5. A superiority relation on R is an acyclic relation $>$ on R. $r1 > r2$ expresses that r1 may override r2 if they lead to conclusive decisions that contradict each other.

A conclusion in a defeasible theory D is a tagged litteral with one of the following modalities:

1. $+\Delta q$ means that q is definitely provable in D. q is proved with only facts and strict rules.
2. $-\Delta q$ means that q is not definetly provable in D. q is not proved with only facts and strict rules.
3. $+\delta q$ means that q is defeasibly provable in D. A defeater to contradict q can exist. The superiority relation leads to conclude q before $\neg q$.
4. $-\delta q$ means that q is not defeasibly provable in D. No set of rules leads to q or for every set of rules that leads to q, the superiority relation leads to $\neg q$ first.
5. $+\delta_{ap} q$ means q is provable with ambiguity in D. There is a set of rules that can lead to q. The superiority relation is not taken into account.
6. $+\delta_{ap} q$ means that q is not provable even with ambiguity in D. There is no set of rules that can lead to q.

However, we didn't look any deeper to confirm the matching between these two structures. This observation was the result of intuition. In our experiments, we focused on defining structures and testing their efficiency. Translation of n-opposition structures into logic is one of our future tasks.

For the second experiment (ECAP), we had defined six judgments:

- **I agree, I don't agree** to judge a statement on its purpose.
- **Correct syntax, incorrect syntax** to judge a statement on its syntax.
- **This confirms, this contradicts** to add supporting references or contradictory references to the statement.

Figure 4: Structure of Opposition for a Defeasible Logic

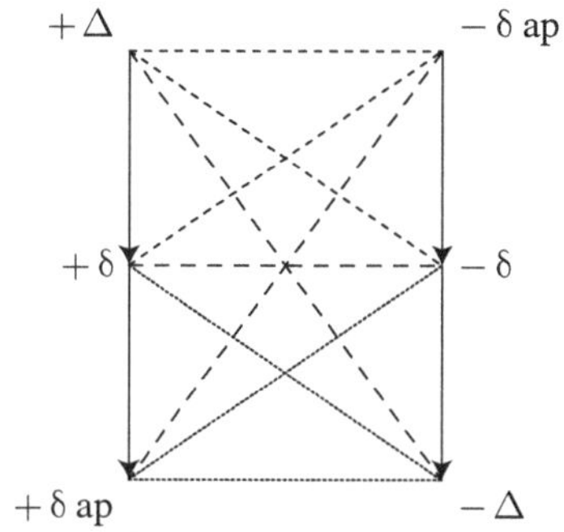

Judgments were opposed by contradiction. The Aim of this experiment was to test if users would accept to rephrase their selections and justify their judgments and we didn't propose a complete structure.

For the third experiment (Intermed), we proposed three levels of judgments:

1. Proved
2. Unclear
3. Incorrect

For each annotation, students had to judge with a combination of an affirmative or negative judgment per level (yes it is proved or no it is not proved, yes it is unclear or no it is not unclear, yes it is incorrect or no it is not incorrect). We tested relations of sub-alternation of the Figure 5.

Figure 5: Structure of Opposition for the Intermed Experiment

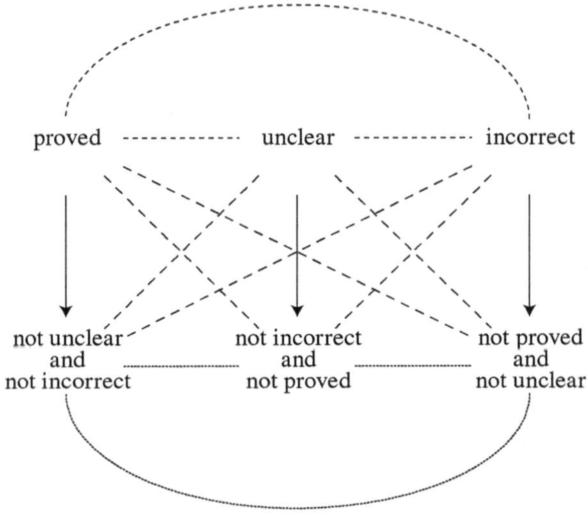

3. Experiments

3.1 Conditions of Experiments

Before dealing with experiments' results and experiments' observations, we present experiments' context. For the first experiment, we worked with the development team of betapolitique. Betapolitique is a blog of politics news. The experiments' duration was three months (February to may 2007) during the French presidential elections. No support was provided to users. There was no objective nor a subject of debate defined. Users of betapolitique are anonymous to each other and were not forced to participate. They could read their arguments on the right of the article (Figure 6) and use a treemap (developed by the society Pikko www.pikkosoftware.com) to see which article was the most annotated (Figure 7).

For the second experiment, we collaborated with the organization committee of ECAP congress. The web site of ECAP was designed to read and discuss abstracts of the speakers of the congress. The experiment begun few days before the congress and ended with the congress. A tutorial is accessible. Participants of the congress were unknown to each other. Their

Figure 6: Screenshot of Annotations for Betapolitique

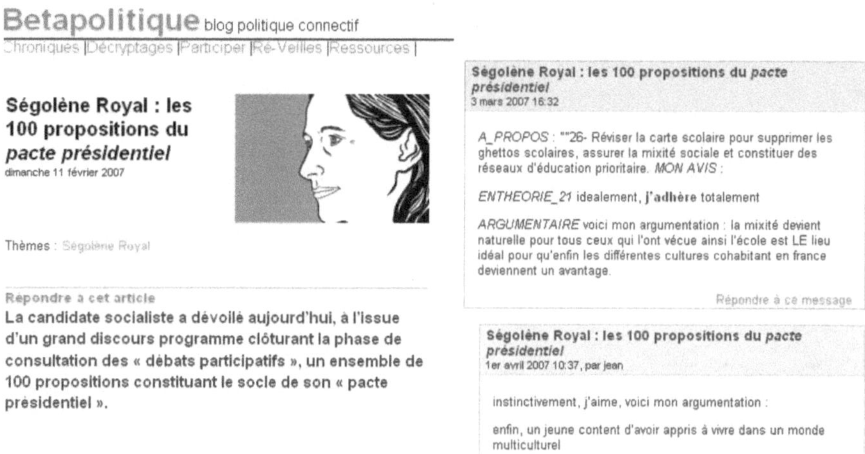

Figure 7: A Treemap to Visualize Annotated Documents

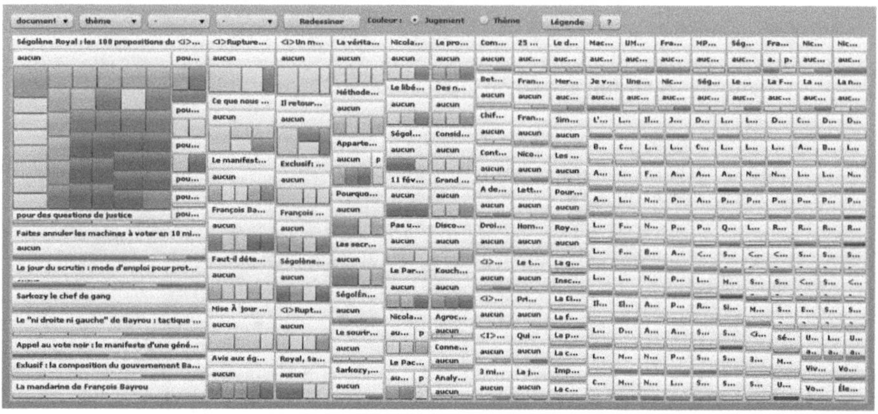

arguments were displayed on the right of the text and corresponding selections were highlighted into the text (Figure 8). The goal of the debate was to enhance the quality of presentations by taking into account remarks from others.

Figure 8: Screenshot of Annotations for ECAP

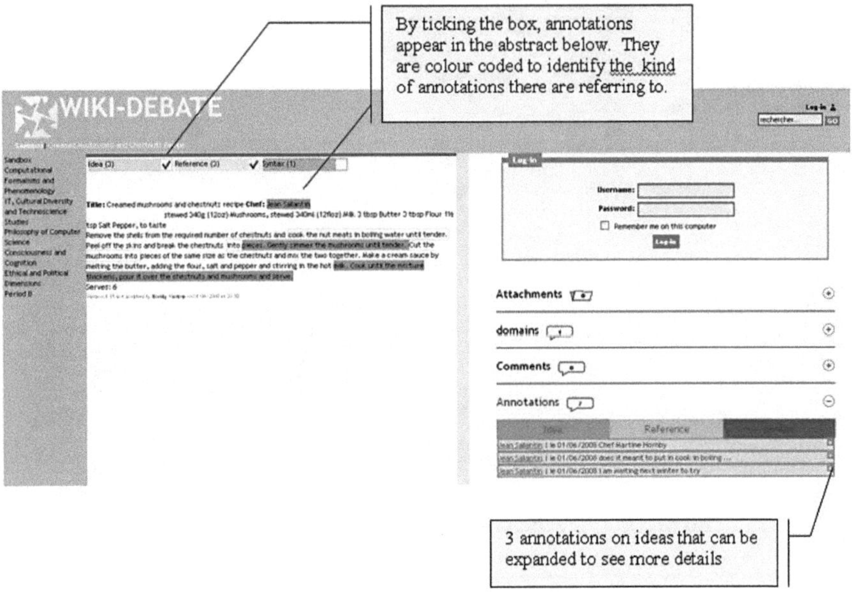

Figure 9: A Graph Based on Opposition to Cartography Debates

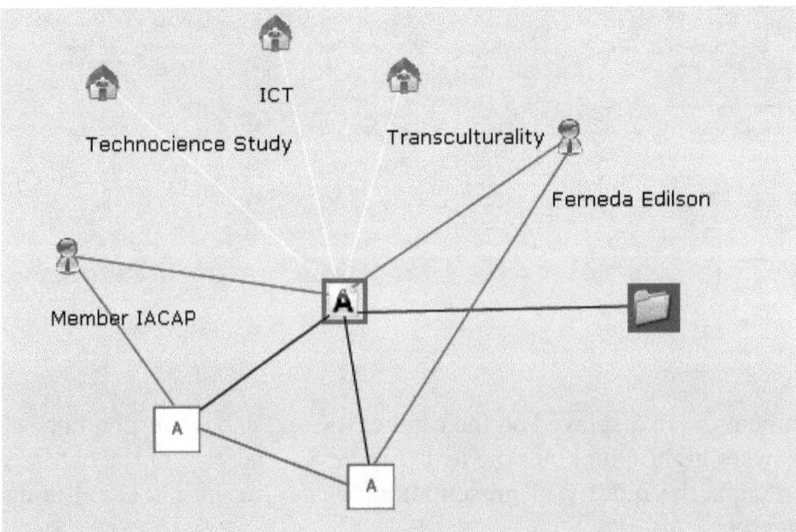

For the Intermed experiment, a group of experts of nanotechnologies domain redacted a set of documents to answer the following questions:

1. Do nanotechnologies bring innovations ?
2. Is nanotechnology a buzzword for already well known technologies?
3. Are nanotechnologies a chance for the developping countries?
4. Will nanotechnologies accentuate the division between north and south ?
5. Is it sacrilege or progress to alter human nature through nanotechnologies?
6. Do nanosciences and nanotechnologies contains unknown risks?
7. Should we adopt a precautionary principle ?
8. Should we be confident that further research will identify the risks and provide means to control them?
9. Is the lack of proof that nanotechnologies are harmful a justification for pursuing production and research, or a justification for stopping them?

These texts were the documents that students have annotated. The subject of the debate was: Should a moratorium on nanosciences and nanotechnologies be necessary? And the goal was to produce a synthesis of argumentations. The experiment was done during two class sessions. Our team was present to assist them. Their annotations were listed on the left of the texts. The parts of text discussed were highlighted by a background color (Figure 10).

Each of these experiments evolved our experimental protocol. We decided to define a subject of debate for the third experiment. Maybe, to not clearly define a subject is a reason of the lack of discussions. In the third experiment, each argument was discussed and it's the only one having a subject.

It appeared clearly that judgments have to be defined depending on the community. For the third experiment, we defined judgments depending on the subject of the debate. We are now convinced that judgments must be defined with respect to the categories of documents. People don't judge scientific statements the same way as philosophical statements.

With the second experiment we introduced the obligation to rephrase the selection. This enhances comprehension between users. A judgment is associated to the new statement (rephrasing the selection), not to the selected part of document. So, even if there is a misunderstanding of the selected part of document, users precise the statement they judge.

Figure 10: A Screenshot of Annotations for Intermed

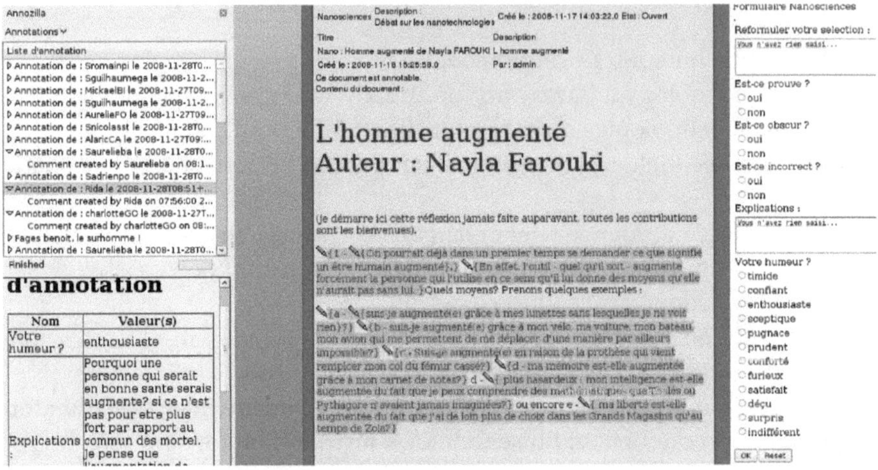

Figure 11: Comparison of the Conditions of the Experiments

	Betapolitique	ECAP	Intermed
kind of documents	political articles	scientifical and philosophical acts	legal or scientific texts of experts
user support	none	tutorial	support of our team
duration	3 months	a week (the congress was on the last 3 days)	2 meetings in 2 hours
goal	none	to improve presentations	to write a synthesis
subject	none	none	Should a moratorium on nanotechnology be necessary?
visualization	treemap	annotated text was highlighted, cartography of oppositions	annotated text was highlighted

3.2 *Observations and Results*

We present in this section our three experiments' conclusions.

The first experiment confirmed the use of judgments and the choice of annotation as a solution for implementing debate. We had 870 annotations for three months of experiment. Judgments were all adopted (Figure 12).

Figure 12: Repartition of Judgments in Betapolitique

		percent
I like	147	16.90
I don't like	127	14.60
I agree	120	13.79
I don't agree	86	9.89
it works	115	13.22
it doesn't work	162	18.62
others	113	12.99
Total	870	100

The second experiment validated the need of an ergonomic interface and of a gratification function. The goal of the experiment was not taken into account by participants. We did interviews as well as polls and the general feeling was that the idea is interesting but the tool is too complicated to be used. We analysed frequencies of visit on the site through Google analytics tool (www.google.com/analytics). The Congres was held from the 16th to the 18th of June 2008. The highest number of visits was the 16th of June, 147 visitors for 80 participants registered to the congress. We had 306 visitors from the 15th to the 19th of June. The average time of visit during this period was four minutes. Though we had visitors, we had only 95 annotations, most of them done before 19th June. The most annotated text was annotated 7 times (there were 70 texts). Nearly nobody attempted to discuss through the tool.

The third experiment validated the interest of rephrasing the annotated statement. The most interesting observation is the fact that there seems to be patterns in justifications for judgments. Users adopt the same kind of locution for a same combination of judgments to justify their choice. Satisfaction, interest and evolution of the opinions were evaluated with a survey.

Figure 13: Statistics of Visits of ECAP Website

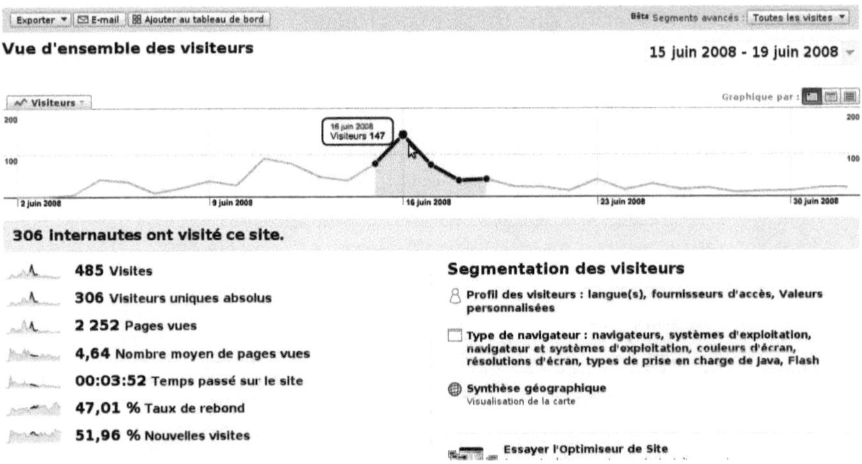

Most of students have appreciated the quality of debates using our tool. They suggested enhancements for the tool.

Most of the arguments are of ethical content. They are on managing risks or on managing conflicts between nanotechnologies potential danger and public health.

Opinions become more precise with controversies. Users made 39 arguments during the first session and 19 during the second one. Each argument was systematically discussed.

The content of justifications leads to the following conclusions:

- If a statement is tagged **proved**, the justification explains the statement.
- If a statement is tagged **incorrect**, the justification is a negative reason.
- If a statement is tagged **with two criteria** (proved and unclear or proved and incorrect), the justification shows a contradiction and suggests a new direction.
- If a statement is tagged **not incorrect, unclear and not proved**, the justification is expression of a belief or of a conviction.

If these results are confirmed by more experiments, it means that we can guide users in justification of their choices of judgment. 41 annotations were well-formed (i.e a judgment per level), the Figure 14 presents the repartition of these annotations with respect to relations of sub-alternation of our structure.

Figure 14: Sub-Alternation in Intermed

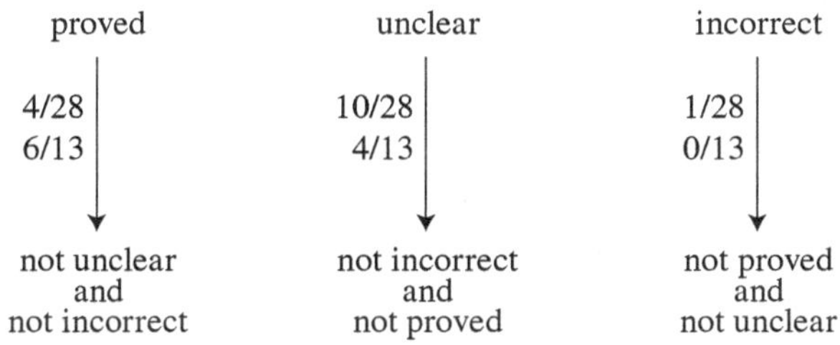

3.3 Desire and Pleasure

Notions of desire and pleasure have appeared during our experiments. Desire is an impulsion for users to participate. Pleasure is a satisfaction, a gratification. The third experiment was the most representative of the need to not neglect these notions. Every frustration (misunderstanding of the software, technical issues, lack of information) can block the participation but an ergonomic interface (nice and easy to use) can enhance desire to debate.

Pleasure depends on quality and goal of the debate. Reasoning and arguing, clarifying one's own opinion and producing a synthesis of one's argumentation are pleasurable activities. These two notions (desire and pleasure) deserve more attention in our future works. But we have seen we could mesure desire and pleasure with poll. Students of the Intermed experiment attributed an average mark of 4/5 for the quality of the debate and of 3,5/5 for the ergonomy and the interest of the website.

4. Conclusion

Facing the lack of web 2.0 solutions to make debates of quality, we propose to create web tools founded on logic to enhance structure and comprehension of debates. In our solution, arguments are justified judgments on statements. The experiments we presented show that our model of argumentation is natural and easy to adopt. The n-opposition theory is a geometrical tool we use to define relations between judgments and to categorize relations between arguments. That's a first step to describe reasonings. The third experiment showed that it is possible to define for each judgment an "illocutionary-act". We can associate an interpretation to each judgment. But, is it natural to represent a judgment with a logical modality? What kind of logic can express relations between judgments?

In our future experiments, we suggest the implementation of an help system that will suggest to users a form of justification for their judgments. We have to define notions of desire and pleasure in the context of the debate. We want to develop an agent founded on n-opposition and logic for helping users when reasoning.

Acknowledgements

We thank J. B. Soufron for its help during the first experiment. J. B. Soufron was responsible of the website betapolitique. Thank to V. Pinet and P. Thomaso and their students for testing our tools. Thank to N. Desquinabo and all the team of the Intermed project. This research was sponsored by the Intermed-ANR project. Thank to Pikko-software society and specially its developers for technical help. Thank to J. Divol, N. Rodriguez, A. Gouaich for their help at LIRMM. And thanks to M. Hornby for her technical and engineering support.

References

Antoniou, G., Billington, D., Governatori, G., & Maher, M. J. (2000). A flexible framework for defeasible logic. *American National Conference on Artificial Intelligence*. AAAI.

Desquinabo, N. (2005). *Caractéristiques et contraintes des genres interlocutifs dans les émissions de plateaux télévisées: analyses pragma-discursives et en réception*. PhD thesis, Université Paris III-Sorbonne Nouvelle.

Desquinabo, N. (2008). Dynamiques et impacts des propositions politiques dans les webforums partisans. *Réseaux*, *26* (150), .

Hugues, B., Schmidt, D., & Smith, A. (2006). *Towards interoperable secondary annotations in the e-social science domain*.

Luzeaux, D., Sallantin, J., & Dartnell, C. (2007). Logical extensions of aristotle's square. In *Handbook of the Square of Opposition*.

Moretti, A. (2005). Geometry for modalities? yes: through n-opposition theory. In Elsevier BV (Ed.), *Aspects of Universal Logic*. Vol. 6. J-Y. Beziau, A. Costa-Leite, A. Facchini.

Nute, D. (1994). Defeasible logic. In D. M. Gabbay, C. J. Hogger, J. A. Robinson, *Handbook of Logic in Artificial Intelligence and Logic Programming*. (Vol. 3, pp. 353–395). Oxford: Oxford University Press.

Vignaux, G. (1988). *Le discours acteur du monde*. Ophrys.